FRACTOGRAPHY OF MODERN ENGINEERING MATERIALS: COMPOSITES AND METALS

A symposium
sponsored by
ASTM Committees E-24
on Fracture Testing and
D-30 on High Modulus Fibers
and Their Composites
Nashville, TN, 18–19 Nov. 1985

ASTM SPECIAL TECHNICAL PUBLICATION 948
John E. Masters, American Cyanamid Co.,
and Joseph J. Au, Sundstrand Corp.,
editors

ASTM Publication Code Number (PCN)
04-948000-30

 1916 Race Street, Philadelphia, PA 19103

Library of Congress Cataloging-in-Publication Data

Fractography of modern engineering materials.

(ASTM special technical publication; 948)
"Papers presented at the Symposium on Fractography of
Modern Engineering Materials"—Foreword.
"ASTM publication code number (PCN) 04-948000-30."
Includes bibliographies and index.
1. Fractography—Congresses.   2. Composite materials—
Fracture—Congresses.   3. Metals—Fracture—Congresses.
I. Masters, John E.   II. Au, Joseph J.   III. ASTM
Committee E-24 on Fracture Testing.   IV. ASTM Committee
D-30 on High Modulus Fibers and Their Composites.
V. Symposium on Fractography of Modern Engineering
Materials (1985: Nashville, Tenn.)   VI. Series.
TA409.F683   1987   620.1′126   87-14970
ISBN 0-8031-0950-4

Printed in Baltimore, MD
September 1987

Foreword

This publication, *Fractography of Modern Engineering Materials: Composites and Metals*, contains papers presented at the Symposium on Fractography of Modern Engineering Materials, which was held in Nashville, Tennessee, 18–19 Nov. 1985. The symposium was sponsored by ASTM Committees E-24 on Fracture Testing and D-30 on High Modulus Fibers and Their Composites. John E. Masters, American Cyanamid Co., and Joseph J. Au, Sundstrand Corp., presided as symposium chairmen and were editors of this publication.

Related
ASTM Publications

Composite Materials: Fatigue and Fracture, STP 907 (1986), 04-907000-33

Delamination and Debonding of Materials, STP 876 (1985), 04-876000-33

Short Fiber Reinforced Composite Materials, STP 772 (1982), 04-772000-30

Fracture Mechanics of Composites, STP 593 (1976), 04-593000-33

A Note of Appreciation
to Reviewers

The quality of the papers that appear in this publication reflects not only the obvious efforts of the authors but also the unheralded, though essential, work of the reviewers. On behalf of ASTM we acknowledge with appreciation their dedication to high professional standards and their sacrifice of time and effort.

ASTM Committee on Publications

ASTM Editorial Staff

Contents

Metals II: Nonferrous Alloys

Overview

Fractography is the detailed analysis of a fracture surface to determine the cause of the fracture and the relationship of the fracture mode to the microstructure of the material. Fractography techniques are used, for example, to identify the origin of a crack and to determine what type of loading caused the crack to initiate. They are also used to establish the direction of crack propagation and the local loading mode which drove the crack. Fractography is, perhaps, the most important analytical approach used by materials scientists in attempting to establish structure-property relationships involving strength and failure of materials.

The science of fractography has been developed into a tool which can be applied to all phases of material investigation. It plays a key role in the failure analysis of established materials. Fractography can also provide useful information in evaluating new materials and in defining their response to mechanical, chemical, and thermal environments.

The technological challenges of recent developments in the aircraft, power generation, and communications industries have led to the development and application of numerous advanced, high-performance materials. These include carbon fiber-reinforced composites, high-strength steel and nonferrous alloys, and superalloys. In addition to design requirements which demand increased strength and stiffness, these new materials are often subjected to harsh fatigue and environmental conditions which may limit their lives. Unfortunately, the development of the fractographic data required to interpret the failure modes of these materials has not always kept pace with their rapid advancement.

The objective of the Symposium on Fractography of Modern Engineering Materials was to provide a forum for presentation and discussions of results of fractographic investigations of these emerging materials. An excellent overview of this topic was presented by Professor R. W. Hertzberg of Lehigh University in his keynote address entitled "Fracture Surface Micromorphology in Engineering Solids."

In addition to the keynote address, 22 papers were presented at the symposium, which was jointly sponsored by ASTM Committees E24 on Fracture Testing and D30 on High Modulus Fibers and Their Composites. These presentations were divided equally between the fractography of metallic and composite materials. The bulk of the work presented on the latter topic discussed failure of continuous fiber graphite/epoxy material systems. This included several presentations on delamination propagation in these

1

materials. The effects of impact and radiation damage on these systems and the fracture surface characterization of notched laminates were also reviewed. The fractography of ferrous alloys comprised the majority of the papers presented on metallic materials. This included discussions of the fractography of steam turbine blading steels, steel weldments in pressure vessels, fatigue damage in carburized steel, and ductile-brittle transition in austenitic stainless steels. In addition, papers were also presented on fatigue crack growth in Inconel and high-strength aluminum alloys and on hydrogen-assisted cracking in titanium alloys.

The expanded use of all of these materials and the increasingly severe loading and service environments to which they are exposed necessitates a thorough knowledge of their fracture behavior. This is particularly true, for example, of composite materials since they are now being used in primary structural applications in the aircraft industry. The ability to correctly interpret the material's fracture mechanisms and to define component failure modes is critical in determining air worthiness and in investigating in-service component failure. The situation in this case is further complicated because these materials exhibit failure modes not normally encountered in metallic materials due to their anisotropic and heterogeneous nature.

It is hoped that the papers presented in this volume will aid investigators conducting failure analyses of advanced composite and metallic materials. It is also hoped that additional symposia will be held as this body of knowledge continues to be developed.

The editors would like to gratefully acknowledge the many contributions provided to this volume by the authors, the reviewers, and the ASTM staff.

John E. Masters

Senior Research Engineer, American Cyanamid Co., Stamford, CT 06904; symposium cochairman and coeditor.

Keynote Address

Richard W. Hertzberg[1]

Fracture Surface Micromorphology in Engineering Solids

REFERENCE: Hertzberg, R. W., **"Fracture Surface Micromorphology in Engineering Solids,"** *Fractography of Modern Engineering Materials: Composites and Metals, ASTM STP 948,* J. E. Masters and J. J. Au, Eds., American Society for Testing and Materials, Philadelphia, 1987, pp. 5–36.

ABSTRACT: The macroscopic and microscopic features of fracture surfaces in metal alloys, polymeric solids, and their composites are examined with particular attention given to the appearance of fatigue damage micromorphology. Macroscopic features such as chevron markings, shear lips, and fatigue clam shell bands and ratchet lines are observed in metal and polymer alloys and are useful in identifying the crack origin, direction of crack growth, and stress state. Certain micromechanisms are common to both metals and plastics (for example, microvoid coalescence, interfacial separation, river markings, and fatigue striations), whereas other fracture markings are unique to a particular material as a result of different microstructures and deformation processes (for example, craze fracture markings and discontinuous growth bands). In general, fracture surface features are seen to depend on the magnitude of the crack driving force, scale of the microstructure relative to the crack tip plastic zone size, and the viscoelastic state of the material.

KEYWORDS: fractography, fatigue, polymers, metals, composites, microstructure, deformation mechanisms, crazing

Much can be learned from an assessment of one's past mistakes. This is particularly true when one examines the details found on a fracture surface. For example, fracture surface markings often reveal the crack origin, the direction of crack propagation, and the stress state that prevailed during the fracture event. Observations of this kind have been reported in various journal articles and conference proceedings, and are contained in several handbooks devoted to the subject of fracture surface analysis [1–7].

Though a detailed fractographic examination in either the transmission or scanning electron microscope (SEM) provides a wealth of useful information pertaining to the failure event, this author firmly believes that

---

[1]Professor, Department of Materials Science and Engineering, Lehigh University, Bethlehem, PA 18015.

the initial step in any fractographic analysis should involve a macroscopic examination of a given fracture surface. In this manner, the investigator gains an overview of the progression of the crack(s), and this enables him/her to focus attention on the critical initiation site instead of on the fast fracture region that usually accounts for much of the total fracture surface area generated. Accordingly, some discussion will be focused on an analysis of several commonly observed macrofractographic features that provide information concerning the stress state, crack origin(s), and direction of crack growth in metal and plastic alloys, respectively.

Since much has been written elsewhere on the subject of electron fractography of metal alloys [1–6], brief mention of common fracture surface markings will suffice; the remainder of the paper is devoted to a study of fracture processes in emerging structural materials such as engineering plastics and their composites. Particular attention is given to an analysis of the fatigue fracture surface micromorphology of metals, plastics, and their composites since most service failures involve cyclic stresses and/or strains.

Preparation Procedures

In preparation for an electron fractographic examination of a particular fracture, one should conduct both a macroscopic examination of the fracture surface and a detailed examination of the underlying microstructure of the material in question. Surely, the fracture surface markings reflect the underlying microstructure to the extent that the crack path either prefers or avoids certain microconstituents. Chestnutt and Spurling [8] demonstrated effectively that one could simultaneously examine the fracture surface and identify the underlying microstructure. Their procedure involved masking part of the fracture surface with a lacquer and then electropolishing the exposed portions of the fracture surface. After removing the lacquer, the specimen is analyzed in the SEM so as to reveal *both* microstructural and fractographic features. An example of this technique is shown in Fig. 1, which reveals elongated microvoid coalescence in an aluminum (Al)–nickel aluminite (Al_3Ni) eutectic composite. The microvoids were nucleated by rupture of Al_3Ni whiskers (see arrows).

To be sure, a fractographic examination must be preceded by careful cleaning of the fracture surface. Zipp [9] has cited several cleaning procedures that include benign techniques such as: dry air blasting to remove loosely adhering particles; the use of various cleaning fluids such as acetone and toluene to remove greases and oils; and repeated stripping of the fracture surface with cellulose acetate replicating tape for removal of tenacious films. When such procedures are found wanting in the case of badly corroded fracture surfaces, more aggressive techniques are needed. In this regard, Zipp

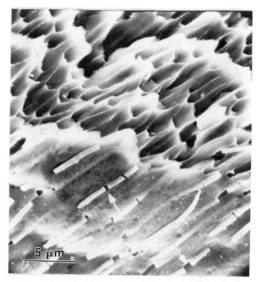

FIG. 1—*Fracture surface (top) and microstructural (bottom) appearance of Al-Al₃Ni eutectic composite. Stopoff lacquer applied to fracture surface, and specimen electropolished in 0.5% HF + H₂O solution. Arrows indicate fractured Al₃Ni whiskers. (Photograph courtesy of K. Vecchio.)*

has recommended that a degraded fracture surface be placed in a bath containing an Alconox cleaning solution (15 g Alconox powder + 350 cm³ water), which is then heated to 95°C and agitated ultrasonically. Vecchio and Hertzberg [10] have found that the sample should not be exposed to this bath for more than 15 to 30 min to prevent chemical attack of the fracture surface.

It should be noted that the use of cleaning solvents may be suitable for metal alloys, but are unsuitable for use with most plastics and their composites since these organic fluids either dissolve or severely attack the polymer structure. Furthermore, the mechanical action of stripping replica tapes (water soluble polyacrylic acid in this instance) from a polymer fracture introduces undesirable artifacts [11]. Problems may also be encountered when polymer fracture surfaces are examined in the SEM. Depending on the thermal stability of the polymer and the thickness of the metal coating, the electron beam may alter the fracture surface micromorphology and even cause localized melting. Figure 2 shows a region on the fatigue fracture surface of polymethyl methacrylate (PMMA) that was damaged during reduced area imaging for the purpose of focus optimization [12]. In some instances, when excessive beam potential or current is applied, fissures may develop in conjunction with the presence of internal stresses [13].

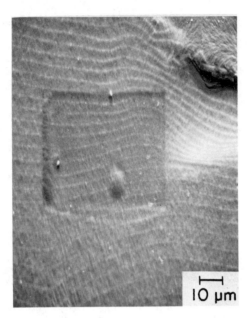

FIG. 2—*Beam damage zone created on PMMA fatigue fracture surface. Damage generated during reduced area imaging* [12].

Macroscopic Features

Chevron Markings

Chevron markings are particularly useful in assessing the direction of crack propagation and, by inference, the location of the crack origin (Fig. 3). The curved lines which radiate in from the opposing free surfaces form an arrow which points back towards the origin of the metal fracture (Fig. 3a), while the radiating pattern of lines on the polystyrene fracture surface reveals the crack origin (Fig. 3b). Another example of a chevron pattern is shown in Fig. 4, which reveals the fatigue fracture surface of a welded web and flange assembly. (Note the lack of welding along the bottom edge of the web section.) The chevron-like pattern at the lower right side of Fig. 4A points back to the porosity origins located in the toe welds on either side of the web (Fig. 4B and 4C).

When a component shatters into many pieces, the chevron markings on the various fragments point in different directions. It should be recognized that while the chevron markings on individual portions of the fracture surface may point in different absolute directions, the *relative* direction of crack growth is the same—that is, all chevron patterns point back to the common initiation site for the component.

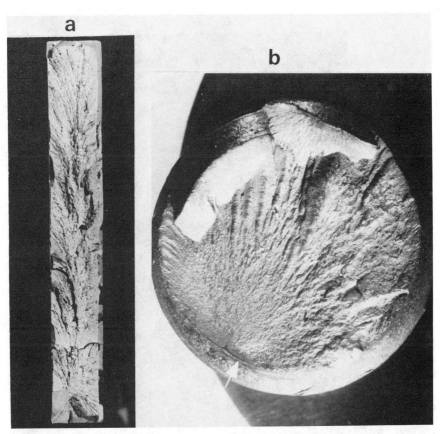

FIG. 3—*Chevron markings point back to origin at bottom of sample. (a) Steel plate; (b) polyvinyl chloride (PVC) threaded bolt. Arrow indicates crack origin (X7). (Photo reduced 10% for printing purposes.)*

Shear Lips

Shear lips which reflect the size of the crack tip plastic zone provide important information concerning the stress state in the specimen at any particular crack length. As a crack grows under cyclic loading conditions, the stress intensity factor increases according to a relationship of the form

$$K = \sigma Y \sqrt{a} \qquad (1)$$

where

K = stress intensity factor,
σ = applied stress,
Y = geometrical correction factor $[f(a/w)]$
a = crack length, and
w = panel width.

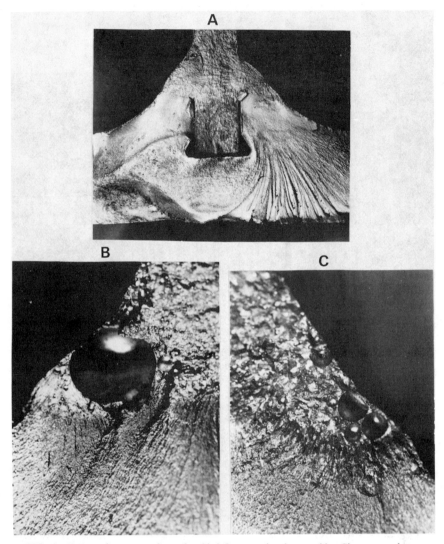

FIG. 4—*Fatigue fracture surface of welded flange and web assembly. Chevron markings at lower right of* (A) *point upward to porosity origins* (B *and* C). (*Photographs courtesy of R. Jaccard, Alusuisse Corp.*)

The increase in K contributes to an increase in the crack tip plastic zone as noted in Eq 2

$$r_y = D(K/\sigma_{ys})^2 \qquad (2)$$

where

   r_y = plastic zone size,
   K = stress intensity factor,

σ_{ys} = yield strength, and

D = constant ranging from $\frac{1}{2\pi}$ to $\frac{1}{6\pi}$.

For conditions of plane stress, the size of the plastic zone is large compared with the thickness of the component and $D = \frac{1}{2\pi}$. Under this condition, the fracture surface assumes a $+/-45°$ orientation with respect to the loading direction and the specimen thickness. This may be rationalized in terms of failure occurring on those planes containing the maximum resolved shear stress. Since plane stress prevails at the free surface of a component, shear lips are always present, albeit small in some instances. When a state of plane strain exists, the plastic zone size is much smaller than the section thickness and $D = \frac{1}{6\pi}$. For this condition, the fracture surface is oriented at 90° to the loading direction and specimen thickness. Examples of plane stress, plane strain, and mixed mode fracture are shown in Fig. 5.

It is possible to relate the size of the plastic zone to the amount of shear lip found on the fracture surface [14]. Since the shear lip forms along a 45° plane, the width of the lip can be equated to the radius of the plastic zone (Eq

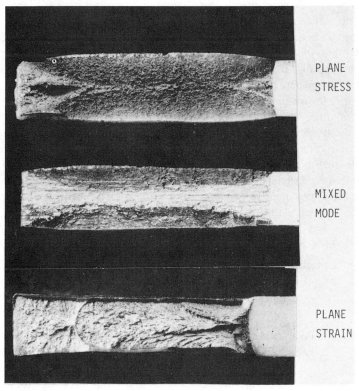

FIG. 5—*Fracture surface appearance corresponding to slant, mixed-mode, and flat fracture in fatigue-precracked fracture toughness specimens.*

FIG. 6—*Double set of shear lips in a steel Charpy sample. Internal shear lips are associated with the formation of center-line delamination. (Photograph courtesy of K. Vecchio.)*

2). Knowing the geometrical correction factor Y for the component and the crack length a where the shear lip was measured, it is possible to estimate the prevailing stress level. This approach to the estimation of stress level must be regarded as highly empirical in that shear lip–inferred estimates of the prevailing stress may be computed with only certain materials such as high-strength aluminum alloys and quenched and tempered steel alloys. By sharp contrast, the size of shear lips in ferritic and pearlitic steels tends to be considerably smaller than that which would be computed based on known stress levels [15].

From this discussion, it is clear that shear lips form at the free surfaces of the specimen where plane stress conditions prevail. A unique exception to this pattern is shown in Fig. 6 where *two* sets of shear lips are found on mating surfaces of a Charpy specimen. The external pair of shear lips correspond to the plane stress condition that normally exists at the specimen's free surfaces. The internal pair of shear lips was produced as a result of the delamination created during the fracture process. Once the delamination was created, two additional free surfaces were created, each corresponding to a plane stress condition. Consequently, a second set of shear lips was produced in the middle of the specimen—a most unusual location for shear lip development.

Fatigue Bands and Ratchet Markings

Fatigue fracture surfaces often reveal a series of parallel bands that eminate from an origin and are oriented parallel to the advancing crack front (Fig. 7a and 7b). These markings, often referred to as "beach markings"[2] (Fig. 7c)

---

[2]The use of the expressions "beach markings" and "clam shell" markings have long been associated with the appearance of certain fatigue fracture surfaces. "Clam shell" markings on the outer surface of bivalve mollusks represent periods of shell growth and are analogous to macroscopic markings on fatigue fracture surfaces in metals and plastics which correspond to periods of crack growth. No such direct analogy exists for the case of "beach markings," which depend on the vagaries of wind and water flow patterns. Nevertheless, reference to "beach markings" provides a mental picture of a pattern of parallel lines that typifies the surface of many fatigue fractures.

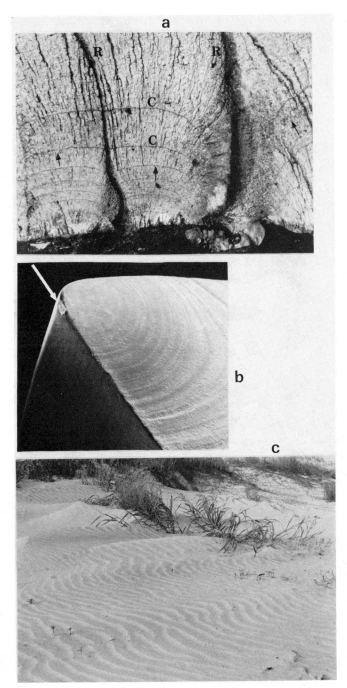

FIG. 7—*Fatigue marker bands.* (a) *Clam shell markings* (C) *and ratchet lines* (R) *in aluminum. Arrow indicates crack propagation direction.* (*Photograph courtesy of R. Jaccard.*) (b) *Fatigue bands in high-impact polystyrene toilet seat* [16]. *Arrow shows crack origin.* (c) *Beach markings in South Carolina.*

or "clam shell markings,"[2] represent different periods of crack extension resulting from intermittent load cycling or from a variable amplitude block loading history. A second set of fracture surface markings are seen in Fig. 7a. In addition to the set of horizontal clam shell markings, one also notes two vertically oriented curved black lines which separate parallel packets of clam

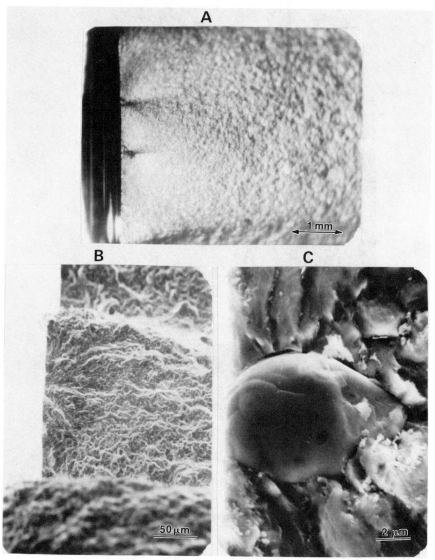

FIG. 8—*Fatigue initiation sites in steel specimen.* (A) *Two ratchet lines separate three crack origins.* (B) *SEM image revealing height elevation difference between three cracks.* (C) *Inclusion at origin of fatigue crack* [18].

shell markings. These lines are called ratchet lines and represent the junction surfaces between the three adjacent crack origins [17]. Since each microcrack is unlikely to form on the same plane, their eventual linkage creates a vertical step on the fracture surface. Once the initial cracks have linked together, the ratchet line disappears. This point is best illustrated by examination of the fatigue fracture surface in Fig. 8. This specimen possessed a polished semicircular notch root from which three fatigue cracks initiated. Note the presence of two small horizontal lines at the edge of the notch root (left side of Fig. 8A); these markings are ratchet lines that separate the three different crack origins. By reducing the depth of field in the SEM, it is possible to clearly demonstrate that the three cracks originated on three separate planes and that the ratchet lines represent the linkage planes that connect the planes containing the original cracks (Fig. 8B). Note the chevron-like pattern on the crack surface in the middle of the photograph. Such markings can be used to locate the crack origin, such as the inclusion which was found at the surface of the polished notch root (Fig. 8C) [18].

Microsocopic Appearance

Fast Fracture and Sustained Loading in Metal Alloys

Since most readers of this review are probably familiar with the basic micromechanisms of fracture in metal alloys, little discussion is given as to their fracture surface markings. For completeness, however, examples of microvoid coalescence, cleavage, and intergranular failure in metal alloys are shown in Fig. 9. Microvoid coalescence occurs by a process involving fracture of inclusions or failure along the inclusion-matrix interface. The associated stress concentration then triggers localized plastic deformation which generates a microvoid within the material. Depending on the nature of the stress state, the microvoid is either spherical (axisymmetric tensile stress) or elongated (shear stress or nonaxisymmetric tensile stress). When a critical number of microvoids have formed along a particular macroscopic plane, they coalesce and cause failure (Fig. 9A). A number of studies have been conducted which reveal that the size of the microvoids generally increases with the fracture toughness of the material and often varies directly with the inclusion spacing (for a recent review, see Ref 19).

Cleavage fracture involves separation along specific crystallographic planes such as the (100)-type planes in body-centered cubic (BCC) steels (Fig. 9B). The relatively flat fracture surface facets are often disrupted by the development of "river markings" which parallel the local direction of crack propagation. Such markings arise from reorientation of the cleavage crack plane as the crack either traverses a grain boundary or encounters a change in the sense of the applied or residual stress field.

When grain boundaries are weakened due to the presence of a brittle phase or preferentially attacked by an aggressive gaseous or liquid environment,

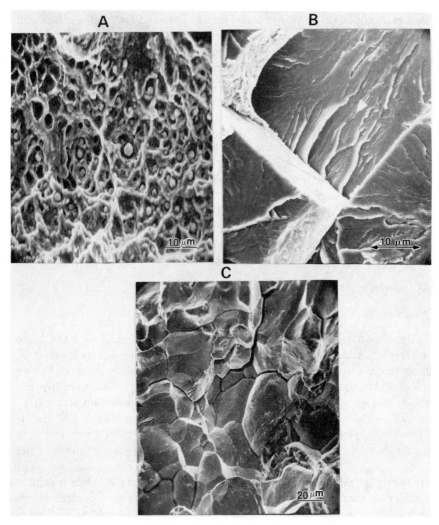

FIG. 9—*Fracture surface markings in metals. (A) Microvoid coalescence in steel. (B) Cleavage in plain carbon steel. Arrow reveals direction of crack growth (river marking). (C) Intergranular fracture associated with environment-assisted cracking in steel.*

failure will occur along a grain boundary surface path (Fig. 9*C*). The observant reader may recognize that the fracture surface of a styrofoam drinking cup also possesses the same interparticle crack path.

Fast Fracture and Sustained Loading in Engineering Plastics

As was noted for the case of metal alloys, the fracture surface appearance of engineering plastics depends on the underlying microstructure [7,14,20].

For example, the fracture surface of a rubber-modified polystyrene [21] reveals the presence of microvoid coalescence with the nucleating phase corresponding to complex spherical particles consisting of a polybutadiene skeletal matrix surrounding discrete particles of polystyrene (Fig. 10a). The duplex structure of these toughening particles is shown clearly in the transmission electron microscope photograph shown in Fig. 10b. In preparing the sample for this image, a thin section was cut from the impact-modified polymer and subsequently exposed to osmium tetroxide, which interacted with the unsaturated double bonds in the polybutadiene phase; the presence of osmium in the polybutadiene phase provides phase contrast in the two-phase particle.

The fracture surface micromorphology of engineering plastics also depends strongly on the deformation mechanism(s) that are active prior to fracture. In most instances, deformation takes place either by shear yielding along discrete bands or by normal yielding (that is, crazing) within narrow lenticular-shaped zones in the polymer. In both instances, bundles of molecules—called fibrils—are aligned parallel to the prevailing stress direction. In the case of the craze, the fibrils become oriented parallel to the applied tensile stress direction and are separated from one another by many microvoids. As such, the overall material volume of the craze zone is reduced by 50% or more from that of the parent matrix. Several different fracture surface markings are produced when the crack passes through or around these crazes. When the crack travels slowly through a preexistent craze,

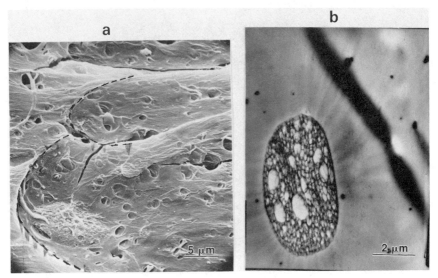

FIG. 10—*Fracture of impact modified polyblend (Noryl). (a) Elongated microvoids nucleated by complex particles (SEM)* [21]. *(b) TEM image revealing polybutadiene skeletal matrix surrounding discrete particles of polystyrene.*

failure occurs along the midplane (midrib) of the craze where prior damage is more extensive (that is, higher void density). As crack speed increases, the crack path jumps back and forth from one craze-matrix interface to the other as shown in Fig. 11a. This fracture surface micromorphology is called "patch" and results from the removal of isolated patches of craze matter from each half of the fracture surface [22–26]. It should be noted that

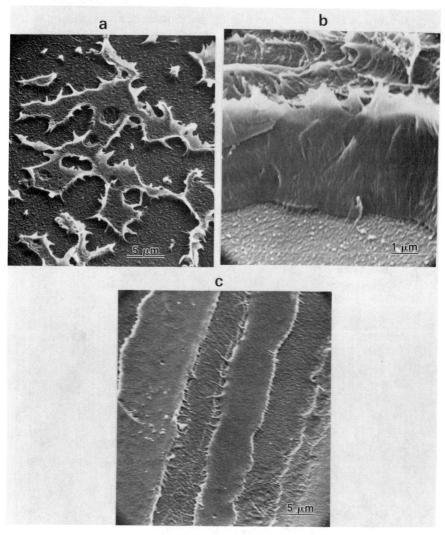

FIG. 11—*Fracture through preexistent crazes in amorphous plastics.* (a) *Patch morphology revealing craze matter attached in segments on craze-matrix interface.* (b) *Edge view of patch. Bottom of photo represents interfacial failure with top portion indicative of midrib failure.* (c) *Mackerel pattern of craze detachment.*

patches of craze matter on one surface dovetail with craze-free zones on the mating fracture surface. Figure 11*b* shows the fracture step associated with passage of the crack from the midrib to the craze interface. Note both the fine-stippled appearance associated with fracture of the fibrils at the craze-matrix interface and the highly elongated nature of the craze matter normal to the plane of the craze. Occasionally, oscillation of the crack path between the two craze-matrix interfaces takes place in stripelike fashion; this results in the alternating presence and absence of parallel bands of craze matter on the fracture surface. The latter pattern has been described as a "mackerel" pattern [24] (Fig. 11*c*).

When a rapidly propagating crack passes through previously uncrazed polymer, the fracture surface often takes on a banded appearance. This pattern arises from the periodic development of bundles of crazes immediately ahead of the advancing crack front and the subsequent passage of the crack along one of the craze-matrix interfaces. Figures 12*a* and 12*b* show the banded appearance of the "hackle" region and the patch morphology on a given hackle band, respectively. Studies by Kusy and Turner [26,27] and Döll and Weidmann [28] have shown that the spacing of these fracture bands decreases with molecular weight and apparent fracture energy. For a more complete discussion of the fracture surface appearance of glassy polymers, the reader is referred to papers by Hull and Murray and Hull [22–25].

The fast fracture surface appearance of semicrystalline polymers has received far less attention and is not clearly understood. In high-density

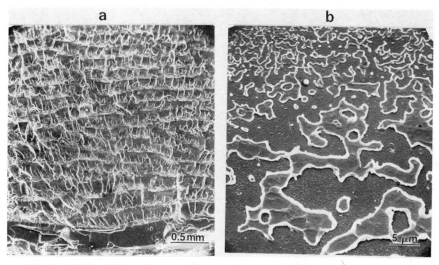

FIG. 12—(a) *"Hackle" markings on fracture surface in amorphous polymer which radiate from crack origin, X30. (Photo reduced 35% for printing purposes.)* (b) *Higher magnification of hackle band revealing patch morphology.*

polyethylene and polyacetal copolymer, the fast fracture surface takes on a plate-like appearance as shown in Fig. 13a. It is not clear as to whether these fracture facets represent failure through specific spherulite regions and whether they represent the consequence of failure through shear bands or crazes; more studies are needed to further clarify these markings.

Under sustained loading conditions and in the presence of water, these two materials undergo slow stable crack extension, resulting in the development of a tufted fracture surface appearance (Fig. 13b). These tufts are believed to represent fibril extension and subsequent rupture. For the case of polyethylene, tufting is observed over more than two orders of magnitude of crack growth rate, with the number and length of tufts decreasing at lower crack velocities in association with lower stress intensity levels [29]. The identification of this fracture feature in polyethylene is often a clue for the presence of environment-assisted cracking in water pipe systems.

Fatigue Mechanisms and Transitions in Metal Alloys

The fatigue striation which represents the successive position of the advancing crack front after each loading cycle is one of the most important features to be found on a fatigue fracture surface (Fig. 14a). Of particular importance, the spacing between adjacent striations represents the increment of crack advance resulting from a single loading cycle. It should be noted that each clam shell or beach marking (recall Fig. 7) represents a period of growth and may contain thousands of individual striae. Much research has

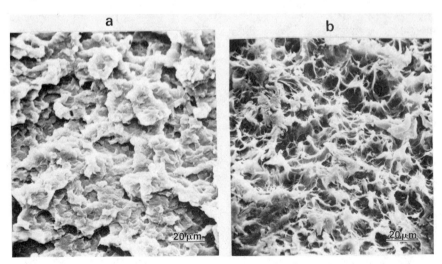

FIG. 13—(a) *Plate-like fracture zones corresponding to fast fracture in polyacetal copolymer (Celcon). (b) Tufted appearance in Celcon associated with sustained loading in water. Tufts represent elongated fibrils that eventually ruptured.*

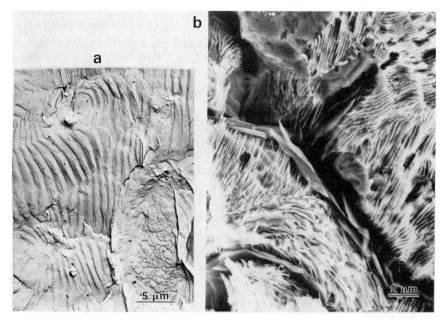

FIG. 14—(a) *Fatigue fracture surface in aluminum alloy. Note localized region of microvoid coalescence.* (b) *Fast fracture through pearlite colony in steel alloy.*

been conducted on the nature of fatigue striations and how their spacing varies with the crack driving force [*14,30*]. To this end, Bates and Clark [*31*] proposed an empirical relationship (Eq 3) between the fatigue striation spacing and the prevailing stress intensity factor range

$$\text{striation spacing} = 6(\Delta K/E)^2 \qquad (3)$$

where

ΔK = stress intensity factor range, and
E = elastic modulus.

Based on past experience, this writer concludes that the transmission electron microscope (TEM) is a better tool than the SEM for use in examining fine-scale details on fatigue fracture surfaces. The use of replicas allows for improved resolution of individual striae that are closely spaced and possess very limited peak to trough depth variation. For those not experienced with replication techniques—a growing sign of the times—it should be recognized that it is possible to change the apparent striation spacing through rotation of the replica in the specimen goniometer holder. Even when this influence is taken into account, one often finds a two- or three-fold variation in the striation spacing for a particular ΔK level. Stofanak et al. [32] concluded that such differences in striation spacing are due primarily to metallurgical factors such as variations in grain orientation and not due to orientation differences between individual fracture surface facets. Regarding the latter, it was found that the carbon replica, containing an array of fracture facets, collapses onto the support grid after being removed from the acetone bath.

It should be recognized that most authors (including the present author) tend to submit with his/her manuscript the very best examples of fracture surface photographs. As a result, the reader is often led to believe that the fracture surface of typical fatigue failure contains only fatigue striations and crisp examples of fatigue striations at that. This is certainly not the case. In fact the typical fatigue fracture surface micromorphology in a metal alloy shows a combination of fatigue striations, microvoid coalescence, cleavage facets, intergranular fracture, and other nondescript regions (Fig. 14a). As such, the spacing between adjacent fatigue striations (a measure of the local microscopic growth rate) usually does not agree with the overall macroscopic fatigue crack propagation rate. Occasionally, the fracture surface of a specimen that did not experience cyclic loading will reveal parallel markings that may be interpreted as being fatigue striations. This may arise when the crack propagates through a microstructural unit that possesses a lamellar microstructure. For example, the parallel markings shown in Fig. 14b reflect fast fracture through several pearlite colonies in a low-carbon steel alloy. As discussed earlier, it is extremely important to identify the underlying microstructure before drawing conclusions about a particular fracture surface micromorphological feature.

Further complicating the picture is the fact that the formation of clearly defined fatigue striations depends on the microstructure of a given material. For example, fatigue striations, when present, are much easier to identify in face-centered cubic (FCC) crystal structures such as in aluminum, copper, and nickel-based alloys. On the other hand, striations are more difficult to identify in BCC alloys such as quenched and tempered steels.

The development of fatigue striations depends strongly on the prevailing stress intensity factor range. At high ΔK levels close to final fracture, the fracture surface often contains microvoids. When the specimen experiences a predominantly plane stress condition associated with the development of a $+/-45°$ fracture plane, the microvoids are elongated parallel to the crack front (that is, the shearing direction on the inclined fracture surface). The author noted earlier that the width of these elongated microvoids (measured in the direction of crack extension) did not vary over more than a two-decade change in the fatigue crack propagation rate [33]. Instead, the width of the elongated microvoids tended to correspond to the dimension of the dislocation cell wall as noted from thin-film studies of the same alloy. At intermediate stress intensity levels, fatigue striations are generated over a two- to three-decade range of fatigue crack propagation rates.

At progressively lower ΔK levels, fatigue striations are no longer observed and the fracture surface assumes a highly faceted appearance reminiscent of the appearance of cleavage facets in BCC metals (Fig. 15*a*) [*34*]. Often these cleavage-like patterns contain river markings as shown in Fig. 15*b*. Faceted fatigue fracture surfaces have been observed in the threshold regime of many

FIG. 15—(a) *Faceted appearance on fatigue fracture surface in Ni-base alloy in threshold regime.* (b) *River markings and faceted appearance in extruded aluminum alloy.* (c) *Intergranular fatigue failure at pearlite colony in VAN-80 steel.*

metal alloys with BCC, FCC, and hexagonal-close packed (HCP) crystal structures. In this ΔK regime, the cyclic plastic zone dimension at the advancing crack tip is found to be smaller than the relevant microstructural unit (for example, the grain diameter in a nickel-based alloy or the Widmanstaetten packet size in a titanium alloy) [35]. Under such conditions, the fatigue crack propagation process is sensitive to metallurgical factors. On the other hand, when the plastic zone size is larger than the microstructural unit, metallurgical considerations are less important and the fracture surface contains either fatigue striations, microvoid coalescence, or other fast fracture micromechanisms. In certain metals (for example, several BCC steel alloys), intergranular fracture is observed at intermediate ΔK levels and is bounded by the threshold regime where faceted growth is noted and higher ΔK levels where fatigue striations are observed. A typical region of intergranular failure during fatigue crack propagation in a pearlitic steel is shown in Fig. 15c. Cooke et al. [36] noted a maximum amount of intergranular fracture when the reversed plastic zone size was comparable to the relevant microstructural dimension in the material.

Fatigue Mechanisms and Transitions in
Engineering Plastics and Composites

The fatigue fracture surface micromorphology of engineering plastics is extremely complex with mechanisms changing as a function of stress intensity level and underlying microstructure. At relatively high K levels, most amorphous and semicrystalline polymers reveal the presence of fatigue striations with the width of each fatigue striation being in excellent agreement with the associated macroscopically determined fatigue crack propagation (FCP) rate [20,37] (Figs. 16A and 16B). This excellent correlation between macroscopic and microscopic growth rates reflects the fact that 100% of the fracture surface in this ΔK regime is striated. As just noted, this finding contrasts with that observed in metal systems. When examined at higher magnification, striations in amorphous plastics reveal little detail with the exception of fine tear lines which are oriented in the crack propagation direction. On the other hand, fatigue striations in semicrystalline polymers often contain two sets of markings: ridges parallel to the FCP direction and many fine lines oriented parallel to the fatigue striation line (Figs. 16C and 16D). Since these fine lines do not correspond to the macroscopic growth rate of the polymer, it is difficult for the investigator to infer the ΔK-level based on the observed fracture surface markings. Without knowledge of the macroscopic growth rate, the investigator might construe the fine deformation lines to be actual striations, thereby inferring a lower crack growth rate than that which actually occurred. The ribbon-like ridges shown in Fig. 16D point in the FCP direction and are believed to represent the original spherulites which had subsequently been drawn out in the direction of crack

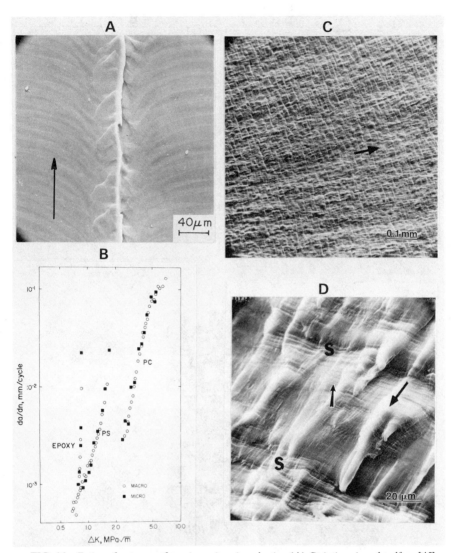

FIG. 16—*Fatigue fracture surfaces in engineering plastics. (A) Striations in polysulfone [45]. (B) Macroscopic versus microscopic growth rates in epoxy, polystyrene (PS) and polycarbonate (PC) [37]. (C) Striations in Nylon 66 + 2.7% H_2O. $\Delta K = 3.3\ MPa\sqrt{m}$. Marker band = 0.1 mm. (D) High magnification of striations in N66 revealing fine lines parallel to striation lines (S) and ridges parallel to crack direction. Arrows indicate crack propagation direction.*

growth [38–41]. The rumpled markings on the fracture surface of an impact-modified Nylon 66 blend reveal another set of fatigue fracture surface lines that are parallel to the advancing crack front position (Fig. 17A). These lines do not represent true striations (that is, fracture lines that correspond to the successive position of the crack front after each loading cycle) [42,43]. The

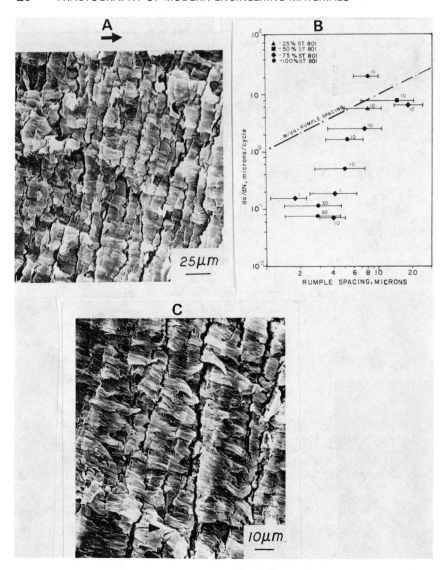

FIG. 17—*Fracture markings in impact-modified N66.* (A) *Rumple markings in fatigue.* $\Delta K = 5.8\,MPa\sqrt{m}.$ *Note crevices on fracture surface and evidence for drawing of polymer in FCP direction.* (B) *Macroscopic growth rate versus rumple spacing.* (C) *Rumple markings in fast fracture zone. Note extensive drawing of polymer* [42].

data shown in Fig. 17*B* reveal that the rumple spacing is generally greater than the average growth increment per loading cycle. Furthermore, rumple spacings varied by only an order of magnitude even though the macroscopic growth rate changed by 1000 times. The fact that the band width of these fracture bands does not correlate with the rate of subcritical flaw growth suggests that these markings are not unique to the FCP process. Indeed, rumple bands have been observed in this material in the fast fracture regime (Fig. 17C).

Though most metallic alloys reveal a second power-dependence of fatigue striation spacing versus the stress intensity factor range, no such correlation has been clearly identified for the case of engineering plastics. Since close agreement is found between fatigue striation spacings and the associated macroscopic growth rates in polymers, the ΔK dependence on polymer striation width is not constant but varies greatly with material. For example, fatigue crack propagation rates have been found to vary with ΔK from the third to twentieth power of the stress intensity range [20]. Whereas striation width-crack opening displacement correlations have been proposed for the case of metallic systems (based on the second power dependence of ΔK on the crack opening displacement), such correlations are suspect for the case of polymer fatigue fracture [20]. A similar conclusion was drawn by Döll et al. [44] based on striation spacing and crack opening displacement measurements.

In many amorphous plastics and in some semicrystalline polymers, an additional set of fatigue fracture bands are found which are oriented parallel to the advancing crack front (Fig. 18a,b); the presence of these markings is seen to depend on the molecular weight of a particular polymer as well as the prevailing ΔK level and test frequency [11,12,20,37,45]. These fracture bands reflect discrete crack advance increments that occur after several hundred to several thousand loading cycles of total crack arrest. Elinck et al. [46] demonstrated that the width of these fracture bands corresponded to the dimension of the crack tip plastic zone as formulated by the Dugdale model [47] (Eq 4)

$$\text{Dugdale plastic zone size} = \frac{\pi}{8}\left(\frac{K}{\sigma_{ys}}\right)^2 \tag{4}$$

As such, the width of each band varies with the second power of the stress intensity factor [11,37]. The present author and coworkers [37,48,49] confirmed that these bands form in two distinct stages. Initially, a craze develops at the leading edge of the crack which then lengthens continuously with repeated load cycling. After the craze reaches some limiting size—the Dugdale plastic zone dimension corresponding to the prevailing stress intensity level—the crack then extends abruptly across the midrib of the craze during the next loading cycle before terminating at or near the craze tip. With continued load cycling, the process is repeated so as to produce a series of parallel bands on the fracture surface. When these discontinuous growth bands (DGB) are examined at higher magnification in the electron microscope, a clearly defined void gradient is noted on each DGB with large microvoids located near the crack tip and small voids found near the craze

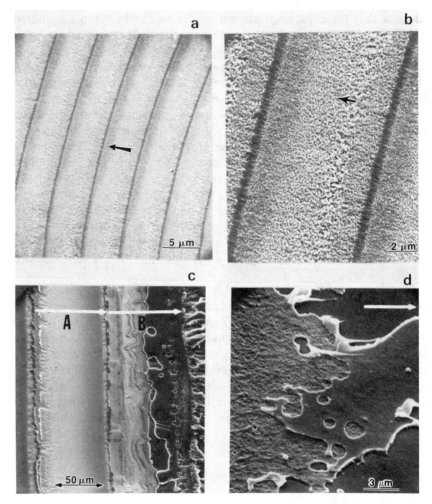

FIG. 18—*Discontinuous growth band formation in polyvinyl chloride.* (a),(b) *Constant load amplitude. Note void gradient.* (c) *Effect of overload on DGB morphology. A is normal band and B is interrupted band* [50]. (d) *Higher magnification of Band B from* (c) *showing transition of crack path from midrib to interface of craze. Arrows indicate crack propagation direction.*

tip (Fig. 18*b*). These microvoids are surrounded by broken fibrils. (Separating each DGB is a narrow stretch zone consisting of elongated microvoids that point back toward the crack origin.) This morphology suggests that the craze breakdown process takes place by void coalescence and fibril rupture, with the void size distribution reflecting the internal structure of the craze just prior to crack extension.

When a given discontinuous growth band is overloaded prior to its anticipated lifetime at a given ΔK level, the fracture surface micromorphology reveals that the crack initially assumes a midrib fracture path (microvoid

formation) but then progresses alternately along both craze-matrix interfaces (patch morphology) (Fig. 18c,d). This change in crack path reflects the
fact that the midrib region near the craze tip had not yet been sufficiently
damaged during cyclic loading to permit passage of the crack [50]. Instead,
the crack chose the next easiest path—the craze-matrix interfaces (recall
Fig. 11).

At relatively low stress intensity levels, the fracture surface in semicrystalline polymers reveals a different set of markings. When viscoelastic deformation processes are restricted, the crack progresses across the spherulites so

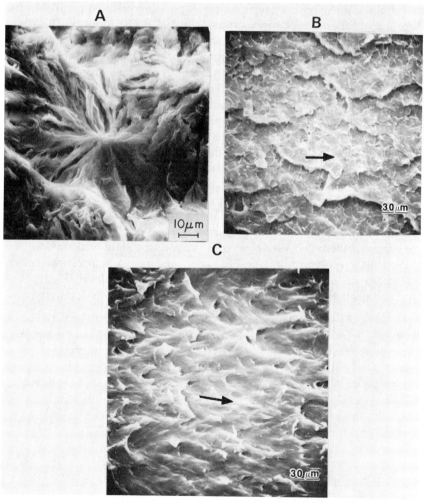

FIG. 19—*Fatigue fracture surfaces in semicrystalline polymers.* (A) *Polyacetal homopolymer*
(Delrin). (B) *Dry Nylon 66* [38]. (C) *Nylon 66 + 2% H$_2$O. In* (B) *and* (C), *$\Delta K = 2.1$ MPa$\sqrt{m}$*
and marker band = 30 μm. Arrows indicate crack direction.

as to reveal circular regions on the fracture surface [38,39,51]. Within each circular zone, one observes a series of radial lines corresponding to the radial pattern of individual crystalline lamellae in the spherulite (Fig. 19A). In other semicrystalline polymers, such as Nylon 66, the fracture surface appearance depends on both the viscoelastic state of the polymer and the stress intensity level. Hahn et al. [42] altered the viscoelastic state of Nylon 66 by the addition of up to 8.5% water to the polymer and by conducting fatigue crack propagation tests at temperatures ranging between −70° and 52°C. Two different fracture surface features were identified: a patchy appearance corresponding to transspherulitic fracture (Fig. 19B) and a highly drawn morphology with the fracture surface lineage being oriented parallel to the direction of fatigue crack propagation (Fig. 19C). The results of this study are summarized in Table 1 and show the dependence of test temperature and moisture on the fractographic appearance of Nylon 66. It is seen that the transspherulitic fracture surface morphology is observed at higher levels of water, the lower the test temperature. This transition in fractographic appearance reflects changes in the viscoelastic state of nylon with different test temperature and water content. By comparison of these results with dynamic modulus data for nylon, it was found that the transspherulitic fracture surface appearance occurred only when the test temperature was below the glass transition temperature for the material (that is, corresponding to the glassy state) [42,52]; alternatively, the drawn fracture surface appearance was noted when the specimen was tested at temperatures in excess of the glass transition temperature. Thus, the change in fractographic appearance as a function of water content and test temperature is really a manifestation of the effects of these two variables on the viscoelastic state of the nylon.

Particulate polymer composites also undergo fracture mechanism transitions as a function of the stress intensity factor level. Figure 20A typifies the fatigue fracture surface appearance in a high-impact polystyrene–

TABLE 1—*Nylon 66 fracture morphology variation with temperature and water content* [42].

Temperature °C	% Absorbed Water			
	0.2%	2.6%	4.0%	8.5%
−70	...	...	...	T
−33	T[a]	T	T	D
0	T	T	D	D
25	T	D	D	D
52	D[b]	D	D	D

[a]T = transspherulitic appearance.
[b]D = drawn appearance.

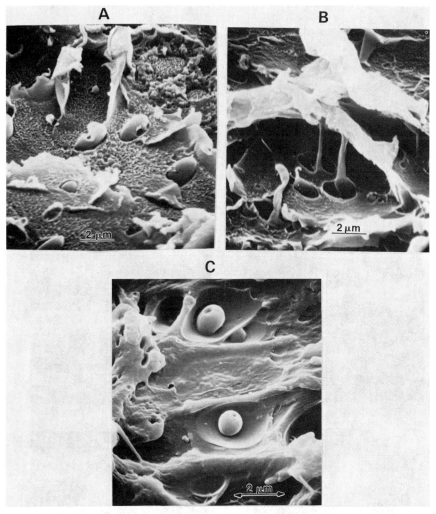

FIG. 20—*Fracture surface morphology in high-impact polystyrene-polyxylenol ether blend* [21]. (A) *Fatigue zone at* $\Delta K = 0.9\,MPa\sqrt{m}$. (B) *Fatigue zone at* $\Delta K = 3\,MPa\sqrt{m}$. (C) *Fast fracture zone. Microvoids associated with hollow polyethylene spheres.*

poly(xylenol ether) blend at low ΔK levels (0.9 MPa$\sqrt{m}$) [21]. Note the presence of incipient particle tearing in the hollow polyethylene (PE) spheres along with interfacial separation of the particles from the matrix; the matrix fracture surface reveals an extensive pattern of microvoid coalescence associated with fracture through crazes in the matrix. At higher ΔK levels (3 MPa$\sqrt{m}$), the polyethylene spheres are almost completely torn from the matrix with the diameter of the associated microvoids being roughly twice that of the undeformed hollow PE beads; at the same time, the fracture

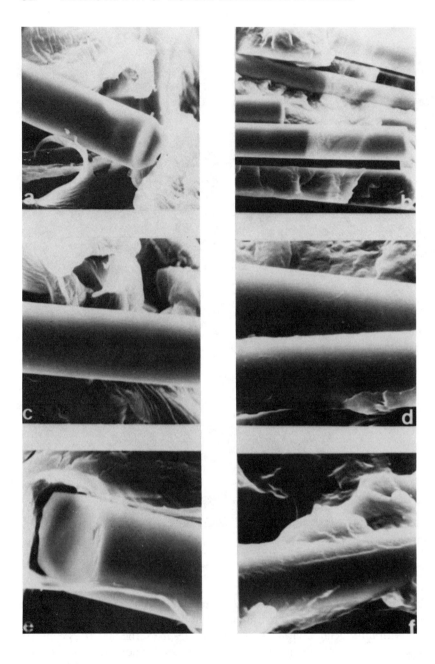

FATIGUE   FAST FRACTURE

FIG. 21—*Fracture surface morphology in glass fiber–reinforced polypropylene in fatigue and fast fracture as a function of low-, intermediate-, and high-bond strength* [54].

surface of the matrix phase appears more drawn with less evidence noted for the presence of microvoid coalescence (Fig. 20B). From these changes in fracture surface appearance, it is concluded that both phases undergo considerably more plastic flow at higher ΔK levels with the matrix phase experiencing more shear yielding than normal crazing [21]. In the fast fracture regime, corresponding to the highest K level experienced by the material, shear yielding was found to be the dominant deformation mechanism as evidenced by the smooth appearance of the matrix fracture surface and large microvoid size relative to the diameter of the PE bead (Fig. 20C).

For the case of a glass fiber–polypropylene (PP) reinforced composite, a fracture mechanism transition occurs between fatigue and fast fracture conditions as a function of the strength of the fiber-matrix interface [53]. Figures 21a, c, and e, corresponding to low, intermediate, and high interfacial strength levels, reveal the same fiber-matrix boundary crack path under fatigue loading conditions, even though differences were noted in fatigue crack propagation rates for a given ΔK level [53]. For the case of fast fracture, the crack path continues to follow the fiber-matrix interface (Figs. 21b and 21d) except when the fiber-matrix interfacial strength is high. Note the presence of PP matrix material along the length of the fiber (Fig. 21f).

An additional example of interfacial fracture between particle and matrix is shown in Fig. 22. Here we see the fatigue fracture surface of an epoxy resin containing hollow glass beads [54]. Figure 22a shows that some of the beads

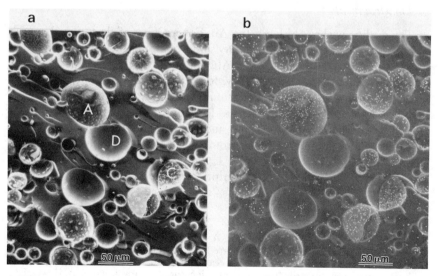

FIG. 22—*Fracture surface in hollow glass sphere-filled epoxy resin.* (a) *Glass particles located at A, detached at D, and cracked at C.* (b) *Silicon X-ray map showing location of glass spheres* [55].

are retained on this half of the fracture surface whereas other particles were either pulled out or shattered. Note the fracture step on the trailing surface of many glass beads, which represents the location where the crack front linked up after passing around opposite sides of the spherical glass particle. The silicon X-ray map shown in Fig. 22b clearly reveals the presence of both undamaged and fragmented hollow glass spheres on the fracture surface.

Conclusions

The fracture surface appearance of metals, plastics, and their composites has been reviewed with particular attention given to mechanisms associated with failure under cyclic loading conditions. Though certain micromechanisms are common to all materials (for example, microvoid coalescence, interfacial separation, river markings, and fatigue striations), other fracture patterns are unique to a particular material as a result of different microstructures and deformation processes (for example, patch and hackle morphology, fibrillar drawing, crystallographic cleavage, and discontinuous growth bands). In general, the development of particular fracture surface markings varies as a function of many external test and internal material variables such as: stress intensity level, crack speed, test temperature, relationship between the size of the crack tip damage zone and dimension of the relevant microstructural unit, viscoelastic state, and nature of the deformation processes that are active prior to the fracture event.

Acknowledgments

The author wishes to thank the Air Force Office of Scientific Research (Grant 83-0029), the Polymer Division, National Science Foundation (Grant DMR 8106489), the Swiss Aluminium Ltd. Co., and the Materials Research Center, Lehigh University, for support during the preparation of this manuscript. Special appreciation is given to K. S. Vecchio, J. S. Crompton, R. Jaccard, B. Braglia, R. Stofanak, C. M. Rimnac, R. S. Vecchio, M. T. Hahn, M. Carling and M. Breslaur for use of previously unpublished photomicrographs from the MRC Mechanical Behavior Laboratory fracture library, and to L. Valkenburg and J. Phillips for their assistance in preparing this manuscript.

References

[1] Phillips, A., Kerlins, V., and Whiteson, B. V., *Electron Fractography Handbook*, AFML TDR 64-416, Wright-Patterson Air Force Base, Dayton, OH, 1965.
[2] *Metals Handbook*, Vol. 9, American Society for Metals, Metals Park, OH, 1974.
[3] *Fracture Handbook*, S. Bhattacharyya, V. E. Johnson, S. Agarwal, and M. A. Howes, Eds., Illinois Institute of Technology, Chicago, 1979.
[4] *Electron Fractography*, ASTM STP 436, American Society for Testing and Materials, Philadelphia, 1968.

[5] *Fractography and Materials Science, ASTM STP 733*, L. N. Gilbertson and R. D. Zipp, Eds., American Society for Testing and Materials, Philadelphia, 1981.

[6] *Fractography of Ceramics and Metal Failures, ASTM STP 827*, J. J. Mecholsky, Jr. and S. R. Powell, Jr., Eds., American Society for Testing and Materials, Philadelphia, 1984.

[7] Engel, L., Klingele, H., Ehrenstein, G. W., and Schaper, H., *An Atlas of Polymer Damage*, Prentice-Hall, Englewood Cliffs, NJ, 1981.

[8] Chestnutt, J. C. and Spurling, R. A., *Metallurgical Transactions*, Vol. 8A, 1977, p. 216.

[9] Zipp, R. D. in *Scanning Electron Microscopy I*, SEM, Inc., AMF O'Hare, Chicago, IL, 1979, p. 355.

[10] Vecchio, R. S. and Hertzberg, R. W. in *Fractography of Ceramic and Metal Failures, ASTM STP 827*, J. J. Mecholsky, Jr. and S. R. Powell, Jr., Eds., American Society for Testing and Materials, 1984, p. 267.

[11] Skibo, M. D., Hertzberg, R. W., and Manson, J. A., *Journal of Materials Science*, Vol. 11, 1976, p. 479.

[12] Janiszewski, J., Hertzberg, R. W., and Manson, J. A., *Fracture Mechanics—13th Conference, ASTM STP 743*, 1981, p. 125.

[13] White, J. R., *Institute of Physics Conference, Series No. 36*, Chap. 10, 1977, p. 411.

[14] Hertzberg, R. W., *Deformation and Fracture Mechanics of Engineering Materials*, 2nd ed., Wiley, New York, 1983.

[15] Hertzberg, R. W. and Goodenow, R., *Proceedings, Microalloying 1975*, Oct. 1975, Washington, DC.

[16] Hertzberg, R. W., Hahn, M. T., Rimnac, C. M., Manson, J. A., and Paris, P. C., *International Journal of Fracture*, Vol. 23, 1983, p. R57.

[17] Wulpi, D. J., *How Components Fail*, American Society for Metals, Metals Park, OH, 1966.

[18] Braglia, B. L., Hertzberg, R. W., and Roberts, R., *Fracture Mechanics, ASTM STP 677*, C. W. Smith, Ed., American Society for Testing and Materials, Philadelphia, 1979, p. 290.

[19] Van Stone, R. H., Cox, T. B., Low, Jr., J. R., and Psioda, J. A., *International Metallurgical Review*, Vol. 30, 1985, p. 157.

[20] Hertzberg, R. W. and Manson, J. A., *Fatigue of Engineering Plastics*, Academic Press, New York, 1980.

[21] Rimnac, C. M., Hertzberg, R. W., and Manson, J. A., *Polymer*, Vol. 23, 1982, p. 1977.

[22] Hull, D., *Polymeric Materials*, American Society for Metals, Metals Park, OH, 1975, p. 487.

[23] Murray, J. and Hull, D., *Polymer*, Vol. 10, 1969, p. 451.

[24] Murray, J. and Hull, D., *Journal of Polymer Science, A-2*, Vol. 8, 1970, p. 583.

[25] Hull, D., *Journal of Materials Science*, Vol. 5, 1970, p. 357.

[26] Kusy, R. P. and Turner, D. T., *Polymer*, Vol. 18, 1977, p. 391.

[27] Kusy, R. P. and Turner, D. T., *Journal of Biomedical Materials Research Symposium*, No. 6, 1975, p. 89.

[28] Döll, W. and Weidmann, G. W., *Journal of Materials Science Letters*, Vol. 11, 1976, p. 2348.

[29] Chan, M. K. V. and Williams, J. G., *Polymer*, Vol. 24, 1983, p. 234.

[30] Laird, C., *Metallurgical Treatises*, J. K. Tien and J. F. Elliott, Eds., American Institute of Mining, Metallurgical, and Petroleum Engineers, Warrendale, PA, 1981, p. 505.

[31] Bates, R. C. and Clark, W. G., *Transactions Quarterly*, American Society of Metals, Metals Park, OH, Vol. 62, No. 2, 1969, p. 380.

[32] Stofanak, R. J., Hertzberg, R. W., and Jaccard, R., *International Journal of Fractography*, Vol. 20, 1982, p. R145.

[33] Hertzberg, R. W., *Fatigue Crack Propagation, ASTM STP 415*, American Society for Testing and Materials, Philadelphia, 1967, p. 205.

[34] Hertzberg, R. W. and Mills, W. J., *Fractography-Microscopic Cracking Processes, ASTM STP 600*, C. D. Beachem and W. R. Warke, Eds., American Society for Testing and Materials, Philadelphia, 1976, p. 220.

[35] Yoder, G. R., Cooley, L. A., and Crooker, T. W., *Titanium 80*, H. Kimura and O. Izumi, Eds., Vol. 3, American Institute of Mining, Metallurgical, and Petroleum Engineers, Warrendale, PA, 1980, p. 1865.

[36] Cooke, R. J., Irving, P. E., Booth, C. S., and Beevers, C. J., *Engineering Fracture Mechanics*, Vol. 7, 1975, p. 69.
[37] Hertzberg, R. W., Skibo, M. D., and Manson, J. A., *Fatigue Mechanisms, ASTM STP 675*, J. T. Fong, Ed., American Society for Testing and Materials, Philadelphia, 1979, p. 471.
[38] Bretz, P. E., Hertzberg, R. W., and Manson, J. A., *Journal of Materials Science*, Vol. 16, 1981, p. 2070.
[39] Bretz, P. E., Hertzberg, R. W., and Manson, J. A., *Polymer*, Vol. 22, 1981, p. 1272.
[40] Teh, J. W. and White, J. R., *Journal of Polymer Science* (Polymer Letter Edition), Vol. 17, 1979, p. 737.
[41] White, J. R. and Teh, J. W., *Polymer*, Vol. 20, 1979, p. 764.
[42] Hahn, M. T., Hertzberg, R. W., and Manson, J. A., *Journal of Materials Science*, Vol. 21, 1986, p. 39.
[43] Flexman, Jr., E. A., Paper 14, Int. Conf. on Toughening of Plastics, Plastics and Rubber Institute, London, July 1978.
[44] Döll, W., Konczol, L., and Schinker, M. G., *Polymer*, Vol. 24, 1983, p. 1213.
[45] Skibo, M. D., Hertzberg, R. W., Manson, J. A., and Kim, S. L., *Journal of Science*, Vol. 12, 1977, p. 531.
[46] Elinck, J. P., Bauwens, J. C., and Homes, G., *International Journal of Fracture Mechanics*, Vol. 7, No. 3, 1971, p. 227.
[47] Dugdale, D. S., *Journal of the Mechanics and Physics of Solids*, Vol. 8, 1960, p. 100.
[48] Hertzberg, R. W. and Manson, J. A., in *Materials, Experimentation and Design in Fatigue*, F. Sherratt and J. B. Sturgeon, Eds., Westbury House, Warwick, England, 1981, p. 185.
[49] Hertzberg, R. W. in *Advances in Fracture Research*, Vol. I, S. R. Valluri, D. M. R. Taplin, P. Rao, J. F. Knott, and R. Dubey, Eds., Pergamon Press, Oxford, England, 1984, p. 163.
[50] Rimnac, C. M., Hertzberg, R. W., and Manson, J. A., *Journal of Materials Science*, Vol. 19, 1984, p. 1116.
[51] Hertzberg, R. W., Skibo, M. D., and Manson, J. A., *Journal of Materials Science*, Vol. 18, 1983, p. 3551.
[52] Kohan, M. I., *Nylon Plastics*, Wiley-Interscience, New York, 1973.
[53] Carling, M. J., Manson, J. A., Hertzberg, R. W., and Attalla, G., *Proceedings, 43rd Annual Technical Conference*, Society for Plastics Engineers, Brookfield Center, CT, 1985, p. 396.
[54] M. Breslaur, Lehigh, private communication.

Composites

Composites I: Delamination

Lelia Arcan,[1] *Mircea Arcan,*[2] *and Isaac M. Daniel*[3]

SEM Fractography of Pure and Mixed-Mode Interlaminar Fractures in Graphite/Epoxy Composites

REFERENCE: Arcan, L., Arcan, M., and Daniel, I. M., **"SEM Fractography of Pure and Mixed-Mode Interlaminar Fractures in Graphite/Epoxy Composites,"** *Fractography of Modern Engineering Materials: Composites and Metals, ASTM STP 948*, J. E. Masters and J. J. Au, Eds., American Society for Testing and Materials, Philadelphia, 1987, pp. 41–67.

ABSTRACT: The objective of this study was to characterize the morphology of interlaminar fractures in unidirectional graphite/epoxy composites. The specific objective was to correlate characteristic fractographic patterns for different pure and mixed-mode loadings with the loading parameters and fracture mechanisms. Two graphite/epoxy composite materials were investigated, one with a brittle matrix and the other with a rubber-toughened matrix. The tests were conducted with a specially designed loading system which allows the application of pure Mode I or Mode II loadings or any combination thereof. SEM fractographs, including stereo pairs, were taken of the specimens after failure at various points across the width of the crack and at various distances from the initial crack front. In the case of the brittle matrix composite, the fibers are, in general, covered with a film of matrix. The matrix fracture surface consists of hackles resulting from interfacial tensile or shear failure and tensile failure normal to the maximum tensile stress. The hackle density increases from Mode I, to Mode II, to mixed Mode I and II. In the case of the toughened matrix composite, the fibers are covered with matrix for Mode I and mixed Mode I and II fractures, whereas they are clean under Mode II and mixed Mode II with compression loading. Matrix fracture is granular or consists of hackles with a low degree of orientation and regularity increasing from Mode I to Mode II. Failure mechanisms include interfacial shear and high ductile or plastic elongation of the matrix.

KEYWORDS: graphite/epoxy, delamination, Mode I fracture, Mode II fracture, mixed mode fracture, test methods, fractography, failure mechanisms, toughened composites

[1]Research physicist, Volcani Research Center, Beit Dagan, Israel.
[2]Professor, Department of Solid Mechanics, Materials and Structures, Tel Aviv University, Tel Aviv, Israel.
[3]Professor, Department of Civil Engineering, Northwestern University, Evanston, IL 60201.

Delamination or interlaminar cracking in laminated fiber-reinforced composite materials is considered one of the major failure modes. The initiation and growth of delamination results in progressive stiffness degradation and eventual failure of composite structures. Interlaminar crack propagation can occur under opening, forward shearing, tearing, or a combination thereof. Therefore, delamination fracture toughness is usually characterized by critical stress intensity factors or critical strain energy release rates in Modes I, II, and III.

The global material characterization in terms of fracture toughness is well complemented by local fractographic characterization. Fracture morphology in composites for various failure modes has been studied by several investigators [1–6]. Fracture morphology sheds light on the micromechanisms of failure and, in addition, can lead to the development of mixed-mode failure criteria [3–6].

Most of the delamination testing is done in Mode I employing the double cantilever beam (DCB) specimen [7]. A limited amount of data on mixed Mode I-II delamination has been obtained by using a cracked lap shear (CLS) specimen [8] or an end notch flexure (ENF) test [9]. Most of the mixed mode testing to date has been done by using an off-axis unidirectional specimen with a central crack parallel to the fibers [1,4]. Although the induced failure surface is normal to the plane of the lamina (1-3 plane), it is judged to be equivalent to delamination failure (1-2 plane). This type of specimen suffers from limitations. The coupling between the applied normal stress and the in-plane shear stress requires high aspect ratios, especially for small angles between the loading and fiber directions. The specimen is best suited for a high ratio of Mode I to Mode II. It is not suitable for high Mode II or pure Mode II loading.

Pure shear behavior has been studied by using the three-rail shear test with a vertical center crack in one of the gage sections [4]. In addition to the pure shear in the test section, this test produces high transverse (in the horizontal direction) tensile stresses in the exposed edges, which tend to cause premature failure. Reinforcing the edges against this failure produces the desired failure in the notch but makes the state of stress more indefinite and affects the uniformity of shear stress distribution. However, it may have little effect on the measured fracture toughness.

Recently, a new mixed mode fracture specimen was proposed, which is most suitable for cases of predominant Mode II deformation [10]. It is based on a specimen proposed by Arcan et al. [11] for producing various uniform states of plane stress. This specimen was developed into a pure Mode II specimen by introducing a crack at one edge [12]. It was further adapted to the study of interlaminar fracture in composite materials under opening, shearing, and mixed mode conditions [13].

The present paper deals with the fractography of delamination fractures using an edge-cracked composite specimen loaded in the Arcan test fixture under pure and mixed mode loadings.

Experimental Procedure

Two materials were investigated, AS4/3501-6 (Hercules, Inc.), having AS4 graphite fibers in a relatively brittle epoxy matrix (3501-6), and T300/F-185, having Thornel T 300 fibers (Union Carbide) in an elastomer-modified epoxy resin (F-185, Hexcel Corp.). Twenty-four ply unidirectional plates were fabricated from each of these two materials following standard layup and bagging procedures. The specimens were coupons 1.91 cm (0.75 in.) wide and 3.82 cm (1.5 in.) long with an initial, 1.91-cm (0.75-in.)-long inter-laminar crack at one end (Fig. 1). Several techniques were tried for introduc-ing the initial crack at the midplane, including embedment of Teflon film, machining the crack with a thin saw blade, and extending a natural crack by means of a wedge. The first one of these techniques was found to be most practical and was used throughout, except when noted otherwise.

The specimen was bonded on both faces to two aluminum parts of a specimen holder (Fig. 2). The specimen holder was in turn mounted by pins unto the Arcan loading fixture (Fig. 2). The latter was equipped with loading holes on the periphery for introduction of pure Mode I, Mode II, and mixed Mode I and II loadings. A special attachment has been designed for finer adjustment of the loading angle.

Specimens of both materials were loaded to failure in the fixture described in pure Mode I ($\alpha = 90°$), pure Mode II ($\alpha = 0°$), and a mixed Mode I and II ($\alpha = 20°$). Scanning electron microscopy (SEM) fractographs, including stereo pairs, were taken of the specimens after failure at various points across

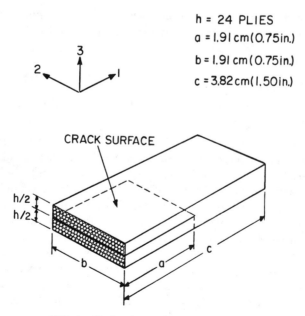

h = 24 PLIES

a = 1.91 cm (0.75 in.)

b = 1.91 cm (0.75 in.)

c = 3.82 cm (1.50 in.)

CRACK SURFACE

FIG. 1—*Single-edge notch specimen geometry.*

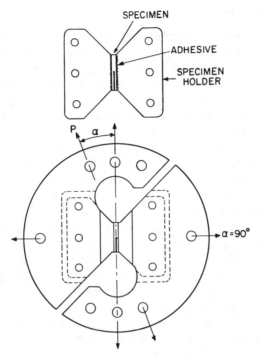

FIG. 2—*Specimen holder and loading fixture for pure and mixed mode loading.*

the width of the crack and at various distances from the initial crack front. The resulting various fracture morphologies are discussed in paragraphs that follow.

Fracture Morphology

Mode I Fracture (Tension)

The first material tested was the AS4/3501-6 with a relatively brittle epoxy matrix. The strain energy release rate for Mode I delamination of this material, measured by means of a DCB test is $G_{Ic} = 0.200$ kJ/m²(1.14 in · lb/in.²).

Typical fractographs for the AS4/3501-6 material are shown in Figs. 3 and 4. Several mechanisms can be recognized. Generally the matrix fails in tension near the fiber/matrix interface with many fibers appearing clean, with occasional areas of matrix covering the fiber. Fiber breaks occur as a result of fiber bridging. The matrix directly under the fiber shows segments of fiber debonding several diameters long interrupted by deep matrix cracks inclined towards the direction of crack propagation (Fig. 4). Since the matrix is brittle, failure occurs where the tensile stress is maximum. At the crack front,

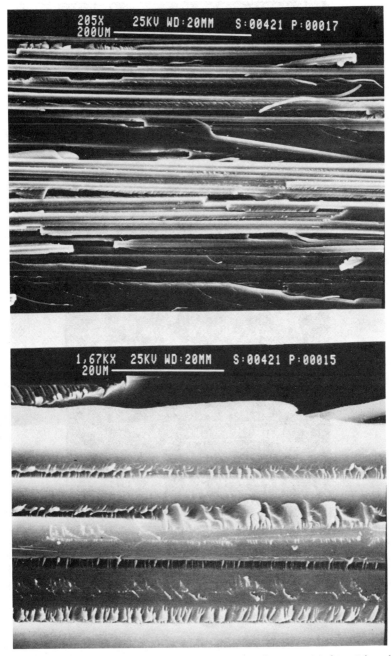

FIG. 3—*Mode I fracture in AS4/3501-6 graphite/epoxy (crack propagation from right to left).*

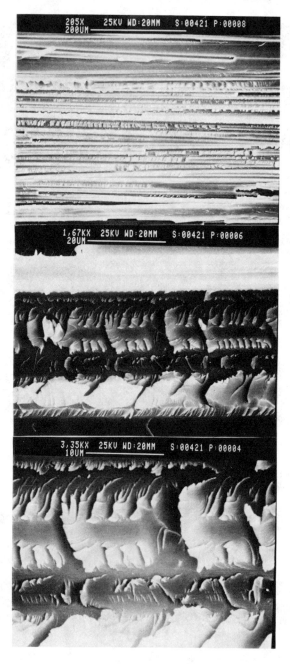

FIG. 4—*Mode I fracture in AS4/3501-6 graphite/epoxy (crack propagation from left to right).*

the maximum tensile stress occurs at the interface normal to the fiber. Away from the crack front, the presence of shear produces a principal stress inclined towards the direction of crack propagation. Since the loading system has very low compliance, the high energy release rate causes crack bifurcation along the fiber matrix interface and along the normal to the maximum tensile stress ahead of the crack tip. A physical model explaining this phenomenon is shown in Fig. 5. In the regions between the fibers, the matrix is under a more complex state of stress, producing smaller and denser hackles as seen in Fig. 4.

The strain energy release rate for the toughened-matrix composite, measured with a DCB test, is

$$G_{Ic} = 1.88 \text{ kJ/m}^2 \ (10.71 \text{ in} \cdot \text{lb/in.}^2)$$

which is approximately an order of magnitude higher than for the AS4/3501-6 material.

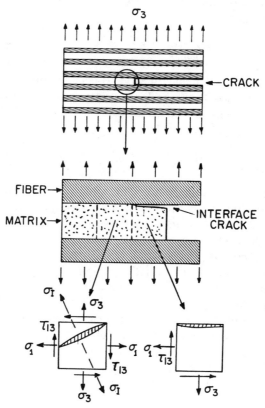

FIG. 5—*Physical model explaining crack branching under Mode I loading.*

Typical fractographs for the T300/F-185 graphite/epoxy with the rubber-toughened matrix are shown in Fig. 6. The low magnification photograph shows several fiber fractures. The fracture surface in general appears rough with all fibers covered with resin. It is apparent that this type of fracture absorbed more energy than in the case of the brittle 3501-6 matrix. Although the hackles are irregular, they appear to have some orientation away from the direction of crack propagation. This is consistent with a maximum tensile stress failure criterion and the contribution of the shear stress ahead of the crack as illustrated in Fig. 5.

Another phenomenon observed in the toughened composite is the presence of slow energy release and fast energy release zones associated with stable and unstable crack propagation. The stable propagation zone appears whitish in the low magnification photograph of Fig. 7. This is usually referred to as a "stress whitening" zone and corresponds to large matrix deformation. The fractured tops of the spikes give the whitish appearance in the low magnification photograph. The white zone is brighter at the beginning of the zone and gradually turns to black. The fracture appearance changes from an oriented one in the white zone to a granular nonoriented one in the black zone.

Mode II Fracture (Pure Shear)

The strain energy release rate for Mode II delamination of the AS4/3501-6 material was obtained from the load versus shear deformation curve to failure of the test coupon. The average value of the Mode II strain energy release rate determined is

$$G_{IIc} = 1.07 \text{ J/m}^2 \text{ (6.1 in} \cdot \text{lb/in.}^2)$$

which is 5.35 times the Mode I value. This value may be higher than it should be because of the high strain rate of fracture and the positive rate dependence of G_{Ic} for brittle matrix composites. However, it is in agreement with the extrapolated value obtained by Bradley and Cohen using asymmetrically loaded DCB specimens [5].

Figure 8 shows the fracture initiation region near the crack front for the AS4/3501-6 graphite/epoxy. It is seen that the characteristic features of this type of fracture develop within a very short (approximately 10 μm) distance from the crack front. Two stereo pairs of a characteristic fracture surface are shown in Fig. 9. The general appearance of the fracture surface is flat on a macroscopic scale. Fracture occurs in the matrix phase with an occasional fiber break. The fibers generally remain coated with a sheath of matrix epoxy. The fracture surface is rather uniform with regularly oriented smooth hackles spaced approximately 50% more closely than in the case of pure tension. Figure 10 shows magnifications of the hackle morphology. Directly

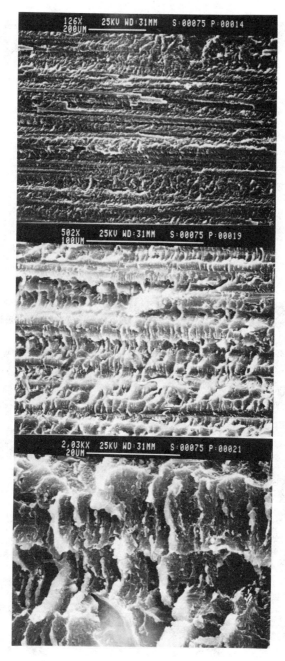

FIG. 6—*Mode I fracture in T300/F-185 graphite/epoxy (crack propagation from right to left).*

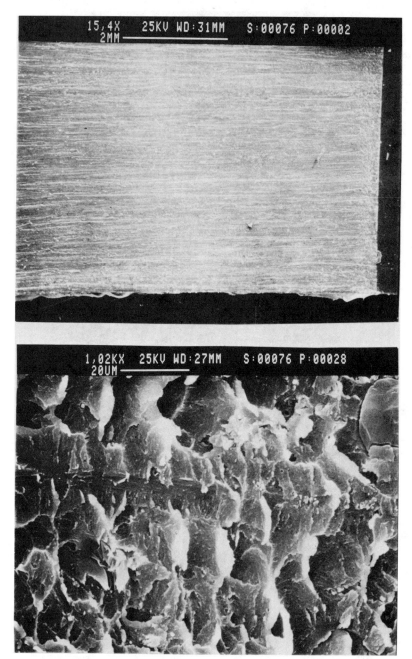

FIG. 7—*"Stress whitening" zone and detail for Mode I fracture of T300/F-185 graphite/epoxy* (*crack propagation from right to left*).

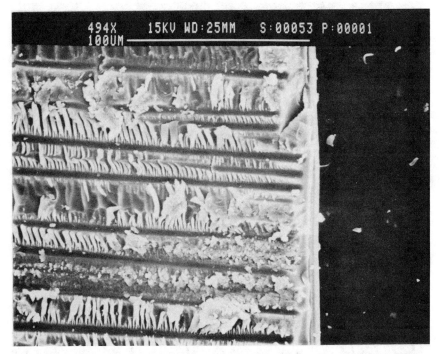

FIG. 8—*Fracture initiation region in unidirectional AS4/3501-6 graphite/epoxy under Mode II loading (crack propagation from right to left; shear stress direction from left to right).*

under the missing fiber the matrix fails in tension at approximately 45° with the fiber direction. These cracks, occurring at intervals of approximately two fiber diameters, are inclined away from the direction of applied shear stress, that is, they are normal to the maximum tensile stress. A physical model explaining this phenomenon is shown in Fig. 11. The matrix area between fibers shows hackles which are normal to the fibers and two to three times as dense as those directly under the missing fibers and are the result of a more complex state of stress (Fig. 12).

The Mode II strain energy release rate for the T300/F-185 material, determined in the same manner as for the AS4/3501-6 material, is

$$G_{IIc} = 1.58 \text{ kJ/m}^2 \ (9.02 \text{ in} \cdot \text{lb/in.}^2)$$

which is higher than the corresponding value for the AS4/3501-6 material but somewhat lower than the G_{Ic} value for the same material. The result is in agreement with the extrapolated value obtained by Bradley and Cohen [5]. The reduction in G_{IIc} may be related in part to the dynamic nature of fracture and the negative rate dependence of the toughened composite material.

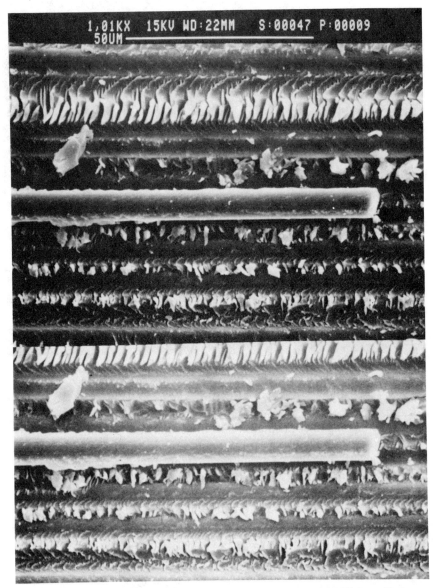

FIG. 9—*Stereo pair of Mode II fracture surface in unidirectional AS4/3501-6 graphite/epoxy (crack propagation and shear direction right to left).*

Fracture initiation in the T300/F-185 graphite/epoxy appears to be characterized by local slips on three different ply levels as illustrated in Fig. 13. The length of this initiation region is approximately 200 μm, much longer than in the case of the brittle matrix composite (AS4/3501-6). In one case, the initial crack front was produced by a natural crack obtained by Mode I

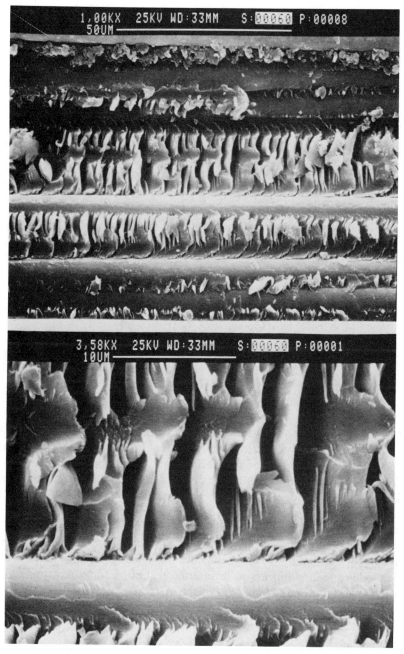

FIG. 10—*Mode II fracture in unidirectional AS4/3501-6 graphite/epoxy (crack propagation and shear direction from right to left).*

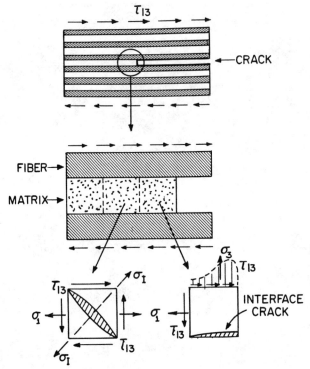

FIG. 11—*Physical model explaining crack formation under Mode II loading.*

crack extension. The transition from this natural crack front to Mode II shear fracture is characterized by longitudinal interpenetration of smooth fibers and regions of grooves and hackles with rough appearance (Fig. 14). At some distance from the initial crack front, an intermediate region exists where hackles start developing with increasing orientation at 45° with the fiber direction. The characteristic features of clean fibers and regular elongated and oriented hackles are illustrated in Fig. 15. The hackles in the grooves between fibers tend to assume a "V" shape pointed in the shearing direction of the fiber above them, the sides still clinging to the adjacent fibers. The mode of failure appears to be interfacial shear failure following a high ductile, or even plastic, elongation of the matrix.

Mixed Mode I and II Fracture (Tension and Shear)

Mixed mode fracture was obtained for one case by loading the specimens at an angle $a = 20°$ with the fiber direction, corresponding to a ratio of normal to shear stress $\sigma_3/\tau_{13} = 0.36$.

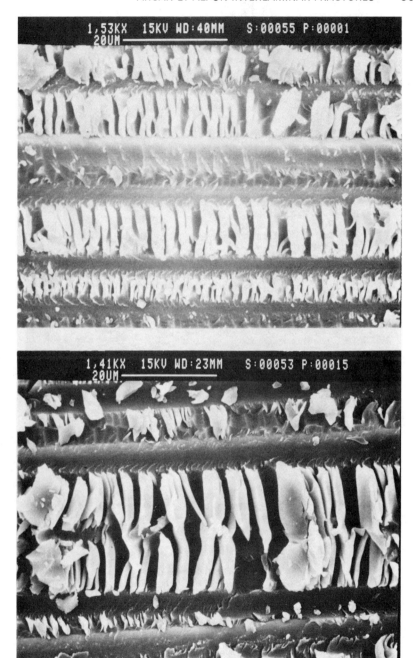

FIG. 12—*Mode II fracture in unidirectional AS4/3501-6 graphite/epoxy (crack propagation direction from right to left; shear direction from left to right).*

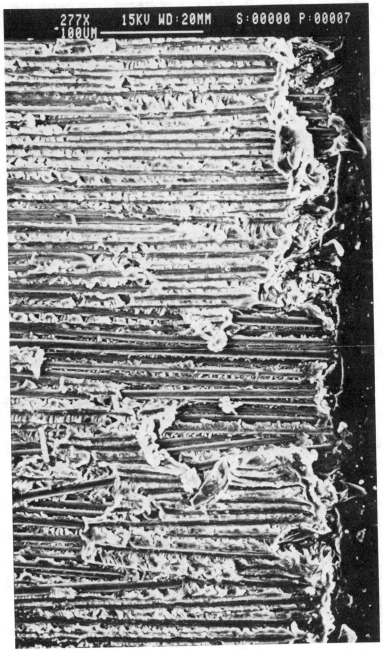

FIG. 13—*Fracture initiation region in unidirectional T300/F-185 graphite/epoxy under Mode II loading (crack propagation and shear direction from right to left).*

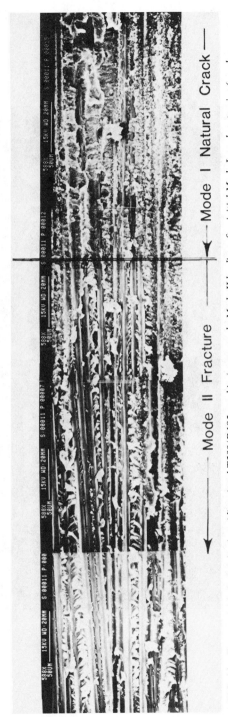

FIG. 14—*Fracture initiation region in unidirectional T300/F-185 graphite/epoxy under Mode II loading after initial Mode I crack extension (crack propagation and shear direction from right to left).*

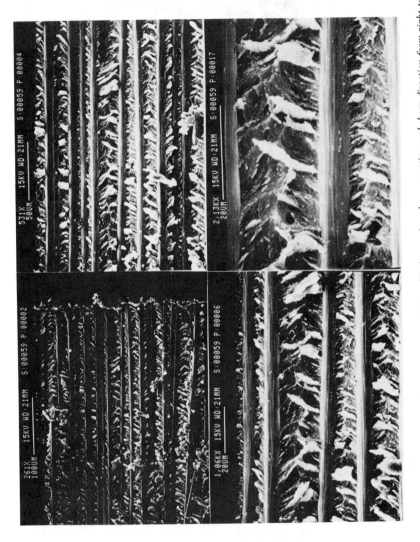

FIG. 15—*Mode II fracture in unidirectional T300/F-185 graphite/epoxy (crack propagation and shear direction from right to left).*

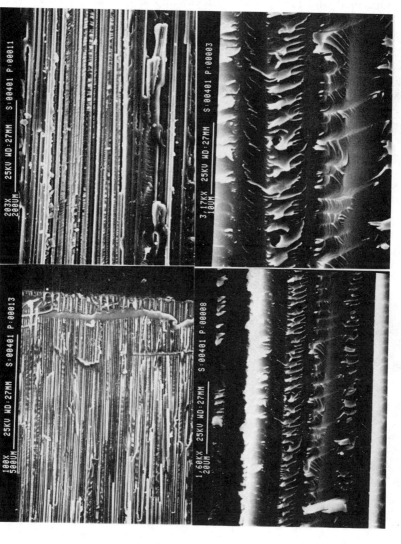

FIG. 16—*Mixed Mode I and II, $K_I/K_{II} = 0.36$, fracture in unidirectional AS4/3501-6 graphite/epoxy (crack propagation direction from right to left; shear direction from left to right).*

Figure 16 shows fractographic features of this type of fracture for the AS4/3501-6 graphite/epoxy. There are many broken fibers due to the bridging effect. In general, the fibers are covered with epoxy matrix. The fracture surface appears corrugated with valleys and plateaus running along the fiber direction. Small and closely spaced hackles appear between the fibers. The

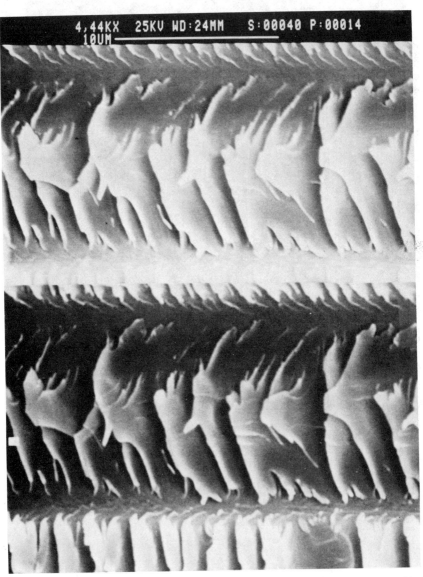

FIG. 17—*Stereo pair of mixed Mode I and II fracture in unidirectional AS4/3501-6 graphite/ epoxy* ($K_I K_{II}$ = 0.36) (*crack propagation direction from right to left; shear direction from left to right*).

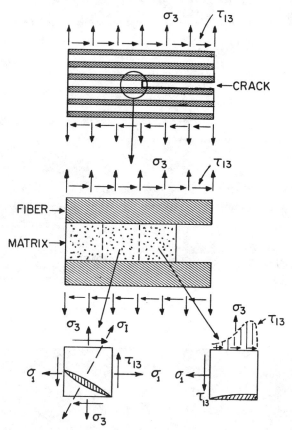

FIG. 18—*Physical model explaining crack formation under mixed Mode I and II loading.*

hackles appear primarily in the valleys; they are triangular in shape, very thin-like flakes, and curled at their tips (Fig. 17). The stress biaxiality (tension σ_3 and shear τ_{13}) imposed on the specimen results in a maximum tensile stress at an angle of 50° with the fiber direction. Assuming a maximum tensile stress failure criterion, one would predict cracking of the matrix at an angle of 40° with the fiber direction on the 1-3 plane (Fig. 18). This is the case for the tips of the flake-like hackles of Fig. 17; however, away from the tips the state of stress is more complex.

Two zones of fracture were observed in the T300/F-185 graphite/epoxy. The first zone, extending up to a distance of 8 mm from the initial crack front, is characterized by broken fibers and deep longitudinal grooves. The fibers are covered with resin as in the case of Mode I fracture (Fig. 19). Small hackles are visible at the bottom of a groove in Fig. 19. The higher magnification photograph of this figure reveals a granular fracture, suggesting small

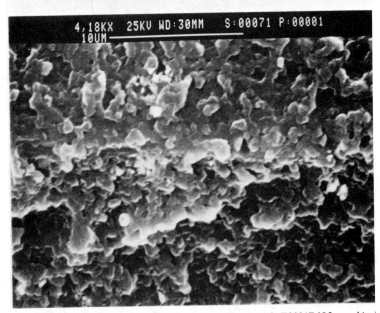

FIG. 19—*Mixed Mode I and II fracture in unidirectional T300/F-185 graphite/epoxy* $(K_IK_{II} = 0.36)$ *(crack and shear direction from right to left).*

FIG. 20—*Mixed Mode I and II fracture in unidirectional T300/F-185 graphite/epoxy* ($K_I K_{II} = 0.36$) (*crack propagation and shear direction from right to left*).

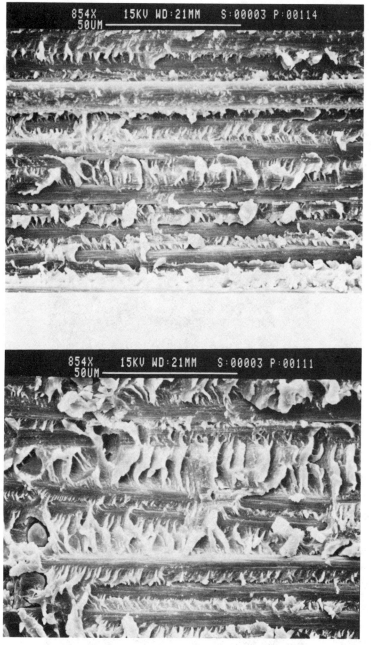

FIG. 21—*Mixed mode shear and compression fracture in unidirectional T300/F-185 graphite/ epoxy* ($K_I K_{II} = 0.36$) (*crack propagation and shear direction from right to left*).

deformation to failure. In the second zone beyond the first, groups or bundles of covered fibers alternate with deep grooves (Fig. 20). The matrix layer covering the fibers appears to have deformed in a more ductile fashion. The hackles in the grooves are thick and random.

Mixed Mode Fracture—Shear and Compression

The influence of superimposed compression on the pure shear mode was investigated in the case of the T300/F-185 graphite/epoxy. Specimens were loaded at an angle $\alpha = -20°$ with the fiber direction in the fixture of Fig. 2, corresponding to a stress ratio of $\sigma_3/\tau_{13} = -0.36$.

Typical fractographs of this mode of fracture are shown in Fig. 21. The fibers appear clean due to the interfacial shear failure of the matrix and the absence of tensile matrix stress normal to the interface. The hackles are nearly normal to the fibers and more widely spaced than in the case of the shear and tension mode.

Summary and Conclusions

A single-edge notch compact specimen was used to study the delamination fracture morphologies of two types of graphite/epoxy materials under Mode I, Mode II, mixed Mode I and II, and Mode II combined with compression. The specimen type and loading system were specially designed to allow the application of any desired mode of loading using only one specimen type. The materials investigated were AS4/3501-6, having AS4 graphite fibers in a relatively brittle and low-toughness epoxy (3501-6), and T300/F-185, having T300 graphite fibers in a rubber-toughened epoxy matrix (F-185).

Under Mode I loading, the AS4/3501-6 material shows some clean fibers and some covered with matrix. The general appearance is rough with longitudinal grooves. Many fibers are broken due to bridging. The matrix under removed fibers shows areas of interfacial failure due to the tensile interfacial stress, interrupted by deep inclined cracks attributed to bifurcation of the interfacial crack and the shear stress ahead of the crack. These large hackles are branched at their base.

Similar results have been observed by Sinclair and Chamis, who used an off-axis specimen [1], Donaldson, who used a rail shear fixture [4], and Bradley and Cohen, who used DCB specimens [5]. However, these investigators noted more clean fibers than have been observed here. This may be related to the higher fiber volume ratio and higher interfacial bond strength of the material used.

Under Mode I loading, the T300/F-185 material is characterized by matrix-covered fibers and irregular hackles tending towards orientation consistent with the maximum tensile stress failure criterion. The material is further characterized by stepwise release of energy with a "stress whitening" zone corresponding to slow energy release and large matrix deformation and a fast energy release zone.

Bradley and Cohen, who used the same matrix resin (F-185), observed a characteristic scalloped fracture surface with most fibers covered with resin [5]. Similar results were observed by Donaldson in a PEEK matrix composite [4].

Under Mode II loading, the AS4/3501-6 material shows fibers generally coated with a sheath of matrix. The fracture surface shows regularly oriented smooth hackles more densely spaced than under Mode I loading. The tops of the hackles correspond to interfacial shear failure and their 45° inclination results from tensile failure across the maximum tensile stress direction. The hackles are branched at their base.

Under Mode II loading, the T300/F-185 material is characterized by clean fibers and regular oriented hackles at 45° with the fiber direction. The hackles in the grooves between fibers tend to assume a V-shaped form. The failure mechanism appears to be interfacial shear debonding and a high ductile or even plastic elongation of the matrix.

The characteristic features of clean fibers and regularly spaced hackles oriented in the shear direction have been observed by the investigators just mentioned and by Russell and Street [9].

Under mixed Mode I and II loading, the AS4/3501-6 material shows many fiber breaks due to bridging. Fibers in general are covered with epoxy matrix. The fracture surface appears corrugated with valleys and plateaus in the fiber direction. Hackles, appearing primarily in the valleys, are triangular flakes with curled tips. In general, as observed also by Russell and Street [9], the surface roughness and degree of matrix microcracking increase with increasing percentage of Mode II contribution.

Under mixed Mode I and II loading, the T300/F-185 material displays many broken fibers and deep longitudinal grooves. The fibers are covered as in the case of Mode I fracture. Matrix fracture is granular or consists of thick random hackles.

Under shear and compression loading, the fibers of the T300/F-185 material appear clean. The hackles are nearly normal to the fibers and more widely spaced than in the case of the shear and tension (Mode I and II) mode.

The problem of determining the direction of crack propagation from fractographs remains unresolved. As described by Hibbs and Bradley, crack propagation and hackle formation is a result of coalescence of sigmoidal microcracks in the resin-rich regions ahead of the crack tip [14]. Depending on where the separation occurs, the hackles may be oriented in the direction of crack propagation or in the opposite direction.

Acknowledgment

The work described in this paper was conducted at the Illinois Institute of Technology while the second author (Mircea Arcan) was a visiting professor. The assistance of Scott Cokeing in specimen preparation is greatly appreciated.

References

[1] Sinclair, J. H. and Chamis, C. C., "Fracture Surface Characteristics of Off-Axis Composites," *Proceedings of the 14th Annual Meeting of the Society of Engineering Science,* Bethlehem, PA, Nov. 1977, Society of Engineering Science, Blacksburg, VA, pp. 247–249.

[2] Purslow, D., "Some Fundamental Aspects of Composites Fractography," *Composites,* Vol. 12, No. 4, Oct. 1981, pp. 241–247.

[3] Hahn, H. T. and Johannesson, T., "A Correlation Between Fracture Energy and Fracture Morphology in Mixed-Mode Fracture of Composites," *Proceedings of the 4th International Conference on the Mechanical Behaviour of Materials,* Stockholm, Sweden, 1983, pp. 431–438.

[4] Donaldson, S. L., "Fractography of Mixed Mode I-II Failure in Graphite/Epoxy and Graphite/Thermoplastic Unidirectional Composites," AFWAL-84-TR-4186, Air Force Wright Aeronautical Laboratories, Wright-Patterson Air Force Base, OH, June 1985.

[5] Bradley, W. L. and Cohen, R. N., "Matrix Deformation and Fracture in Graphite-Reinforced Epoxies," *Delamination and Debonding of Materials, ASTM STP 876,* W. S. Johnson, Ed., American Society for Testing and Materials, Philadelphia, PA, 1985, pp. 389–410.

[6] Johannesson, T. and Blikstad, M., "Fractography and Fracture Criteria of the Delamination Process," *Delamination and Debonding of Materials, ASTM STP 876,* W. S. Johnson, Ed., American Society for Testing and Materials, Philadelphia, PA, 1985, pp. 411–423.

[7] Whitney, J. M., Browning, C.E., and Hoogsteden, W., "A Double Cantilever Beam Test for Characterizing Mode I Delamination of Composite Materials," *Journal of Reinforced Plastics and Composites,* Vol. 1, Oct. 1982, pp. 297–313.

[8] Wilkins, D. J., Eisenmann, J. R., Camin, R. A., Margolis, W. S., and Benson, R.A., "Characterizing Delamination Growth in Graphite/Epoxy," in *Damage in Composite Materials, ASTM STP 775,* American Society for Testing and Materials, Philadelphia, PA, 1982, pp. 168–183.

[9] Russell, A. J. and Street, K. N., "Moisture and Temperature Effects on the Mixed-Mode Delamination Fracture of Unidirectional Graphite/Epoxy," *Delamination and Debonding of Materials, ASTM STP 876,* American Society for Testing and Materials, Philadelphia, PA, 1985, pp. 349–370.

[10] Banks-Sills, L., Arcan, M., and Bortman, Y., "A Mixed Mode Fracture Specimen for Mode II Dominant Deformation," *Engineering Fracture Mechanics,* Vol. 20, No. 1, 1984, pp. 145–157.

[11] Arcan, M., Hashin, Z., and Voloshin, A., "A Method to Produce Uniform Plane-Stress States with Applications to Fiber-reinforced Materials," *Experimental Mechanics,* Vol. 18, No. 4, Apr. 1978, pp. 141–146.

[12] Banks-Sills, L. and Arcan, M., "An Edge-cracked Mode II Fracture Specimen," *Experimental Mechanics,* Vol. 23, No. 3, Sept. 1983, pp. 257–261.

[13] Jurf, R. A. and Pipes, R. B., "Interlaminar Fracture of Composite Materials," *Journal of Composite Materials,* Vol. 16, Sept. 1982, pp. 386–394.

[14] Hibbs, M. F. and Bradley, W. L., "Correlations Between Micromechanical Failure Processes and the Delamination Toughness of Graphite/Epoxy Systems," this publication.

Michael F. Hibbs[1] and Walter L. Bradley[1]

Correlations Between Micromechanical Failure Processes and the Delamination Toughness of Graphite/Epoxy Systems

REFERENCE: Hibbs, M. F. and Bradley, W. L., **"Correlations Between Micromechanical Failure Processes and the Delamination Toughness of Graphite/Epoxy Systems,"** *Fractography of Modern Engineering Materials: Composites and Metals, ASTM STP 948*, J. E. Masters and J. J. Au, Eds., American Society for Testing and Materials, Philadelphia, 1987, pp. 68–97.

ABSTRACT: A combination of macroscopic delamination fracture toughness, scanning electron microscope (SEM) real-time fracture observations, and postfracture morphology were used to study the micromechanical processes of delamination failure in several graphite/epoxy systems.

Strain energy release rate G_{Ic}, G_{IIc}, and mixed mode $G_{I\&IIc}$ were obtained from unidirectional double cantilever beam specimens. Comparisons of these energy release rates with the resulting fracture surface mophology are used to clarify the relative importance of the formation of hackles and the fiber matrix interface adhesion to delamination toughness under Mode I, Mode II, and mixed Mode I & II loading conditions.

Real-time fracture observations of composite delamination in the SEM revealed the microprocesses of microcrack formation and coalescence. These observations coupled with G_{Ic} for the neat material, determined from compact tension specimens, provide insight on how resin toughness can be translated into composite delamination toughness.

The implications of hackle formation and their importance in the failure analysis of graphite/epoxy systems are discussed.

KEYWORDS: graphite/epoxy composites, delamination, fracture, toughness, failure analysis

Because of their very high strength-to-weight and stiffness-to-weight ratios, graphite/epoxy composite materials are now used in a wide variety of aerospace, automotive, and sporting goods applications. The first generation resins developed for use in graphite/epoxy systems optimized stiffness and high glass transition temperatures (Tg) by using a very high cross-link

[1]Graduate research assistant and professor, respectively, Mechanical Engineering Dept., Texas A&M University, College Station, TX 77843.

density. Unfortunately, such resins are quite brittle. Recent developments have centered on how to increase resin toughness with a minimum penalty in stiffness and Tg. As tougher resins have been developed, it has been shown [1] that increased resin toughness does not lead to a commensurate improvement in composite toughness. Three possible reasons for this are: (1) premature failure due to weak resin/fiber adhesion; (2) fibers providing constraint which changes the local state of stress and limits the ductility of the resin [2]; and (3) fibers acting as rigid fillers and thereby reducing the volume of material available for deformation in the yield zone around the crack tip.

In the case of delamination of a graphite/epoxy composite, the fracture surface is characterized by the amount of fiber pullout, interfacial debonding, and the extent of the resin deformation or microcracking. One of the most common fractographic features that results from microcracking of the resin are "hackles." These hackles appear as lacerations or scallops in the resin that form regular saw tooth or wave-like patterns. Several attempts have been made to relate the orientation of the hackles to the direction of crack propagation [3]. Results to be presented in this paper clearly show that the hackles slant in opposite directions on the top and bottom surfaces of the crack face, making it impossible to infer crack growth direction from the hackle slant alone.

To develop an accurate interpretation of fractographic results observed on delaminated components, a systematic experimental program of delamination fracture of composite coupons with varying percentages of Mode I and Mode II loading is needed. This paper summarizes the results of such a study in which the effects of state of stress, resin ductility, and resin/fiber adhesive strength have been correlated with resulting fracture surface characteristics. In particular, the formation of hackles and their potential use as a reliable predictor of crack growth direction has been studied. The fractographic information has been collected using scanning electron microscope (SEM) observations of the delamination fracture surfaces of unidirectional split laminate specimens (DCB). These split laminate specimens have been fractured under Mode I, Mode II, and Mixed Mode I & II loading conditions. To clarify the interpretation of the postmortem fractography, in-situ SEM observations of delamination fracture of small DCB specimens loaded in Mode I and Mode II conditions have been made.

Experimental Methods and Materials

Resin and Composite Systems

The four graphite/epoxy composites used in this study were comprised of Hercules AS4 graphite fibers combined with resins having both different degrees of toughness and interfacial bonding strength. The first system used the brittle Hercules resin 3501-6. The composite systems AS4/Dow-P6 and AS4/Dow-P7 were comprised of a somewhat more ductile Novolac epoxy resin. The AS4/Dow-P7 resin was modified with rubber particles to increase resin toughness. The fourth system, AS4/Dow-Q6, uses a much tougher cross-linkable thermoplastic epoxy. Unidirectional 14-ply laminate panels were made from the prepreg of these systems provided by Dow Chemical. A Teflon strip 0.03 mm in thickness was inserted along one edge of the panel between the midplies to provide a starter crack. The panels were then cured in an autoclave press at Texas A&M University using the temperature/pressure/time cycle specified by Dow Chemical for each system.

Fracture Toughness

Delamination fracture toughness measurements for opening mode (G_{Ic}), shear mode (G_{IIc}), and mixed mode were made by using 25 by 2.5 cm double cantilever beam specimens (DCB), as shown in Fig. 1. These specimens were tested using a Material Test System (MTS) at room temperature (24°C) in stroke control at a rate of 0.0085 cm/s. A partial unloading compliance measurement was made after each approximate 1 cm of crack growth. Stroke, displacement, and load were recorded continuously as a function of time, while crack length was measured discretely using visual measurements on the edge of the specimen.

Analysis of Fracture Toughness

A complete description of the analytical procedures used for calculating the composite Mode I, mixed mode, and Mode II energy release rates for these systems can be found in Tse et al. [4]. In summary, the Mode I critical energy release rate G_{Ic} was calculated from the measured data by three methods: linear beam theory as described by Devitt et al. [5], the change of area under the load-displacement curve method suggested by Whitney et al. [6], and the change in compliance method using an analysis suggested by Wilkins et al. [7]. All three methods gave similar results. Thus, an average value obtained from these three methods is reported in this paper. For the mixed mode loadings, the total energy release rate was calculated by combining the Mode I and Mode II energy release rates, each calculated assuming linear beam theory. Because of the large loads and resulting large deflections needed for Mode II crack propagation, the requirements needed for a linear beam theory analysis were not met. Thus, the Mode II energy release rate was determined by measuring the area under the load-displacement curve bounded by the load-crack extension-unload and dividing this energy by the increase in crack area which occurred during this load-unload cycle. It is worth noting that the area method used for the Mode II calcula-

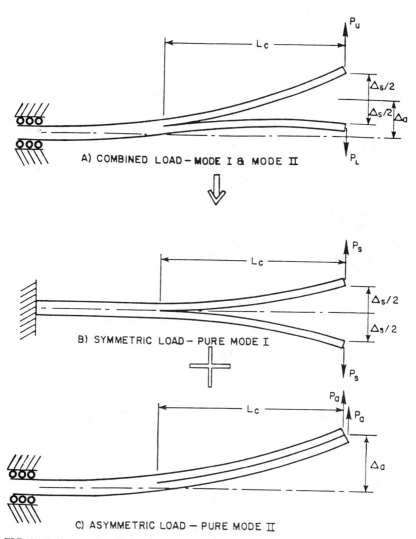

FIG. 1—*Schematic showing how asymmetric loading of a split laminate (DCB) can introduce a mixed (Mode I and Mode II) state of stress at the crack tip.*

tions of G_{IIc} assumes that all damage occurs in the crack tip damage zone rather than in the far field. Thus, it is an upper bound estimate of Mode II energy release rate which equals G_{IIc} when no significant far field damage occurs. The fracture toughness of the neat resin material was determined using compact tension (CT) specimens tested according to ASTM Method for Plane-Strain Fracture Toughness of Metallic Materials (E399-83). Because of the relative brittleness of the resin studied, it was not possible to fatigue-precrack the specimens. Thus, a razor blade was used to pop in a very sharp crack at room temperature.

Fractography

Specimens of suitable size for observation in the SEM were prepared by sectioning the delaminated DCB specimens with a jeweler's saw. Cutting debris on the fractured surface was minimized by cutting the samples before the delamination surfaces were separated. Any loose debris that did occur was blown away using compressed air. The surface was then coated with an approximately 200-Å thick layer of gold-palladium to avoid charging and improve imaging. The photographs were made on tri-X film on a JEOL JMS-25 SEM at accelerating voltages of 12.5 and 15 kV.

Real-time Mode I and Mode II delamination tests of the AS4/3501-6, AS4/Dow-P7, and AS4/Dow-Q6 systems were observed in JEOL JMS-35 SEM using a JEOL 35-TS2 tensile stage. Mode I delamination was achieved by pushing a wedge into the precracked portion of a small DCB-type specimen. The wedge tip was sufficiently blunt to ensure that the wedge remained well behind the crack tip, allowing Mode I opening rather than wedge cutting of the composite. A three-point bend fixture was used to provide the Mode II loading in an end notch flexure (ENF) test arrangement [8]. For both loading conditions, the crack propagation was observed by viewing a polished edge of the specimen in the region of the crack tip. These observations were recorded with video tape and tri-X film. The edges of the specimens to be observed during delamination were carefully polished using standard metallographic techniques and coated with a 200-Å layer of gold-palladium.

Experimental Results

Fracture Toughness Results

A summary of the Mode I (G_{Ic}), mixed mode (G_{IIc}), and Mode II (G_{IIc}) composite fracture toughness are presented in Table 1 along with the fracture toughness values for the neat resin material obtained from the compact tension specimens. The delamination fracture toughness for Mode I loading was found to be greater than the neat resin for the brittle system AS4/3501-6 and similar to the neat resin fracture toughness for the tougher systems (AS4/Dow-P6, AS4/Dow-P7, and AS4/Dow-Q6).

For all of the systems of this study an increase in the fraction of Mode II loading resulted in an increase in the total energy release rate for crack growth, with the most dramatic increase in fracture toughness occurring in the more brittle systems, namely, AS4/3501-6 and AS4/Dow-P6.

TABLE 1—*Resin and composite energy release rates.*

System	Resin, G_{Ic}	Composite G_c			Composite G_{IIc}/G_{Ic} Ratio
		Mode I	43% Mode II	Mode II	
AS4/3501-6	0.07	0.137	0.695	1.292	9.43
AS4/Dow-P6	0.14	0.165	0.334	1.806	10.95
AS4/Dow-P7	0.32	0.340	0.629	1.325	3.90
AS4/Dow-Q6	0.73	0.848	0.969	2.836	3.34

NOTE: Energy release rates given in kJ/m^2.

Fractographic Results: Brittle and Moderately Ductile Composite Systems (AS4/3501-6, AS4/Dow-P6, and AS4/Dow-P7)

Postmortem fractography of the delaminated DCB specimens (Figs. 2, 3, and 4) indicates that the primary Mode I fracture mechanism for the AS4/3501-6 and AS4/Dow-P6 systems was debonding at the fiber/resin interface or fracture through the interphase region. High magnification fractographs of these systems clearly reveal the texture of the fibers, indicating at most a thin sheathing of resin adhering to the fibers. The resulting fracture surface was similar to a flat corrugated roof which has the effect of increasing the surface area compared to a flat, or planar, fracture surface. In the resin rich regions between fibers, a smooth, brittle, cleavage-type fracture was exhibited with a limited number of shallow hackles. Occasional fiber pullout and breakage are also seen. The fracture surface of the AS4/Dow-P7 system reveals much more resin deformation and damage with some resin adhering to the fibers, indicating better interfacial bonding. The surface also is highly pockmarked. These holes (approximately 1 to 5 μm in diameter), which give rise to the pockmarked appearance, are apparently the result of volatiles trapped during the fabrication of the laminate panel (see Fig. 4b). The AS4/Dow-P7 system also exhibited areas where large sheets of resin and fibers were pulled out as the delamination crack jumped back and forth between adjacent fibers (see Fig. 4a).

In all of these systems, qualitative observations indicate the number of hackles and their angle of orientation with reference to the macroscopic fracture plane increased as the percentage of Mode II or shear loading increased. This increase in hackle number and orientation is most conspicuous for Mode II loading greater than 43% (see Figs. 2, 3, and 4). At near pure Mode II loading, the hackles are observed to be nearly perpendicular to the plane of the plies, and their edges sometimes appeared drawn and "tuffed." For the most part, the orientation or slant of the hackles remains constant on a given fracture surface (that is, for a given ratio of Mode I to Mode II loading). They appear to slant towards the direction of crack

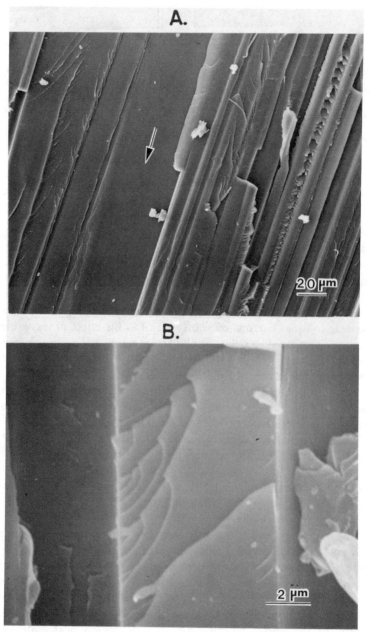

FIG. 2—*Fractographs of AS4/3501-6. Note, as the percentage of Mode II loading is increased, both the number of hackles and their angle with respect to the fracture surface is increased. Arrows indicate the direction of crack growth. (A) pure Mode I loading; (B) 12% Mode II loading; (C) 43% Mode II loading; (D) and (E) pure Mode II loading.*

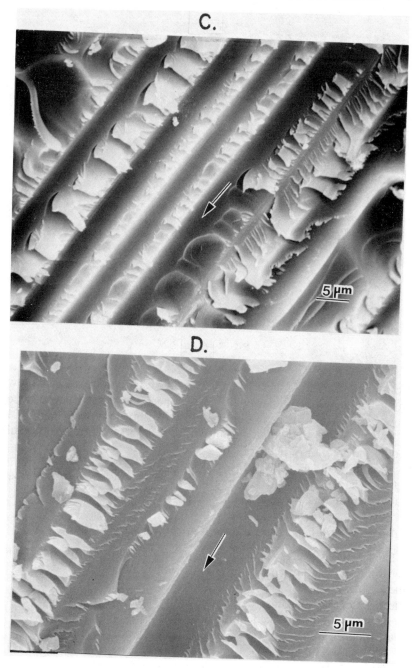

FIG. 2—*Continued.*

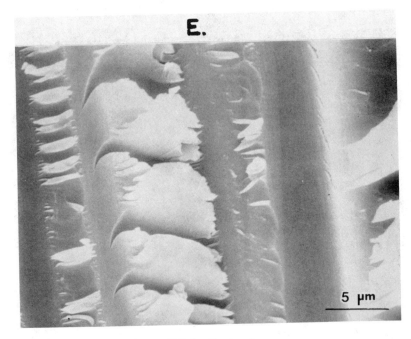

FIG. 2—*Continued.*

growth on one fracture surface and in the direction opposite of crack growth on the matching fracture face (Fig. 5). Comparisons of the Mode I fractography to the Mode II fractography for the two brittle systems AS4/3501-6 and AS4/Dow-P6 show that the amount of resin/fiber debonding as exhibited by the amount of resin adhering or coating the fibers does not seem to be affected by the increase in Mode II loading (Fig. 6). This would seem to indicate that interfacial debonding for these systems is insensitive to the loading conditions.

Fractographic Results: Ductile Resin System (AS4/Dow-Q6)

The fracture surfaces of the ductile system AS4/Dow-Q6 were characterized by resin fracture and deformation, with resin completely coating the fibers, indicating good fiber/resin adhesion. Fractography revealed no distinct hackles (Fig. 7). Only at near pure Mode II loading conditions were any even ill-defined, hackle-like features seen. These features are chaotic with no clearly defined orientation.

Real-Time Delamination Fracture Observations

Mode I Loading—In-situ observations in the SEM of fracture of the composite systems are seen in Fig. 8. For Mode I loading of the brittle

AS4/3501-6, the primary crack is seen to proceed by interfacial debonding. A limited amount of microcracking is seen to occur in the resin behind the primary crack tip and is normally associated with fiber bridging and eventual pullout.

The mechanisms for crack advancement in the delamination fracture of AS4/Dow-P7 were seen to be both resin/fiber debonding and resin deformation and microcracking. In the Mode I fracture of the ductile AS4/Dow-Q6 system, crack propagation occurs primarily by resin deformation and fracture, with only occasional interfacial debonding. Considerable resin deformation and microcracking is seen in the regions outside the resin-rich area between the plies.

Mode II Loading—In-situ Mode II delamination observations were made for the AS4/3501-6 and AS4/Dow-P7 systems. In these systems a series of sigmoidal-shaped microcracks are seen to develop in the resin-rich region between the plies well ahead of the crack tip (see Fig. 9). The primary crack extension occurs by the growth and coalescence of these microcracks. Normally this coalescence occurs near the fiber/resin interface at the upper or lower extent of the resin-rich region between plies.

Discussion

Fracture Surface Characteristics of Moderately Ductile and Brittle Resin Systems and the Formation of Hackles

In Mode I failure, the in-situ observations indicate that the primary micromechanism of delamination is by interfacial debonding for the AS4/3501-6 and AS4/Dow-P6 systems. Because the delamination crack takes the path of least resistance, which in these systems was along the resin/fiber interface, resin fracture in the regions between fibers was only observed occasionally. Significant resin deformation or microcracking leading to hackle formation was limited to the few isolated areas where good resin adhesion and/or fiber pullout occurred. Thus, only a relatively few, shallow hackles are seen in the Mode I fractography. The composite fracture toughness of the brittle AS4/3501-6 system was seen to be twice that of the neat resin material. This may be explained [9] in terms of a greater fracture area due to the corrugated roof fracture surface in the composite as compared to a mirror smooth fracture surface in the neat resin. This greater fracture area, along with fiber bridging and subsequent fiber breakage, apparently compensates for the premature failure by resin debonding. However, the AS4/Dow-P6 system, whose neat resin has twice the toughness of Hercules 3501-6, had a very similar delamination toughness. Assuming the interfacial bonding in each system is similar, this might be expected, since the failure mechanism for the AS4/Dow-P6 system was also principally resin/fiber debonding. Therefore, fiber bridging and breakage along with increased

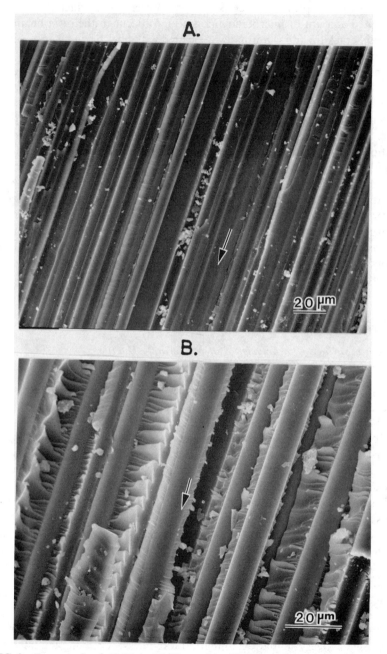

FIG. 3—*Fractographs of AS4/Dow-P6. The number of hackles and their angle with respect to the fracture surface increase as the percentage of Mode II loading is increased. Under pure Mode II loading conditions the hackles appear nearly vertical. Arrows indicate the direction of crack growth.* (A) *pure Mode I loading;* (B) *43% Mode II loading;* (C) *and* (D) *pure Mode II loading.*

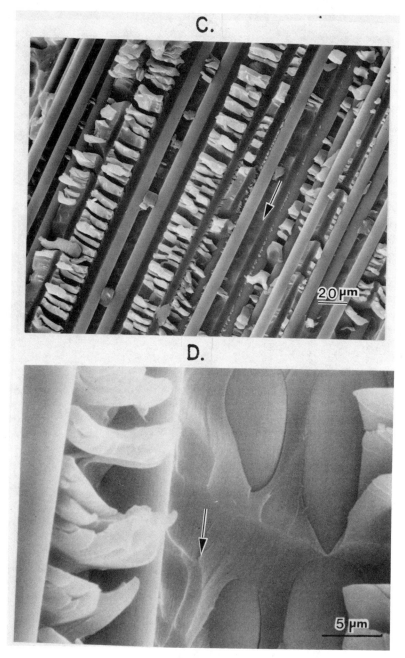

FIG. 3—*Continued.*

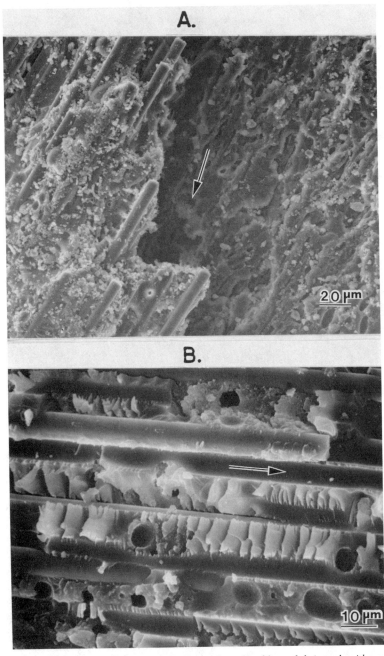

FIG. 4—*Fractographs of AS4/Dow-P7. The number of hackles and their angle with respect to the fracture surface increase as the percentage of Mode II loading is increased. Under pure Mode II loading conditions the hackles appear nearly vertical. The "pock marked" surface may be the result of trapped volatiles during the curing of the laminate panel. Arrows indicate the direction of crack growth. (A) pure Mode I loading—large sheets of fibers and resin are seen to be pulled out; (B) 43% Mode II loading; (C) and (D) pure Mode II loading.*

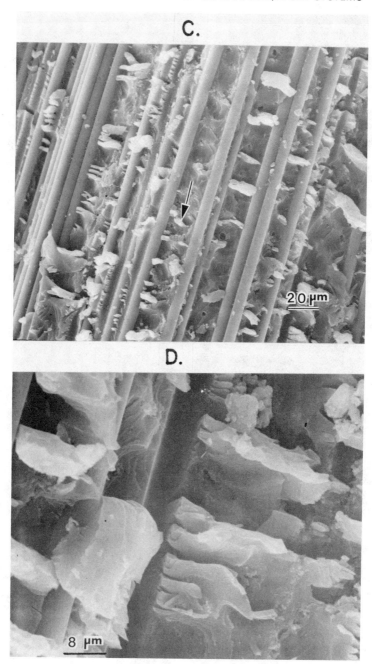

FIG. 4—*Continued.*

FIG. 5—*Hackles slant in the direction of crack growth on the upper fracture surface and in the opposite direction on the lower fracture surface. Arrows indicate the direction of crack growth.*

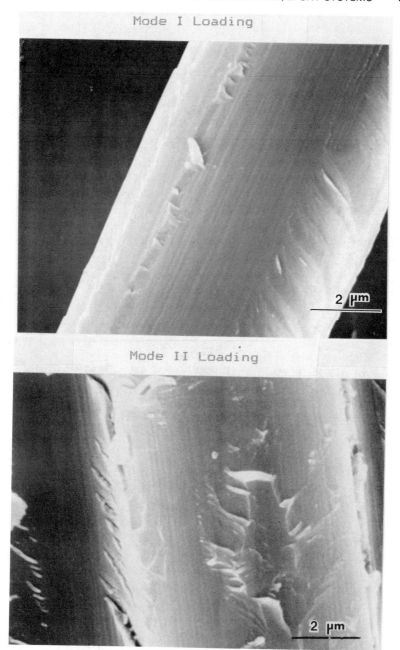

FIG. 6—*Clean fibers indicating fiber/matrix debonding in both Mode I and Mode II loading conditions for the ASA/3501-6 system.*

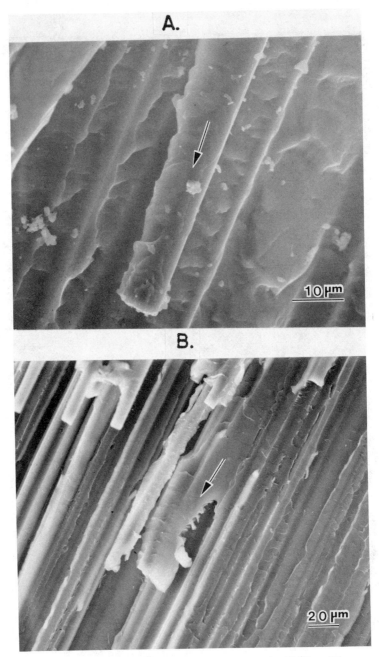

FIG. 7—*Fractographs of delamination fracture of the ductile AS4/Dow-Q6 system. In the AS4/Dow-Q6 system the fibers are completely coated with resin, indicating good resin/fiber adhesion. Note that in this ductile system distinct hackles are not observed even at pure Mode II loading conditions. Arrows indicate the direction of crack growth. (A) Mode I loading; (B) 43% Mode II loading; (C) and (D) pure Mode II loading.*

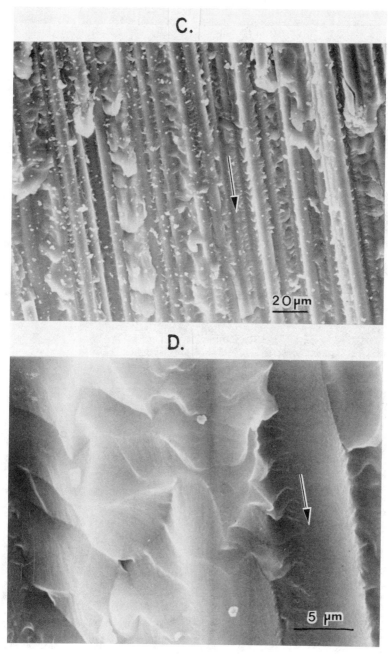

FIG. 7—*Continued.*

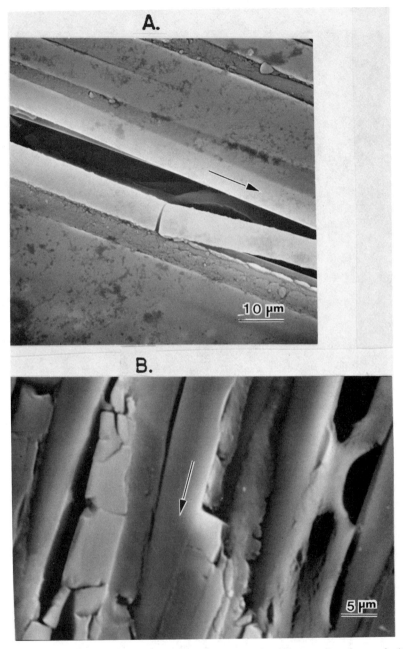

FIG. 8—*Mode I in-situ delamination. Arrows indicate the direction of crack growth.* (A)
AS4/35061-6 failure is seen to be primarily by debonding at the fiber/matrix interface; (B) *in the
Mode I delamination failure of the AS4/Dow-Q6 system, extensive resin yielding is observed;* (C)
and (D) *in the AS4/Dow-P7, system, Mode I delamination failure occurs by a combination of
fiber/resin debonding and resin yielding. Partial resin coating of the fibers indicates improved
resin/fiber adhesion.*

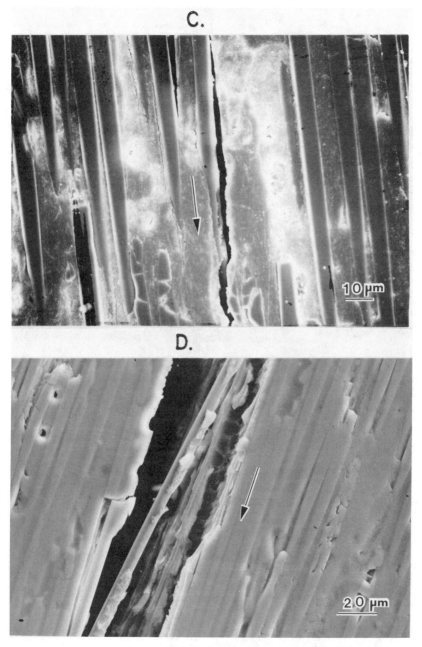

FIG. 8—*Continued.*

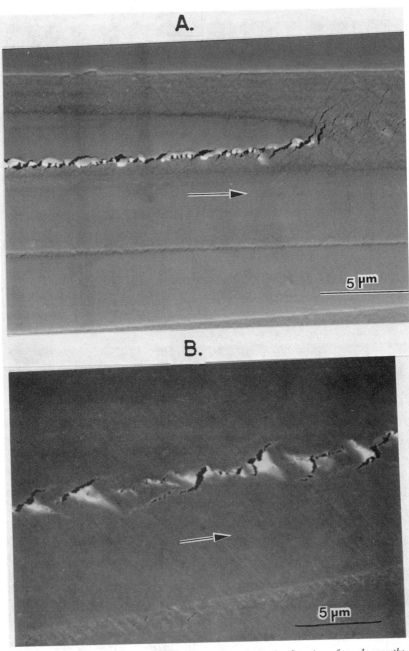

FIG. 9—*Mode II in-situ delamination. Arrows indicate the direction of crack growth:* (A) *development of the sigmoidal microcracks in the resin-rich region between fibers in front of the crack tip in the AS4/3501-6 system;* (B) *sigmoidal microcracks in the AS4/Dow-P7 system under pure Mode II loading conditions. Note: the fiber/resin interfaces are not clearly distinguishable in this fractograph;* (C) *and* (D) *hackles resulting from the coalescence of the microcracks developed in Mode II loading of the AS4/3501-6 system.*

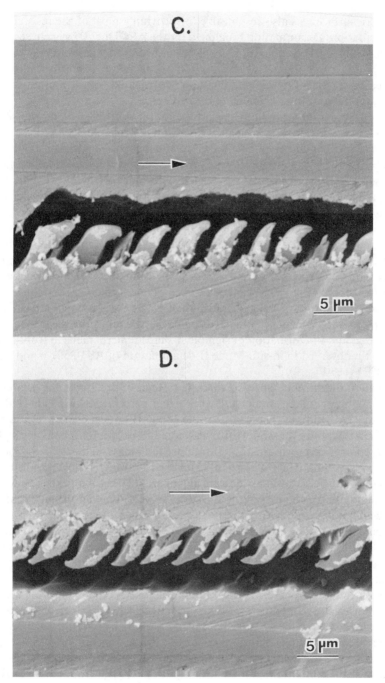

FIG. 9—*Continued.*

fracture surface was also responsible for providing most of the resistance to delamination. Delamination toughness for the AS4/Dow-P7 system seems to be derived from a combination of the better resin toughness, a better extraction of this resin toughness resulting from the increased resin/fiber adhesion, increased fracture surface area due to a "corrugated roof" fracture surface topography, and fiber bridging and breakage.

The fractography for both the brittle system AS4/3501-6 and the somewhat more ductile resin systems, AS4/Dow-P6 and AS4/Dow-P7 clearly shows that the number of hackles as well as their orientation or angle with respect to the ply plane increase as the percentage of shear loading is increased (see Figs. 2, 3, and 4). As the percentage of Mode II loading is increased, the angle of the principal normal stress monotonically increases from being parallel to the ply plane in Mode I to 45° to the ply plane for pure Mode II loading. The hackles, which form only in relatively brittle systems, are apparently the result of microcrack nucleation in advance of the crack tip and the subsequent propagation in a brittle fashion through the resin-rich region on the principal normal stress plane until they are stopped by the graphite fibers (Fig. 10). The coalescence of these microcracks constitutes growth of the primary crack. Confirmation of this explanation is seen in both the fractographic evidence of increasing hackle angles as the shear loading is increased and the formation of sigmoidal-shaped microcracks in front of the crack tip observed in in-situ Mode II delamination of AS4/3501-6 and AS4/Dow-P7 (see Fig. 9).

In all of the graphite/epoxy composite systems of this study, the delamination toughness is seen to increase with higher percentages of Mode II shear loading (see Table 1). The value for the Mode II critical energy release rate found in this study for AS4/3301-6 was noted to be two to three times higher than reported in previous Mode II fracture toughness studies of AS4/3501-6 [8,10]. No comparisons were possible for the four other systems studied, but the large G_{IIc}/G_{Ic} ratios may indicate an apparent larger than expected Mode II delamination toughness in these systems as well. Whether the higher Mode II delamination toughness values reflect differences due to such variables as processing or if they are an artifact of the experimental or analytical procedure is still unclear. However, the general trends between mode of loading and delamination toughness indicated by Table 1 are consistent with other published data and will be used to describe qualitatively the relationship between failure mechanism and fracture toughness. The most dramatic increase in fracture toughness as the percentage of Mode II loading was increased was seen in the systems that failed primarily by interfacial debonding under Mode I loading conditions. For these brittle systems, AS4/3501-6 and AS4/Dow-P6, the increase in delamination toughness appears to be associated with the formation of the hackles and the change from continuous to discontinuous crack growth it represents. Both the increase in fracture surface area generated through the formation of these hackles and the more

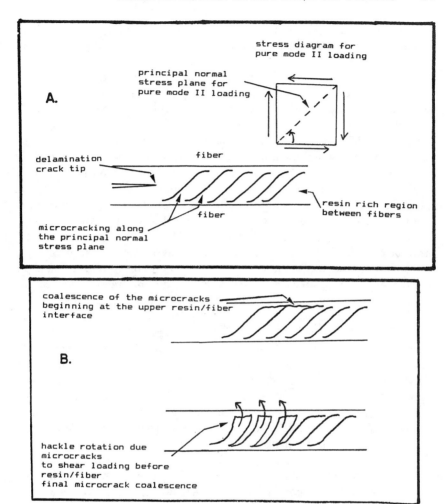

FIG. 10—(A) *The principal normal stresses that developed ahead of a crack tip in the resin-rich region between plies under mixed mode loading conditions. (B) The development of sigmoidal microcracks along the principal normal stress plane and the rotation of the hackles due to the shear loading before final coalescence of the microcracks. The direction in which the hackles slant depends on whether the microcracks coalesce at the upper or lower boundary.*

tortuous path of crack growth (sigmoidal-shaped microcracks impingement on fibers, etc.) would suggest greater resistance to crack growth for increasing Mode II loading. When interfacial adhesion is improved, as in the AS4/ Dow-P7 system the resistance to crack growth in Mode I loading is improved by resin microcracking and deformation. Therefore, the effective increase in delamination toughness under high Mode II loadings that result from the development of regular microcracks and the formation hackles is greatly reduced. It is also worth noting that the more ductile AS4/Dow-Q6 system had no hackle formation and therefore a G_{IIc}/G_{Ic} ratio which was much lower than for the more brittle systems that failed by debonding under Mode I conditions (Table 1).

While the idea that Mode II or mixed mode loading of relatively brittle resin composites gives microcracking on the principal normal stress plane and that subsequent coalescence gives rise to hackle formation seems qualitatively correct, there must be more to the story. The postmortem and in-situ results for the pure Mode II delamination fracture (Mode II fractography in Figs. 2, 3, and 4) clearly indicate a final hackle orientation much greater than the expected 45° to the plane of the ply one would predict based on a simple stress analysis. The more nearly vertical overall orientation of the hackles for pure Mode II loading could result from hackle rotation due to shear loading just prior to microcrack coalescence (see Fig. 10). This rotation would not only orient the microcrack more vertically but could also open up the microcrack near the boundary where coalescence occurs. This opening of the microcracks gives the impression of material lost, which might correspond to matching hackles on the opposite surface. However, the mating fracture surfaces are found to be flat, dish-shaped regions which correspond to microcrack coalescence on that surface.

The Mode II in-situ results of the AS4/3501-6 (Fig. 9) clearly show that the slant direction of the hackles depends on whether the coalescence of the microcracks occurs at the upper or lower boundary. Since the number of hackles pointing in the direction of crack growth on one fracture surface appears to be the same as the number of hackles pointing in the opposite direction on the other fracture surface, no preference as to whether coalescence takes place at the top or bottom of these microcracks is indicated. Once the microcrack coalescence begins on either the upper or lower boundary of the resin-rich region between plies, it will continue on that same boundary. The stress redistribution that accompanies the coalescence of the first two microcracks will favor continued coalescence on this same boundary.

Bascom [11] has recently indicated that resin flow or yielding may be an important mechanism in the formation of hackles. The amount of resin flow in the development of hackles in the AS4/3501-6 and AS4/Dow-P7 under 43% Mode II loading was evaluated by noting changes in the fracture surface details before and after annealing the fractured specimen 15°C above the resin Tg for 4 h to see the extent of recovery (Fig. 11). No visible recovery in terms of change of shape or size of the hackles in these systems was observed, indicating that resin flow is not an important mechanism in the formation of the hackles. However, careful examination suggests that there may have been some change of hackle orientation. After annealing, the angle with respect to the ply plane seems to have decreased very slightly, supporting the idea that rotation of the hackles, which would increase the hackle inclination, may take place due to the Mode II component of loading on the upper and lower boundaries of the sigmoidally microcracked region before final coalescence occurs. One might expect the greatest amount of hackle rotation to occur under pure Mode II loading conditions, and, therefore, a significant change in the hackle inclination with respect to the ply plane after annealing would be observed. At the time of this writing, test to evaluate the effects of annealing on the Mode II fracture specimens have not yet been conducted.

Implications for Failure Analysis

The just-mentioned observations have important application to failure analysis. First, a very hackled surface on a unidirectional laminate clearly indicates a relatively brittle resin with significant Mode II loading. The more nearly vertical the hackles, the greater was the Mode II component of the service load which caused the failure. Delamination of multiaxial composites such as [0, +45, 90] would be expected to show general hackling even for nominally macroscopic Mode I loading due to the development of interlaminar shear stresses through the differential Poisson's contraction.

As mentioned before, some authors have suggested that hackles may be useful in determining the direction of crack growth. This study has clearly shown that this is only true for one of the fracture surfaces. On the other fracture face, the hackles point away from the crack growth direction. There does not seem to be any direct way to infer anything about crack growth direction from the hackle orientation. However, a more careful examination of the detailed river patterns on the individual hackles may allow crack growth direction to be determined. Individual river patterns may be misleading if interpreted to be more than the local microcrack growth direction, but the resolved direction of many such river marks should point in the direction of crack extension. We found these river patterns more distinct and easy to map on relatively flat hackles (that is, for relatively low fraction of Mode II loading).

Fracture Characteristics of Ductile Systems

The large amount of resin deformation and the lack of any distinct hackle formation indicates that the processes of fracture in the ductile AS4/Dow-Q6 system are different from those in the more brittle systems. This is confirmed by the in-situ Mode I delamination observations (Fig. 8) that show that failure is by resin yielding and deformation. The good fiber/resin adhesion allowed the shear loads to be effectively transferred to the resin and results in a deformation and damage zone of up to 8 or 10 fiber diameters across. The Mode II fractography of the AS4/Dow-Q6 system showed an increase in resin deformation compared to that seen in Mode I failure, with little or no hackle formation. Because there was no significant change in failure mechanism between Mode I and Mode II delamination (see Fig. 7), only a

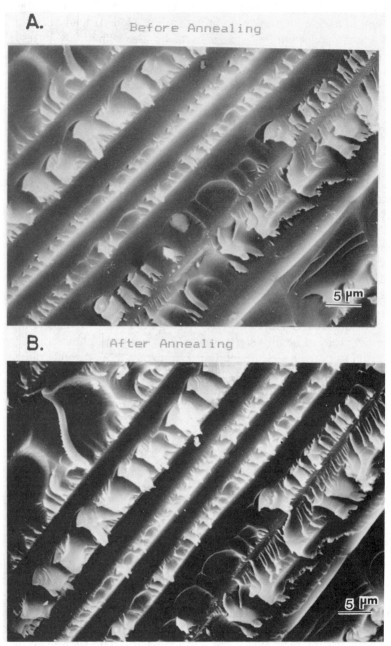

FIG. 11—*The fracture surface before and after annealing 15°C above the resin Tg. There is no observable change in size or shape of the hackles, indicating that resin flow is not an important mechanism in hackle formation. (A) and (B) AS4/3501-6 system—The angle of the hackles with respect to the fraction surface may have decreased slightly after annealing, suggesting that rotation of the hackles due to the shear loading occurred before final coalescence of the microcracks. (C) and (D) AS4/Dow-P7.*

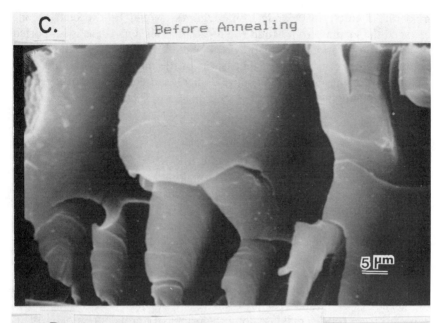

C. Before Annealing

5 µm

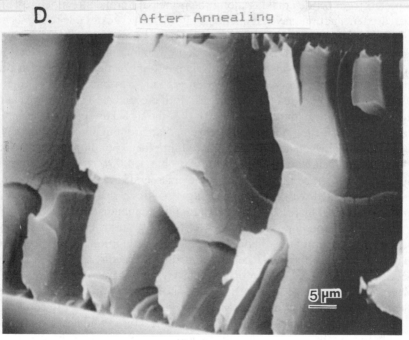

D. After Annealing

5 µm

FIG. 11—*Continued.*

moderate increase in delamination toughness would be expected as one changed the loading from Mode I to Mode II. This hypothesis is supported by a G_{IIc} to G_{Ic} ratio of 3.4, as previously noted. Hackles are not developed because of the resin ductility (that is, no brittle microcracks) failure from yielding occurring in lieu of the development and coalescence of a regular system of microcracks.

Summary

The results of this study show that in brittle AS4/3501-6 (resin $G_{Ic} = 0.07$ kJ/m^2), the slightly more ductile AS4/Dow-P6 (resin $G_{Ic} = 0.14$ kJ/m^2) and AS4/Dow-P7 (resin $G_{Ic} = 0.32$ kJ/m^2) composite systems, the number of hackles and their angle with respect to the ply plane monotonically increase with the percentage of Mode II loading. In-situ delamination observations under Mode II loading show that these hackles develop from the coalescence of sigmoidal-shaped microcracks that form in a brittle fashion on the principal normal stress plane in the resin-rich region ahead of the crack tip. However, under high percentage Mode II loading conditions, the final orientation of the hackles with respect to the ply plane is often much greater than the 45° expected from brittle cracking along the principal normal stress plane. This increased inclination may result from rotation of the hackles due to the shear loading before final coalescence of the microcracks occurs. In the AS4/3501-6 and AS4/Dow-P6 systems where Mode I delamination was dominated by interfacial debonding, the increase in delamination fracture toughness as the loading conditions are changed from pure Mode I to pure Mode II corresponds to the formation of these hackles. Both the increased surface area and the more tortuous path of crack growth that these hackles represent seem to be responsible for the observed increase in fracture toughness. In the AS4/Dow-P7 system where there was enhanced interfacial bonding, at least some of the Mode I delamination toughness was obtained through resin deformation and microcracking. Therefore, the microcracking and formation of hackles associated with higher precentages of Mode II loading produced a much less drastic increase in the delamination toughness.

In the tougher composite system AS4/Dow-Q6 (resin $G_{Ic} = 0.73$ kJ/m^2), hackles do not form under high Mode II loading conditions. This is because in this ductile resin system, yielding and ductile fracture occur before the development and coalescence of a regular system of microcracks can take place. Since no change in failure mechanism (that is, no formation of hackles) takes place as the loading conditions change from Mode I to Mode II, a less substantial change in the delamination fracture toughness occurs.

In failure analysis, hackles can be used to identify a state of shear loading at the crack front during failure. However, care must be taken in using hackles to identify the macroscopic state of the applied loading, since in multidirectional laminates, shear loading at the crack tip can result from differential Poisson's contractions where nominally pure Mode I loading conditions exist macroscopically. Since there is no preference on whether the coalescence occurs at the upper or lower boundary of the microcracks, the direction of crack growth cannot be determined from the direction of hackle slant alone.

References

[1] Hunston, D. L. and Bullman, G. W., "Characterization of Interlaminar Crack Growth in Composites: Double Cantilever Beam Studies," in *1985 Grant and Contract Review*, Vol. II, NASA Langley Research Center Materials Division, Fatigue and Fracture Branch, 13–14 Feb. 1985.

[2] Chakachery, E. A. and Bradley, W. L., "A Comparison of the Crack Tip Damage Zone for Fracture of Hexcel F185 Neat Resin and T6T145/F185 Composite," presented at the ACS International Symposium on Non-Linear Deformation, Fatigue and Fracture of Polymeric Materials, Chicago, IL, 8–13 Sept. 1985, American Chemical Society, Washington, DC.

[3] Morris, G. E., "Determining Fracture Directions and Fracture Origins on Failed Graphite/Epoxy Surfaces," in *Nondestructive Evaluation and Flaw Criticality for Composite Materials, ASTM STP 696*, American Society for Testing and Materials, Philadelphia 1979, pp. 274–297.

[4] Tse, M. K., Hibbs, M. F., and Bradley, W. L., "Interlaminar Fracture Toughness and Real-Time Fracture Mechanism of Some Toughened Graphite/Epoxy Deposits," presented at the ASTM Symposium on Toughened Composites, Houston, TX, 13–15 March 1985, in *Toughened Composites, STP 937*, American Society for Testing and Materials, Philadelphia, 1987.

[5] Devitt, D. F., Schapery, R. A., and Bradley, W. L., "A Method for Determining the Mode I Delamination Fracture Toughness of Elastic and Viscoelastic Composite Materials," *Journal of Composite Materials*, Vol. 14, Oct. 1980, pp. 270–285.

[6] Whitney, J. M., Browning, C. E., and Hoogsteden, W., "A Double Cantilever Beam Test for Characterizing Mode I Delamination of Composite Materials," *Journal of Reinforced Plastics and Composites*, Vol. 1, Oct. 1982, pp. 297–313

[7] Wilkins, D. J., Eisenmann, J. R., Cumin, R. A., and Margolis, W. S., "Characterizing Delamination Growth in Graphite/Epoxy," in *Damage in Composite Materials, ASTM STP 775*, American Society for Testing and Materials, Philadelphia, 1982, pp. 168–183.

[8] Russell, A. J. and Street, K. N., "Moisture and Temperature Effects on the Mixed-Mode Delamination Fracture of Unidirectional Graphite Epoxy," in *Delamination and Debonding of Materials*, American Society for Testing and Materials, Philadelphia, 1985.

[9] Bradley, W. L. and Cohen, R. N., "Matrix Deformation and Fracture in Graphite Reinforced Epoxies," in *Delamination and Debonding of Materials, ASTM STP 876*, American Society for Testing and Materials, Philadelphia, 1985.

[10] Jordan, W. M., "The Effect of Resin Toughness on the Delamination Fracture Behavior of Graphite/Epoxy Composites," Ph.D. dissertation, Texas A&M University, College Station, TX, 1985.

[11] Bascom, W. D., "Fractographic Analysis of Interlaminar Fracture," presented at the ASTM Symposium on Toughened Composites, Houston, TX, 13–15 March 1985, in *Toughened Composites, STP 937*, American Society for Testing and Materials, Philadelphia, 1987.

Composites II: Structures

Carol A. Ginty[1] and Christos C. Chamis[2]

Fracture Characteristics of Angleplied Laminates Fabricated from Overaged Graphite/Epoxy Prepreg

REFERENCE: Ginty, C. A. and Chamis, C. C., "**Fracture Characteristics of Angleplied Laminates Fabricated from Overaged Graphite/Epoxy Prepreg,**" _Fractography of Modern Engineering Materials: Composites and Metals, ASTM STP 948_, J. E. Masters and J. J. Au, Eds., American Society for Testing and Materials, Philadelphia, 1987, pp. 101–130.

ABSTRACT: A series of angleplied graphite/epoxy laminates was fabricated from overaged prepreg and tested in tension to investigate the effects of overaged or "advanced-cure" material on the degradation of laminate strength. Results which include fracture stresses indicate a severe degradation in strength. In addition, the fracture surfaces and microstructural characteristics are distinctly unlike any features observed in previous tests of this prepreg and laminate configuration. Photographs of the surface and microstructures reveal flat morphologies consisting of alternate rows of fibers and hackles. These fracture surface characteristics are independent of the laminate configurations. The photomicrographs are presented and compared with data from similar studies to show the unique characteristics produced by the overaged prepreg. Analytical studies produced results which agreed with those from the experimental investigations.

KEYWORDS: overaged graphite/epoxy prepreg, advanced-cure, angleplied laminates, fracture surface characteristics, strength degradation

Graphite epoxy composites are commonly used in the design and manufacture of components for both aeronautical and space structures. Understanding the fracture modes in these graphite-epoxy composites is very important since many of these structures are primary load-bearing components in the design.

[1]Aerospace engineer, National Aeronautics and Space Administration (NASA), Lewis Research Center, Cleveland, OH 44135.
[2]Senior research engineer, Aerospace Structures/Composites, NASA, Lewis Research Center, Cleveland, OH 44135.

NASA Lewis Research Center has developed an integrated program, Composite Fracture Characterization, that incorporates analytical, experimental, and postfailure analysis techniques to study the progressive fracture of fiber composites. Various factors contribute to the damage in fiber composites and therefore affect the progressive fracture. Factors such as laminate thickness, ply configuration, notch sensitivity, and notch-type sensitivity have been included in the program. Early studies on progressive fracture concentrated on a series of four-ply, angleplied $[\pm\theta]_s$ laminates. Both unnotched and notched (slits and holes) laminates have been tested in uniaxial tension to fracture. In conjunction with the experimental investigation, analytical methods incorporated into in-house computer codes were used to predict progressive damage. The integrated approach used in this program as well as the experimental and analytical results are shown and discussed in Ref 1. More detailed results generated from those studies including fracture surface characteristics of notched angleplied laminates [2] and related fracture modes in notched angleplied laminates [3] have been published. Significant results from these early studies show that: (1) unique microstructural characteristics exist for each ply angle orientation; (2) microstructural characteristics can be used as criteria by which to determine accompanying fracture modes; and (3) the predominant fracture mode and thereby the fracture surface characteristics are a function of the ply angle orientation.

All laminates for the work reported herein were fabricated from Fiberite 1034E prepreg (934 resin impregnated with Thornel 300 graphite fibers), which is the same material used previously [1,3]. The material was delivered and stored on a roll at $-17.8°C$ ($0°F$) as recommended by the manufacturer's data sheet to insure a six-month–rated shelf life. The specimen's fabrication and test dates are randomly selected due to the dynamic nature of the composites program. Thus, experience has demonstrated that this prepreg usually meets or surpasses the material properties provided by the manufacturer even after its rated shelf life has been exceeded. Fracture strengths published in Refs 1 to 3 do not reflect any degradation, and, likewise, similar fracture surface characteristics were observed from six-month-old prepreg, one-year-old prepreg, and even two-year-old prepreg.

Recently, there has been considerable interest in postmortem analysis of in-service failures of composite structures based upon their fracture surface characteristics. It is also known that in-service composite structures may undergo the phenomenon of "advanced-cure," whereby the properties of the resin change due to a chemical reaction in an uncontrolled environment. This phenomenon of "advanced-cure" can be initiated by several factors: the thermal and hygral environments as well as the state or condition of the prepreg itself.

While information is limited on the "advanced-cure" of composites and possible subsequent degradation, Clements, in Ref 4, discovered that in a 0° laminate moisture was found to produce an apparent weakening of the interfacial bond. In addition, she reported that when both temperature and moisture content are increased, the interfacial bond is considerably weakened and the epoxy becomes more brittle, breaking up and falling away from the filaments. A study on the failure morphologies of these "elevated temperature wet" specimens suggests a drop in strength.

Clements and Adamson also reported [5] that a reduction in the glass transition temperature might lead to physical aging of an epoxy, which would produce epoxy embrittlement. The graphite-epoxy prepreg on hand, having been stored at $-17.8°C$ ($0°F$), had physically aged by the passage of time (approximately 3.5 years) and will now be referred to throughout as overaged prepreg. Since overaged prepreg is one form of "advanced cure," specimens were fabricated and tested from the available prepreg to (1) determine possible fracture stress degradation and (2) examine the attendant fracture surface characteristics.

The overaged prepreg specimens were fabricated in a $[\pm\theta_2]_s$ configuration to permit direct comparison of the fracture stresses and the fracture surface characteristics associated with the overaged prepreg to those from the earlier studies [1,3], which involved the same batch of prepreg at a younger age. Results of the comparison would immediately indicate any degradation in strength and the effect of the "advanced-cure" on the microstructure. In addition, specimens were also fabricated from a newly purchased (fresh) batch of prepreg, and five replicates of each configuration ($[\pm\theta_2]_s$) were tested. The tensile strength results from these fresh prepreg laminates were also compared to those from the overaged prepreg to determine the severity of degradation in laminate strength for this particular configuration.

Results including fracture stresses, physical fracture characteristics, and microstructural features are presented and compared for: (1) laminates fabricated from a good batch of prepreg in previous studies; (2) laminates fabricated from the same batch of prepreg in an overaged state; and (3) laminates fabricated from a fresh batch of prepreg.

Characteristics of $[\pm\theta]_s$ Laminates

The effect of ply orientation on the progressive fracture of $[\pm\theta]_s$ laminates constituted one phase of the Composite Fracture Characterization program just mentioned. A series of four-ply laminates was fabricated from graphite/epoxy prepreg with $\theta = 0°$, $3°$, $5°$, $10°$, $15°$, $30°$, $45°$, $60°$, $75°$, and $90°$, where each ply is 0.127 mm (0.005 in.) thick. (The same fabrication process was used for all of the laminates discussed. The curing cycle consists of: (1) heating the molded laminate to $176.67°C$ ($350°F$) at a $-15°C$ ($5°F$)/min rate; (2) applying a pressure of 0.689 MPa (100 psi) for 2 h; (3) cooling the laminate under pressure to about $65.56°C$ ($150°F$); and (4) removing the laminate from the molds.) From the panels, 50.8 by 457.2-mm (2 by 18-in.) specimens were cut using a diamond-tipped cutting wheel. For each ply configuration, three specimens were tested in tension. One specimen was unnotched, and the other two contained through-the-thickness notches to investigate composite notch sensitivity. To determine the effect, if any, of notch type, both a slit [6.35 by 1.27 mm (0.25 by 0.05 in.) and hole [6.35 mm (0.25 in.) radius] were utilized.

Four angleplied configurations were selected for comparison and discussion: $[0]_4$, $[\pm 15]_s$, $[\pm 30]_s$, and $[\pm 45]_s$. Each laminate's performance is described in terms of fracture stress, physical fracture characteristics, and microstructural features. The fracture stresses are tabulated. The first photograph in Figs. 1–4 depicts the physical features of the fractured specimens (unnotched, notch/slit, and notch/hole, respectively). The microstructural features are characterized by a set of photomicrographs from *one* of these specimens. The information gathered from these laminates, which were fabricated from a good batch of prepreg, will serve as a baseline for comparisons to be made later.

The fracture stresses shown in Table 1 for these select laminates were obtained by incrementally loading the specimens in uniaxial tension to fracture. All photomicrographs presented herein document the microscopic investigation conducted with an Amray 1200 scanning electron microscope (SEM). All SEM specimens were coated with a 200-Å-thick gold film to enhance the conductivity of the specimen, thereby improving transmission of the SEM.

The fracture characteristics of the $[0]_4$ laminate are shown in Fig. 1. The laminate experienced a catastrophic fracture, leaving very little of the specimen intact. The dominant microstructural characteristic on the fracture surface, represented by the unnotched specimen, is the tiered fibers resulting from fibers breaking at various locations along their lengths. The photomicrographs show that the tiered fiber surfaces are clean and free of matrix residue, indicative of debonding between the fiber and matrix.

Figure 2 displays the fracture surfaces of the $[\pm 15]_s$ laminates. The laminate fracture occurred across the widths of the specimens. For the notched specimens, the fracture progressed through the notch itself. Analysis of the photomicrographs of the notch/slit specimen revealed an irregular fracture morphology consisting of broken fibers, fiber pullout, and some matrix hackles.

TABLE 1—*Fracture stresses of the graphite/epoxy $[\pm \theta]_s$ laminates.*

Specimen Type	Fracture stress, ksi				Fracture stress, MPa			
	$[0]_4$	$[\pm 15]_s$	$[\pm 30]_s$	$[\pm 45]_s$	$[0]_4$	$[\pm 15]_s$	$[\pm 30]_s$	$[\pm 45]_s$
Unnotched	202	93	66	23	1393	641	455	159
Notch/slit	195	72	54	22	1344	496	372	152
Notch/hole	150	83	44	24	1034	572	303	165

NOTE: 1 ksi = 6.895 MPa.

The fractures of the [±30]$_s$ specimens also occur through the notches and across the widths, as shown in Fig. 3. The unnotched specimen fractured in both locations simultaneously at the recorded fracture stress. Photomicrographs of the unnotched specimen are shown to represent the microstructural characteristics for this particular laminate, that is, both the inner (−30) plies and outer (+30) plies exhibit fiber pullout and fiber breakage in different locations. The morphology for the [±30]$_s$ laminate is irregular and similar to that observed for the [±15]$_s$ laminate.

In Fig. 4, photomicrographs of the [±45]$_s$ notch/slit specimens are shown to typify the characteristics of this laminate, which fractured along a plane parallel to the fiber direction and through the notches. The morphology consists of a flat surface on the outer plies, which exhibited an extensive amount of hackles in contrast to the irregular surface on the inner plies, which experienced pulled-out fibers.

These fracture characteristics have been documented for the purpose of comparison and therefore have been depicted somewhat "generically." Details of these studies and resulting conclusions are published in Refs *1* to *3*.

Characteristics of [±θ$_2$]$_s$ Overaged Laminates

The series of laminates was fabricated in an identical fashion to the previous set. Briefly, the specimen dimensions were: 457.2 mm (18 in.) long by 50.8 mm (2 in.) wide; the thickness of each ply was 0.127 mm (0.005 in.); the values for θ were the same; and the notches were centered through the thickness and were the same size as those discussed in the previous section [6.35 by 1.27 mm (0.25 by 0.05 in.) for the slits and 6.35 mm (0.25 in.) radius for the hole]. The laminate configuration was altered to [±θ$_2$]$_s$, which merely doubled the number of plies and therefore doubled the laminate thickness. This parameter was changed so that the effect of laminate thickness on the fracture characteristics could be investigated.

Elementary analysis would indicate that the fracture loads of the [±θ$_2$]$_s$ laminates should be double in magnitude to those of the [±θ]$_s$ configuration and should exhibit similar fracture characteristics. However, aside from the laminate thickness, another important factor affecting the fracture is the age of the material, especially since these specimens were fabricated from 3.5-year-old prepreg. Even though experience has shown that this prepreg performs exceptionally well even after its shelf life has been exceeded, it was suspected (based upon limited previous information) that a degradation in strength associated with overaged prepreg would probably occur. Therefore this study was initiated to investigate the strength degradation of an "advanced-cure" prepreg where, in this case, the "advanced-cure" is the reuslt of old or overaged prepreg.

The specimens were tested in uniaxial tension by incrementally increasing loads until fracture occurred. The data for the overaged laminates, shown in

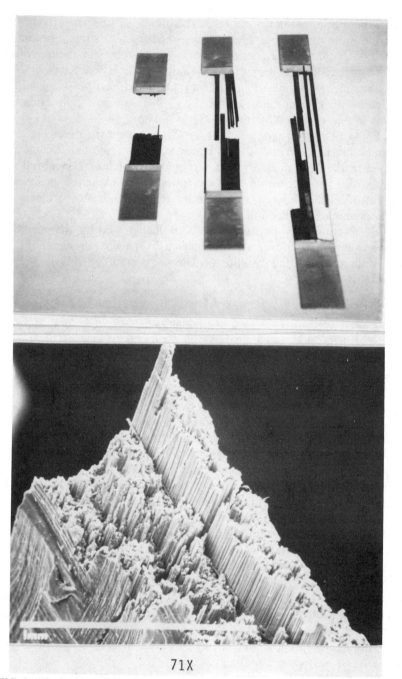

71X

FIG. 1—*The fractured specimens and typical microstructural characteristics for the* [0]₄ *laminate.*

FIG. 1—*Continued.*

FIG. 2—*The fractured specimens and typical microstructural characteristics for the* $[\pm 15]_s$
laminate.

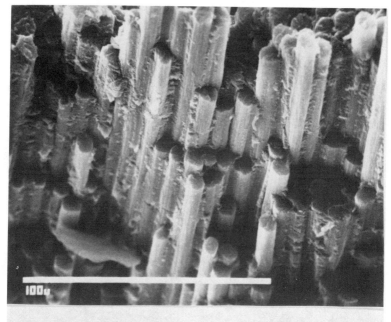

730X

1500X

FIG. 2—*Continued.*

130X

FIG. 3—*The fractured specimens and typical microstructural characteristics for the* [±30]ₛ *laminate.*

260X

290X

FIG. 3—*Continued.*

FIG. 4—*The fractured specimens and typical microstructural characteristics for the* $[\pm 45]_s$
laminate.

600X

1200X

FIG. 4—*Continued.*

Table 2, indicate several trends in the fracture characteristics associated with an overaged prepreg. First of all, since one would expect the fracture loads of the $[\pm\theta_2]_s$ laminates to be about twice those previously recorded for the $[\pm\theta]_s$ laminates, one would then expect to observe approximately the same fracture stresses for each laminate configuration, $[\pm\theta]_s$ and $[\pm\theta_2]_s$. Such is not the case in Table 2. In fact, with the exception of the notch hole/unidirectional laminates, all of the overaged laminates experience a degradation in strength when compared to results [1,3] shown in Table 1. The authors believe the experimental fracture loads used to calculate the fracture stresses for both the unidirectional $[0]_4$ and $[0]_8$ notch/hole laminates do not adequately represent unidirectional composite tensile behavior. Assuming, therefore, that the data points are bad and not to be considered in the comparison, then it can be stated that all of the laminates fabricated from the overaged prepreg degraded in tensile strength.

The second trend observed from the data is that as θ increases, which produces laminates which are matrix/resin dominant, the degradation in strength becomes even more severe. The overaged $[\pm15_2]_s$ laminates registered only 33% (on the average) of the expected fresh prepreg magnitudes while the overaged $[\pm30_2]_s$ and $[\pm45_2]_s$ laminates registered 27% (on the average). This is an indication that the resin and not the fiber is more sensitive to overaging as expected.

Having observed the severe strength degradation in terms of fracture stress, the fracture surfaces were examined using an SEM to document microstructural characteristics associated with overaged prepreg. The first photograph in Figs. 5–7 shows three fractured specimens depicting the physical features of this overaged material. With the exception of the $[\pm45_2]_s$ laminate (notch/slit), all photomicrographs in these figures are for the unnotched specimens. Figure 5 displays the fracture surface and microstructural characteristics for the $[0]_8$ laminate. The fracture resulted in splintery pieces. Under magnification these pieces exhibit an irregular morphology. The fibers are broken in bunches at various locations, forming a patchy pattern on the surface. Higher magnifications reveal extensive matrix hackles and debris.

The $[\pm15_2]_s$ specimens displayed in Fig. 6 fractured at an angle parallel to a fiber direction. For the notched specimen, the fracture split the specimens

TABLE 2—*Fracture stresses of the overaged graphite/epoxy $[\pm\theta_2]_s$ laminates.*

Specimen Type	Fracture stress, ksi				Fracture stress, MPa			
	$[0]_8$	$[\pm15_2]_s$	$[\pm30_2]_s$	$[\pm45_2]_s$	$[0]_8$	$[\pm15_2]_s$	$[\pm30_2]_s$	$[\pm45_2]_s$
Unnotched	163	43	20	⋯	1124	296	138	⋯
Notch/slit	169	25	15	6	1165	172	103	41
Notch/hole	198	14	9	6	1365	97	62	41

NOTE: 1ksi = 6.895 MPa.

into two equal halves. Photomicrographs reveal a flat morphology. Even at lower magnifications, an extensive amount of matrix debris is visible. At higher magnifications, an alternating pattern of rows of fibers and hackles exists. Very few broken fibers were observed. In the presence of a broken fiber, the hackles were larger in size than others in the same view.

Basically, similar fracture characteristics were observed for the $[\pm 30_2]_s$ laminate shown in Fig. 7. Once again the specimen fractured along a line parallel to a fiber direction. For those specimens containing notches, fracture progressed through the center of the machined notches. The morphology of the fracture surface is flat, consisting of rows of fibers and hackles. All hackles in a row appear to be the same size; however, the size of the hackles varies from row to row. Very little fiber fracture was observed on this surface.

In Fig. 8, the characteristics associated with the fracture of the $[\pm 45_2]_s$ laminate are displayed. Again, the specimens fracture in the same manner previously observed. The angle of fracture coincides with the 45° orientation of the fibers. The fracture surface is flat and littered with matrix debris. The fibers and hackles alternate in rows where the sizes of the hackles are varied throughout each row.

The fracture characteristics for the angleplied $[\pm \theta_2]_s$ laminates fabricated from the overaged prepreg are similar regardless of the ply configuration. Since this laminate-independency was not observed before, it is believed that this behavior is due to the material itself. Characteristics such as the extensive amount of matrix debris along with the numerous hackles are indicative of matrix shear fracture. The clean exposed fibers show that debonding has occurred from the resin. The lack of fiber fracture indicates that the loads being applied prior to fracture were relatively low. The characteristics just described will accompany an overaged/"advanced-cure" prepreg. Since the prepreg is decomposing chemically, the ability to transfer sufficient shear stress through the composite laminate is diminishing.

Fracture Characteristics of $[\pm \theta_2]_s$ Fresh Laminates

In documenting the characteristics associated with overaged prepreg, a comparison of $[\pm \theta_2]_s$ characteristics was made to a series of $[\pm \theta]_s$ laminates, which was fabricated from a controlled fresh batch of graphite/epoxy prepreg. As previously mentioned, the age of the material was not the only distinguishable difference, since one set of laminates was twice as thick as the other.

To determine if the laminate thickness as well as the age of the prepreg contributed to the results presented thus far, a new batch of graphite/epoxy prepreg was purchased. From this fresh batch, five unnotched replicates [457.2 mm long by 50.8 mm wide (18 in. by 2 in.)] of each laminate configuration, $[0]_8$, $[\pm 15_2]_s$, $[\pm 30_2]_s$, and $[\pm 45_2]_s$, were fabricated and tested in uniaxial tension to fracture.

81X

FIG. 5—*Fracture characteristics associated with an overaged* [0]₈ *laminate.*

FIG. 5—*Continued.*

FIG. 6—*Fracture characteristics associated with an overaged* [± 15₂]ₛ *laminate.*

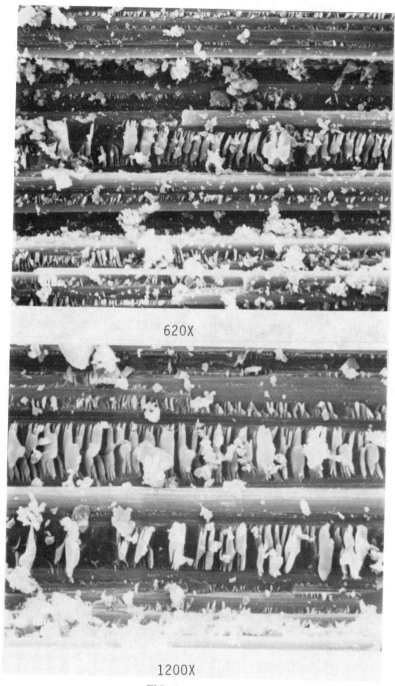

FIG. 6—*Continued.*

FIG. 7—*Fracture characteristics associated with an overaged* $[\pm 30_2]_s$ *laminate.*

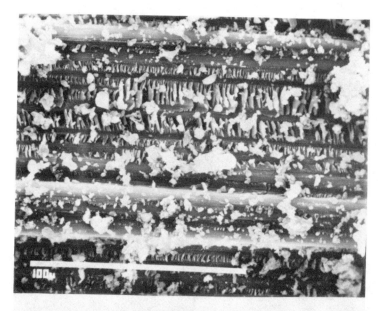

690X

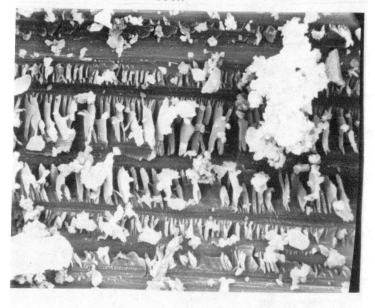

1400X

FIG. 7—*Continued.*

FIG. 8—*Fracture characteristics associated with an overaged* [$\pm 45_2$]$_s$ *laminate.*

500X

2000X

FIG. 8—*Continued.*

TABLE 3—*Fracture stresses of the fresh graphite/epoxy* $[\pm\theta_2]_s$ *laminates.*

Unnotched Replicates	Fracture stress, ksi				Fracture stress, MPa			
	$[0]_8$	$[\pm15_2]_s$	$[\pm30_2]_s$	$[\pm45_2]_s$	$[0]_8$	$[\pm15_2]_s$	$[\pm30_2]_s$	$[\pm45_2]_s$
1	210	140	75	30	1448	965	517	207
2	261	143	75	31	1799	986	517	214
3	206	144	74	29	1420	993	510	200
4	254	128	75	29	1751	882	517	200
5	· · ·	125	75	30	· · ·	862	517	207

NOTE: 1 ksi = 6.895 MPa.

The experimental data (fracture loads), which displayed repeatability, were used to formulate the fracture stresses for the fresh prepreg $[\pm\theta_2]_s$ laminates in Table 3. The thickness effect alone would indicate that the fracture loads of the $[\pm\theta_2]_s$ laminates should be double in magnitude to those observed for the $[\pm\theta]_s$ laminates. If this were the case, then the fracture stresses (which were calculated from the experimental fracture loads) displayed in Tables 1 and 3 for the $[\pm\theta]_s$ and $[\pm\theta_2]_s$ laminates, respectively, should be of the same magnitude. Comparing the fracture stresses for the unnotched specimens in Table 1 to the average fracture stresses in Table 3, it is revealed that the fracture loads for these thicker laminates exceeded the predictions, exhibiting exceptional tensile strength, thereby eliminating laminate thickness as a significant factor in the degradation of strength.

Finally, the physical fracture characteristics of these specimens, shown in Fig. 9, resemble those previously photographed for the $[\pm\theta]_s$ laminates. Primarily, the specimens are breaking at two different locations simultaneously at the recorded fracture stress, producing jagged, irregular surfaces.

Comparisons and Discussion

In comparing the data and results of the $[\pm\theta]_s$ graphite/epoxy laminates to those of the $[\pm\theta_2]_s$ overaged graphite/epoxy laminates, it was shown that the overaged laminates do degrade in strength substantially and that they produce unique microstructural characteristics.

In summarizing the characteristics of the $[\pm\theta]_s$ laminates, photomicrographs reveal distinct microstructural characteristics for each angleplied laminate fracture surface examined. This indicates that the fracture characteristics are a function of the ply angle orientation. The fracture of the specimen itself usually occurred in two places, producing a jagged, irregular surface.

The unidirectional laminate ($[0]_8$) did exhibit some degradation in strength; however, it was slight in comparison to that of the angleplied laminates. As previously mentioned, the resin is found to be more sensitive to overaging. Since the unidirectional laminate is fiber dominant, the degradation was expected to be minimal. This is manifested in comparing the photomicrographs of the two laminates. One would expect the microstructural characteristics of the $[0]_8$ laminate in Fig. 5 to be very similar in nature to those of the $[0]_4$ laminate in Fig. 1 since both laminates are fiber dominant. The unidirectional laminate made from the overaged prepreg does possess the fiber breakage and tiered pattern which was documented for the $[0]_4$ laminate. However, the alternating fiber/hackle pattern is also present in the patchy areas of the surface which, in this paper, is shown to be associated with an overaged prepreg.

The angleplied, overaged $[\pm\theta_2]_s$ laminates fractured along a plane that was parallel to a fiber direction, producing very smooth and flat surfaces. The fracture loads were substantially lower for these laminate configurations. The microstructural characteristics of the angleplied fracture surfaces were not distinct. Each revealed extensive matrix debris along with hackles of various sizes. The hackles surrounding the fibers in the photomicrographs establish a pattern which consists of alternating rows of fibers and hackles. Since these characteristics were observed on all the surfaces regardless of ply angle orientation, it is presumed that these microstructural characteristics are the product of an overaged prepreg.

One other point worth mentioning is that the physical handling of these fractured $[\pm\theta_2]_s$ overaged laminates resulted in a black greasy residue on the skin. The authors have never observed this before even though hundreds of graphite/epoxy specimens have been handled. This is apparently indicative of a chemical breakdown of the old resin and is an additional characteristic of an overaged prepreg.

The fracture loads, surface features, and microstructural characteristics observed and recorded for this particular series of specimens is unique to this batch based upon comparisons made with results from similar studies. Therefore, it is concluded that the overaged graphite/epoxy prepreg does degrade in strength, fractures in a brittle fashion, possesses unique microstructural characteristics, and, for this particular prepreg, attains this state at an age of approximately 2.5 years. This is based on results previously obtained wherein the strengths had not deteriorated significantly for specimens fabricated from 2-year-old prepreg.

These conclusions have been substantiated by an analytical investigation which was conducted in conjunction with the experimental investigation. The ICAN (Integrated Composites Analyzer) computer code [6] was used to model and analyze the eight-ply ($[\pm\theta_2]_s$) laminates using constituent properties associated with a fresh batch of prepreg. Results from the analysis, which include laminate fracture stress predictions as a function of intralaminar shear stress, are shown in Fig. 10 for the $[\pm15_2]_s$ laminate and in Fig. 11 for the $[\pm30_2]_s$ and $[\pm45_2]_s$ laminates.

The fracture stress is plotted as a function of the perturbed intralaminar

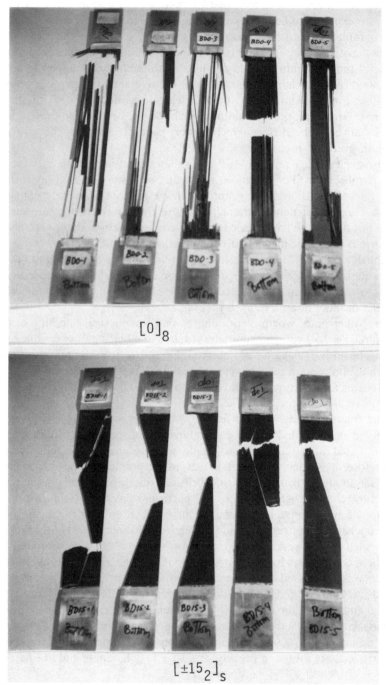

FIG. 9—*Fractured unnotched graphite/epoxy specimens fabricated from a fresh batch of prepreg.*

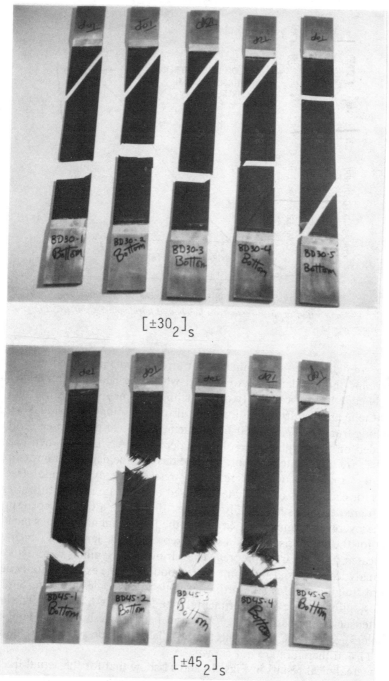

$[\pm 30_2]_s$

$[\pm 45_2]_s$

FIG. 9—*Continued.*

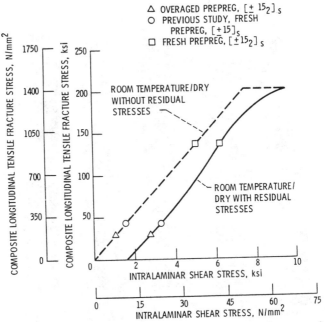

FIG. 10—*Composite fracture stresses as a function of intralaminar shear stress for the laminates where* $\theta = 15°$.

shear strength. The intralaminar shear strength was perturbed in order to determine the intralaminar strength at which the laminate fractured. The ICAN code has the capability of predicting the fracture stress not only for room temperature/dry conditions but also accounting for residual stresses resulting from the 177°C (350°F) cure temperature of the prepreg. Both are included on the graphs. The solid line depicts ICAN results when residual stresses are taken into account, and the dashed line depicts room temperature/dry conditions without residual stresses. The experimental fracture loads for each series of laminates are used to calculate the fracture stresses, which are depicted symbolically. In Figs. 10 and 11, a triangle corresponds to the overaged laminates, a circle for the $[\pm\theta]_s$ laminate data which was used for the comparison, and a square represents the laminates which were fabricated from the fresh prepreg. The various data symbols were placed on the analytical curves at the point corresponding to the experimentally determined fracture stresses (the ordinate). Due to the scale selection, only the $[\pm15_2]_s$ plots include the experimental data for the fresh prepreg. The superior tensile strengths of the $[\pm30_2]_s$ and $[\pm45_2]_s$ fresh prepreg laminates (Table 3) exceed the limits selected for the scale representing the data in Fig. 11 and therefore are not shown.

The analytical results in Figs. 10 and 11 show that for the actual fracture stresses (symbols) the corresponding intralaminar shear stresses associated

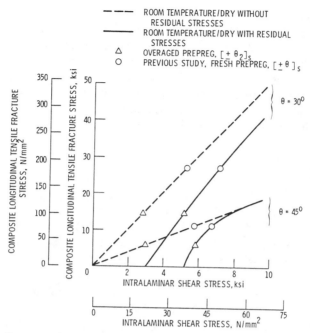

FIG. 11—*Composite fracture stresses as a function of intralaminar shear stress for the laminates where θ = 30° and 45°.*

with the overaged laminates are lower than the thinner [± θ]ₛ laminates and substantially lower than those made with fresh prepreg. As expected, the degradation in shear strength is more severe for the [± 30₂]ₛ and [± 45₂]ₛ laminates, which are more matrix dominant. The only factor separating the data points is the age of the prepreg. Since the intralaminar shear strength is a property of the resin primarily, the analytical results indicate that the degradation in shear strength is the result of a deteriorating resin which is caused by overaging.

Results from the ICAN program did indeed corroborate those resulting from the experimental investigation, namely, that the overaging results in a weakening of the resin, which fractures in shear at a reduced load, producing unique characteristics on the fracture surface associated with the overaged resin.

Summary of Results

Laminates fabricated from the overaged prepreg as expected experience a degradation in strength. This type of prepreg also produces unique microstructural characteristics on the fracture surface. For the case discussed, the "advanced-cure" was the result of overaging. However, "advanced-cure" can also occur in a fresh prepreg when exposed to the environmental conditions of moisture and temperature for some extended period. Note that neither of these conditions needs to be severe to serve as a catalyst for "advanced-cure" leading to degradation. In fact, this phenomenon can occur when a prepreg is exposed to room temperature and normal humidity. The results presented herein provide the experimentalist/analyst with another source of data to consider in a postfailure investigation, especially when anomalies in failure characteristics exist.

References

[1] Irvine, T. B. and Ginty, C. A., "Progressive Fracture of Fiber Composites," NASA TM-83701, National Aeronautics and Space Administration, Washington, DC, 1983.

[2] Ginty, C. A. and Irvine, T. B., "Fracture Surface Characteristics of Notched Angleplied Graphite/Epoxy Composites," NASA TM-83786, National Aeronautics and Space Administration, Washington, DC, 1984.

[3] Irvine, T. B. and Ginty, C. A., "Fracture Modes in Notched Angleplied Composite Laminates," NASA TM-83802, National Aeronautics and Space Administration, Washington, DC, 1984.

[4] Clements, L. L., "Deformation and Failure Mechanisms of Graphite/Epoxy Composites Under Static Loading," (FR-22–3, Advanced Research and Applications Corp., NASA Contract NAS 2-9989), NASA CR-166328, National Aeronautics and Space Administration, Washington, DC, 1981.

[5] Clements, L. L. and Adamson, M. J., "Failure of Morphology of $(0°)_8$ Graphite/Epoxy as Influenced by Environments and Processing," NASA TM-81318, National Aeronautics and Space Administration, Washington, DC, 1981.

[6] Murthy, P. L. N. and Chamis, C. C., "ICAN: Integrated Composites Analyzer," AIAA Paper 84-0974, American Institute of Aeronautics and Astronautics, New York, May 1984.

Charles E. Harris[1] *and Don H. Morris*[2]

A Fractographic Investigation of the Influence of Stacking Sequence on the Strength of Notched Laminated Composites

REFERENCE: Harris, C. E. and Morris, D. H., **"A Fractographic Investigation of the Influence of Stacking Sequence on the Strength of Notched Laminated Composites,"** *Fractography of Modern Engineering Materials: Composites and Metals, ASTM STP 948*, J. E. Masters and J. J. Au, Eds., American Society for Testing and Materials, Philadelphia, 1987, pp. 131–153.

ABSTRACT: A study of the fracture processes in T300/5208 notched laminated composites concentrated on characterizing the influences of laminate stacking sequence on the formation of notch-tip damage zones and the subsequent mode of fracture of each ply of the laminate. The results revealed that laminate notched strength varied considerably with both a specified set of ply fiber orientations as well as the order in which the specific plies were arranged in the laminate. The fractographic examinations identified differences in the laminate fracture process which are believed to be the primary contributor to the differences in notched strength. The twelve laminates exhibited varying degrees of notch sensitivity as a result of the ability of a laminate to develop "subcritical" damage in a manner that relieved the high stress concentration at the notch tip.

The percent of laminate fracture by broken fibers was in direct correlation with the laminate notched strength. Laminates which were not characterized by damage zones containing major delaminations exhibited higher percentages of failure by broken fibers and higher notched strength. Conversely, when delaminations resulted in widespread ply uncoupling, relatively few plies of a laminate failed by broken fibers producing a weaker laminate.

KEYWORDS: composite, graphite/epoxy, notched strength, damage, fracture, delamination

The damage tolerant design requirement for laminated composite structures has provided the motivation for a comprehensive study by the authors

[1]Assistant professor, Aerospace Dept., Texas A&M University, College Station, TX 77843.
[2]Professor, Engineering Science and Mechanics Dept., Virginia Polytechnic Institute and State University, Blacksburg, VA 24061.

of the fracture behavior of laminated composite materials. The emphasis in this research has been on the phenomenological understanding of the actual fracture processes. To facilitate this experimental investigation of laminate damage mechanics, the writers have utilized the standard center-cracked (machined crack-like slit) tension specimen. While the center-cracked tension specimen may not represent a meaningful notched geometry for laminated composite structures, it is a very convenient geometry for the study of fundamental fracture behavior.

The initial phase of the research project had the objective of characterizing the fracture behavior of thick laminates. It was found [1] that laminate thickness significantly influenced the notched strength (far field uniform stress at failure) of T300/5208 graphite/epoxy laminates. This is illustrated in Fig. 1. The notched strength of the $[0/\pm45/90]_{ns}$ ($n = 1, 4, 8, 12, 15$) and $[0/90]_{ns}$ ($n = 2, 8, 16, 24, 30$) laminates decreased with increasing thickness, while the notched strength of the $[0/\pm45]_{ns}$ ($n = 1, 5, 10, 15, 20$) laminate increased with increasing thickness. All three laminate types achieved an approximate plateau value of notched strength beyond about 60 plies. The research revealed that this fundamentally different behavior, as well as the laminate thickness effect, was a result of dramatic differences in the magnitude and content of the load-induced zones of damage that formed at the notch tips prior to catastrophic fracture. In the case of the $[0/\pm45/90]_s$ and $[0/90]_s$ laminates, these notch-tip damage zones provided stress relief in the vicinity of the notch and, therefore, elevated the notched strength. As the laminate thickness increased, the size of the damage zone decreased, result-

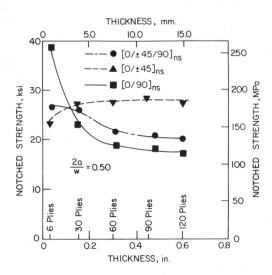

FIG. 1—*Notched strength versus laminate thickness, from center-cracked tension specimens with 2a/W = 0.50.*

ing in less stress relief. Also, damage such as delaminations in the $[0/\pm45/90]_{ns}$ laminate and axial splits (matrix cracks parallel to the 0° fibers) in the $[0/90]_{ns}$ laminate did not occur in the interior of the thick laminates. Therefore, the thicker laminates were more notch sensitive with less stress relief and had a resulting lower value of notched strength. On the other hand, the thin $[0/\pm45]_s$ laminate formed major delaminations which extended from the tips of the notch to the edges of the specimen. These delaminations resulted in complete ply uncoupling which weakened the laminate. Delaminations did not form in the interior of the thick $[0/\pm45]_{ns}$ laminates. Plies which failed by matrix cracks in the thin laminates failed by broken fibers, which requires a higher load, in the thicker ($n \geq 5$) laminates. This resulted in elevated values of notched strength.

Observations of the fracture behavior of laminated composites, such as those just described and by others [2], led to the conclusion that the fracture of laminates was governed by broken fibers in the principal load-carrying plies. (These are the 0° plies in the just-described laminates). The load at which fibers broke and the pattern of broken fibers relative to the notch were influenced by both the constraint provided by adjacent plies and the load redistribution brought about by the formation of the damage zone. In order to more fully understand the role of adjacent ply constraint and the notch-tip damage zone on the fracture of the principal load-carrying plies, the effect of stacking sequence on both notched strength and fracture behavior was investigated. Four basic laminate types were selected for study with three ply stacking order perturbations for each basic type. These twelve laminate stacking sequences are listed in Table 1. The primary emphasis in this investigation was the experimental characterization of the notch-tip damage zone for each laminate and the establishment of the relationship between the notch-tip damage zone, the mode of fracture, and the value of notched strength.

TABLE 1—*Laminate type and stacking sequence.*

Laminate	Stacking Sequence
Type A	$[\pm45/0]_s$ $[45/0/-45]_s$ $[0/\pm45]_s$
Type B	$[\pm60/0]_s$ $[60/0/-60]_s$ $[0/\pm60]_s$
Type C	$[\pm30/90]_s$ $[30/90/-30]_s$ $[90/\pm30]_s$
Type D	$[\pm45/0/90]_s$ $[0/90/\pm45]_s$ $[0/\pm45/90]_s$

The fractographic investigation of the behavior of the notched laminates listed in Table 1 is the subject of this paper. Within the context of this paper, fractography refers to the study of the laminate fracture process on the microscale. This scale of internal damage includes matrix cracks, delaminations, and broken fibers. The fractographic study establishes the initial damage mechanisms and their role in influencing the final fracture of the laminate. The fractographic examinations were conducted utilizing both nondestructive and destructive examination procedures. These procedures and the research results are fully described in the following sections.

Experimental Program

The laminate panels were prepared from graphite/epoxy T300/5208, using a standard tape lay up and autoclave curing process. The basic lamina tensile data were determined experimentally from $[0]_8$, $[90]_8$, and $[\pm 45]_{2s}$ coupons. Using the average of four or five replicate tests, the lamina stiffness properties were: $E_{11} = 138.7$ GPa (20.11×10^6 psi), $E_{22} = 10.77$ GPa (1.56×10^6 psi), $G_{12} = 5.98$ GPa (0.867×10^6 psi), and $v_{12} = 0.318$, where Subscript 1 refers to the direction parallel to the fibers and Subscript 2 refers to the transverse direction. These property values are typical of those reported in the literature for T300/5208 [3].

The influence of ply stacking sequence was investigated using variations in the $[0/\pm 45/90]_s$, $[0/\pm 45]_s$, $[0/\pm 60]_s$, and $[90/\pm 30]_s$ laminates. (All laminate stacking sequences are listed in Table 1.) The $0°$ fiber orientation was perpendicular to the notch. For the [90/30] laminate types the $90°$ fiber orientation was parallel to the notch.

Fracture behavior was studied using the center-cracked tension specimen. The specimen coupon was 51 by 203 mm (2 by 8 in.) with a 25.4-mm (1.0-in.) center notch cut by an ultrasonic cutting tool. The notch tip radius was 0.203 mm (0.008 in); no attempt was made to sharpen the notch tips. The crack length, $2a$, to width, W, ratio was 0.50. In most cases four or more replicate tests were conducted.

All fracture tests were conducted at a constant crosshead displacement rate of 0.02 mm/s (0.05 in./min). The center-cracked tension specimens were held in 51-mm (2-in.)-wide, wedge-action friction grips such that the specimen length between grip ends was approximately 127 mm (5 in.). The thin specimens, six or eight plies, were tested with an antibuckling support to prevent out-of-plane motion. When the center-cracked specimen is loaded in tension, a compression region develops in the central region of the specimen adjacent to the notch but away from the notch tips. The out-of-plane motion of thin specimens in this compression region causes elevated values of apparent fracture toughness. The authors believe that the out-of-plane deflection of thin specimens must be prevented in order to compare the test results to thick specimens where out-of-plane deflections are constrained by thickness.

The crack-opening displacement (COD) was measured using a split-ring–type displacement gage. When compressed, the thin-machined tabs of the gage fit directly into the machined notch in the specimen. As the notch opened, the change in the compression in the gage was continuously monitored by an electrical resistance strain gage attached to the COD gage.

Crack-tip damage that formed in the laminates prior to fracture was investigated using X-ray radiography and the laminate deply technique [4]. Zinc iodide–enhanced X-ray examinations were used to establish the load levels where significant damage developed. Also, the progression of damage as the load approached the failure load was documented using X-ray radiography. X-ray examinations were taken at load intervals of approximately 10% of the anticipated failure load or after discontinuities in the load-COD record. Radiographs will be used throughout the paper to compare notch-tip damage zones.

Using a gold chloride agent to mark delamination regions, the through-the-thickness damage was studied using the destructive laminate deply technique. This technique allowed the study and documentation of the damage in each individual ply. The deply technique was especially useful for studying the details of broken fibers and for establishing the extent of delaminations as well as their exact interface location. Examples of typical results of a deply study are shown in Fig. 2. These photographs were taken at a X20 magnification in a standard optical microscope. Figure 2a shows a delamination marked by gold chloride particles which was illuminated by an external light source. Figure 2b shows a line of broken fibers in 0° ply extending colinearly (90° to fiber direction) from the notch-tip. Figure 2c shows a line of broken fibers in a 0° ply extending in a 45° direction from the notch-tip. Finally, Fig. 2d shows X200 magnification of broken fibers in the 0° ply of a [0/±45/90]$_s$ laminate. (A thorough description of all experimental procedures is given in Ref 1.)

Fractographic Results

Description of the Notch-Tip Damage Zone

As previously stated, the zones of damage that form at the notch tips prior to castastrophic fracture were studied nondestructively by X-ray radiography and destructively by laminate deply. The combined results of these two types of examinations are described in the following paragraphs for each laminate. Load-induced, notch-tip damage zones were developed and examined in several replicate specimens for each laminate type. It was found that the general size and content of the damage zones were the same. Therefore, a generic damage zone is described for each laminate. Only X-ray

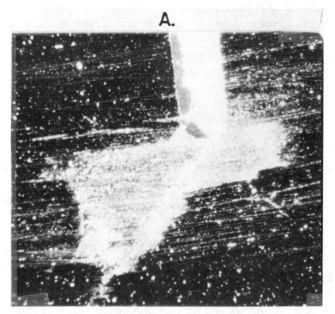

FIG. 2—*Typical optical and SEM photographs of laminate deply results:*
(A) *Delamination marked by gold chloride as viewed in an optical microscope at X20 (0/+45 interface of a [0/±45/90]ₛ laminate).*

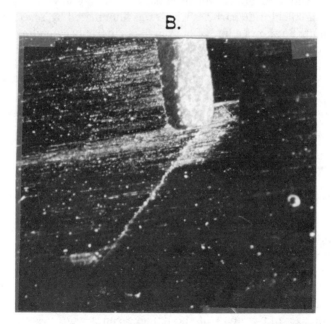

FIG. 2—(B) *Line of broken fibers extending at 45° from the notch as viewed in an optical microscope at X20 (0° ply of a [0/±45/90]ₛ laminate).*

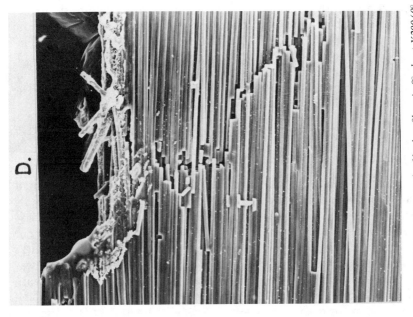

FIG. 2—(D) SEM photograph of broken fibers in 0° ply at X200 (0° ply of a [0/ ± 45]ₛ laminate.

FIG. 2—(C) SEM photograph of broken fibers in 30° ply at X30 (30° ply of a [± 30/90]ₛ laminate).

radiographs will be used to illustrate the damage zones. However, it should be understood that all descriptions of the damage zone regarding delamination interface locations and the details of broken fibers were obtained from the laminate deply technique. In addition, the deply examinations confirmed the symmetry of the damage zones with respect to the laminate midplane. Finally, all directions are specified relative to the 0° fiber direction. In the 90/30 laminates this direction is perpendicular to the 90° fibers. (The notch is perpendicular to the 0° fiber direction.)

Laminate Type A

The damage zones in two specimens of the $[\pm 45/0]_s$ laminate were studied at 91 and 94% of the average failing load of the corresponding replicate specimens for the specific laminate stacking sequence (P_{AVE}). An X-ray radiograph of the damage zones at 94% of P_{AVE} is shown in Fig. 3a. (The dark regions running along the notch are notch-cutting tool damage marks and not load-induced damage and, hence, should be disregarded. It was found that the tool marks had a negligible effect on damage development.) The damage zone is characterized by major matrix cracks at $+45°$ and $-45°$ and a major delamination at the $+45/-45$ interface. (The term *major* will be used to describe delaminations that are quite extensive. The term *minor* will describe those delaminations which are quite localized and typically associated with extensive matrix cracking.) Broken fibers in the 0° ply extended in combinations of 90° and $-45°$ directions. There were no pronounced axial splits (matrix cracks) in the 0° plies. (It has been observed by the authors that broken fibers occur in plies that do not exhibit extensive matrix cracking.)

The damage zones in three specimens of the $[45/0/-45]_s$ laminate were studied at 88, 93, and 94% of P_{AVE}. An X-ray radiograph of the damage zones at 93% of P_{AVE} is shown in Fig. 3b. The primary characteristics of this damage zone are the $+45°$ and $-45°$ matrix cracks along with associated delaminations at the 45/0 and 0/−45 interfaces. Broken fibers in the 0° ply extended in a 90° direction (colinear with the notch). There were no pronounced axial splits in the 0° plies.

The damage zones in two specimens of the $[0/\pm 45]_s$ laminate were studied at 83 and 96% of P_{AVE}. An X-ray radiograph of the damage zones at 96% of P_{AVE} is shown in Fig. 3c. As with the other two laminates, there were major matrix cracks at $+45°$ and $-45°$ along with an associated delamination at the $+45/-45$ interface. Broken fibers extended in a $+45°$ direction from the notch tips of some of the 0° plies while axial splits formed at the notch-tips of the other 0° plies. The presence of the axial splits in the 0° plies distinguishes the damage zone of the $[0/\pm 45]_s$ laminate from the damage zones of the other two Type A laminates.

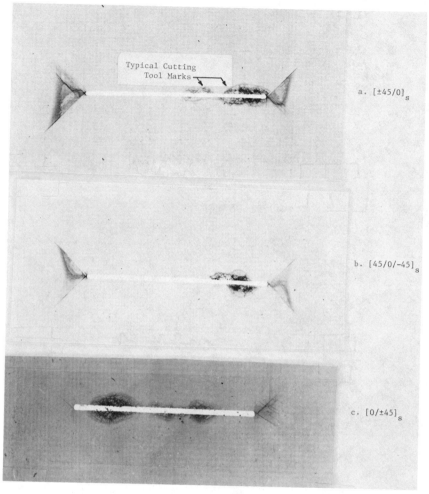

FIG. 3—*X-ray radiographs of damage in Type A laminates.*

Laminate Type B

The damage zones in three specimens of the $[\pm 60/0]_s$ laminate were studied at 94%, 94%, and 101% of P_{AVE}. An X-ray radiograph of the damage zones at 101% of P_{AVE} is shown in Fig. 4a. Axial splits formed in the 0° plies with an associated delamination at the $0/-60$ interface. There were no broken fibers in any plies. Major matrix cracks formed in the $+60°$ plies, but only minor cracks formed in the $-60°$ plies. Also, there was a delamination at the $+60°/-60°$ interface.

The damage zones in two specimens of the $[0/\pm 60]_s$ laminate were both

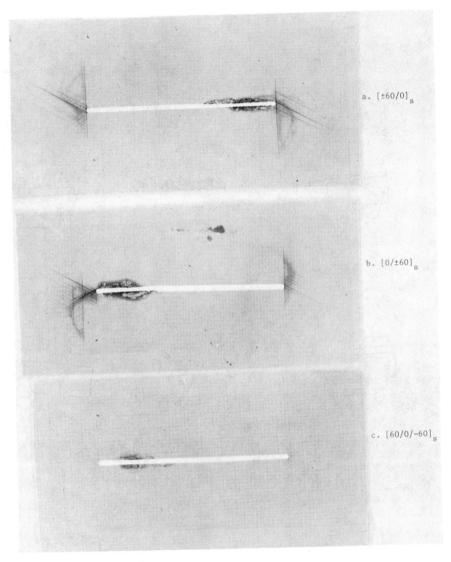

a. [±60/0]$_s$

b. [0/±60]$_s$

c. [60/0/-60]$_s$

FIG. 4—*X-ray radiographs of damage in Type B laminates.*

studied at 85% of P_{AVE}. An X-ray radiograph of the damage zone is shown in Fig. 4b. Major axial splits formed at some of the 0° ply notch-tips, while broken fibers extended in a 60° direction at other 0° ply notch-tips. The axial splits were accompanied by a delamination of the 0/60 interface. Matrix cracks appeared more extensive in the +60 direction than in the −60° direction, and there was a minor delamination at the +60/−60 interface.

The damage zones in three specimens of the $[60/0/-60]_s$ laminate were studied at 90, 99, and 99% of P_{AVE}. An X-ray radiograph of the damage zone at 99% of P_{AVE} is shown in Fig. 4c. The damage zone of the $[60/0/-60]_s$ laminate is obviously quite different from the other two Type B laminates because there is, in effect, no damage zone. The only damage consisted of very minor matrix cracks in the various fiber directions. (The absence of a damage zone should not be interpreted as a function of load level because of the high percentage of the load level at which the examinations were performed and because the load-COD test records did not indicate damage formation.)

Laminate Type C

The damage zones in three specimens of the $[\pm 30/90]_s$ laminate were examined at 82, 98, and 98% of P_{AVE}. An X-ray radiograph of the damage zone at 82% of P_{AVE} is shown in Fig. 5a. A major delamination at the $+30/-30$ interface is bounded by matrix cracks at $+30°$ and by broken fibers in the $-30°$ plies extending at $90°$ from the loading or $0°$ fiber direction. This is the reason for the antisymmetric appearance of the damage zones shown by the radiograph of Fig. 5a.

The damage zones of three specimens of the $[30/90/-30]_s$ laminate were studied at 90, 94, and 104% of P_{AVE}. An X-ray radiograph of the damage zone at 94% of P_{AVE} is shown in Fig. 5b. The damage zone is bounded by major matrix cracks extending at $+30°$ and $-30°$. There are numerous matrix cracks in the $90°$ plies and delaminations at both the $+30/90°$ and $90/-30$ interfaces. Also, there were broken fibers in the $+30°$ plies extending in the $90°$ direction (colinear with the notch).

The damage zones of two specimens of the $[90/\pm 30]_s$ laminate were studied at 79 and 91 of P_{AVE}. An X-ray radiograph of the damage zone at 91% of P_{AVE} is shown in Fig. 5c. The damage zones were quite minor. There were minor matrix cracks in each ply with a minor delamination at the $+30/-30$ interface. There was no evidence of broken fibers in the damage zones.

Laminate Type D

The damage zones in three specimens of the $[\pm 45/0/90]_s$ laminate were examined at 88, 91, and 91% of P_{AVE}. An X-ray radiograph of the damage zone at 91% of P_{AVE} is shown in Fig. 6a. The $0°$ plies exhibited broken fibers extending in the $90°$ direction; there were no pronounced axial splits. There were matrix cracks at $+45°$, $-45°$ and $90°$ and minor delaminations at the $+45/-45$ and $-45/0$ interfaces.

The damage zones in all three specimens of the $[0/90/\pm 45]_s$ laminate were examined at 101% of P_{AVE}. One X-ray radiograph of the damage zones is shown in Fig. 6b. The $0°$ plies exhibited axial splits at most notch-tips with

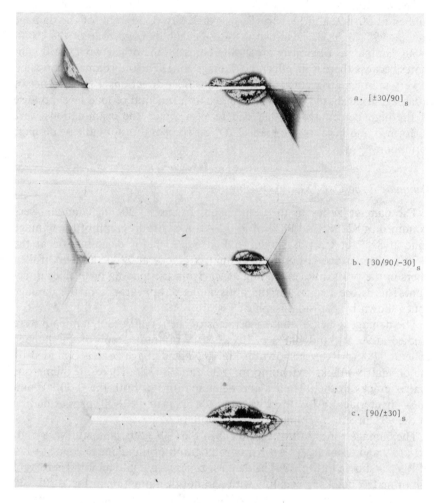

FIG. 5—*X-ray radiographs of damage in Type C laminates.*

an association delamination at the 0/90 interface. Minor broken fibers in the $0°$ plies extending at $90°$ were present at those notch-tips where axial splits did not form. Matrix cracks occurred at $+45°$, $-45°$, and $90°$ along with a delamination at the $+45/-45$ interface.

The damage zones in six specimens of the $[0/\pm45/90]_s$ laminate were examined at 79, 82, 84, 85, 85, and 94% of P_{AVE}. The X-ray radiograph of the damage zone at 85% of P_{AVE} was shown in Fig. 6c. Beyond 84% of P_{AVE} the damage zones were quite extensive. The $0°$ plies exhibited combinations of broken fibers extending at $+45°$ and $90°$ directions as well as axial splits at some notch-tips of the $0°$ plies. There was typically a delamination at the

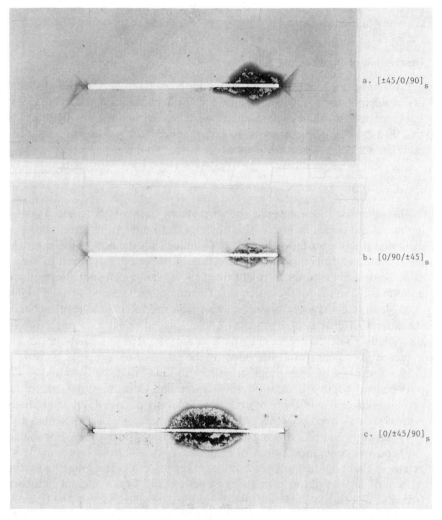

a. [±45/0/90]$_s$

b. [0/90/±45]$_s$

c. [0/±45/90]$_s$

FIG. 6—*X-ray radiographs of damage in Type D laminates.*

0/+45 interface associated with the axial splits. A delamination at the +45/
−45 interface was bounded by major matrix cracks at +45° and −45°.
Also, there was extensive matrix cracking in the 90° plies throughout the
damage zone.

The delaminations in the Type D laminates appeared to be associated with
intersecting matrix cracks. For example, a local delamination may be required to produce sufficient localized compliance that would allow the relative crack-opening displacements of matrix cracks in adjacent 0° and 90°

plies. A similar phenomenon is the local delaminations at crossing cracks in cross-ply laminates [5].

Description of the Fracture Surfaces

It is important to establish the actual mode of fracture of each individual ply of a laminate to fully understand the laminate fracture process. Typical fractured specimens of each laminate were deplied and each individual ply was then examined to determine its failure mode. The ply failure modes for each laminate are described in paragraphs that follow.

Laminate Type A

Photographs of a fractured specimen from each of the three Type A laminates are shown in Fig. 7. It is obvious that none of the three laminates exhibited fracture surfaces that are self-similar with the notch. It is interesting to note the differences in the direction of broken fiber pattern in the $0°$ plies. Broken fibers in the $0°$ ply of the $[0/\pm45]_s$ laminate generally extended at a $45°$ direction from the notch tip. The line of broken fibers in the $[45/0/-45]_s$ laminate was more-or-less colinear with the notch. On the other hand, the pattern of broken fibers in the $[\pm45/0]_s$ was at about a $15°$ direction to the notch. These differences in the fracture of the $0°$ ply are due to the influence of the constraint provided by the adjacent plies.

It is also obvious from the photographs in Fig. 7 that the $+45$ and $-45°$ plies generally failed by matrix cracks extending from the notch-tip to the specimen edge. The fracture of each individual ply of a $[\pm45/0]_s$ laminate is more clearly viewed by the deply results shown in Fig. 8. The $+45°$ plies and $-45°$ failed by matrix cracking and independently of each other. For this to happen, a delamination must form at the $+45/-45$ interface and be so extensive that complete ply uncoupling takes place. On the other hand, the pattern of broken fibers in the $0°$ ply does exhibit some apparent influence from its adjacent $-45°$ plies. This conclusion is reached because the fracture surfaces are at $-30°$ to the notch rather than being colinear with the notch.

The individual ply fracture surfaces of the $[45/0/-45]_s$ laminate are shown in Fig. 9. The $+45°$ and $-45°$ plies failed by matrix cracks extending from the notch-tip to the specimen edge. The broken fiber fracture pattern of the $0°$ ply is essentially colinear with the notch. These fracture surfaces indicate that the plies of this laminate completely uncoupled with each ply failure mode being independent of the other plies.

The fracture of the $[0/\pm45]_s$ laminate was similar to the other Type A laminates with regard to the fracture of the $+45°$ and $-45°$ plies. However, the $0°$ ply and $+45°$ ply failed together as a unit. The $0°$ ply fractured by broken fibers paralleling the $+45°$ matrix crack in the adjacent $+45°$ ply. This is clearly visible in the photograph of the fracture surface in Fig. 7.

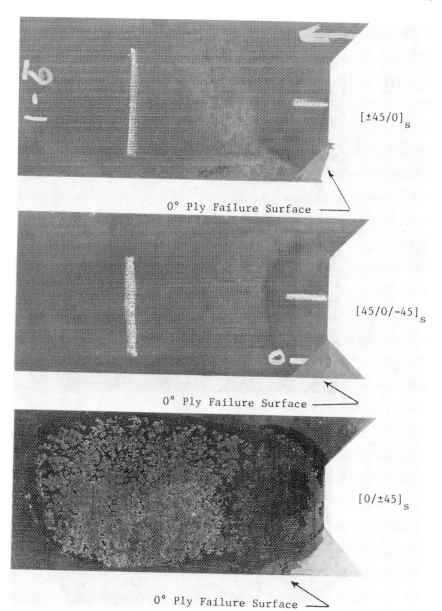

FIG. 7—Comparison of the fracture surfaces of Type A laminates.

FIG. 8—*Individual ply fracture surfaces of a* [±45/0]$_s$ *laminate.*

45° Ply

0° Ply

2@-45° Plies

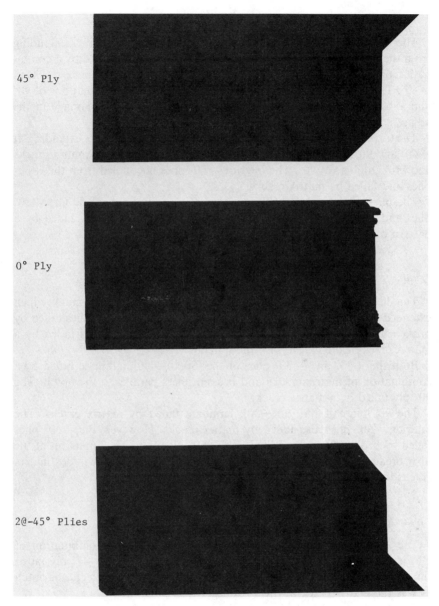

FIG. 9—*Individual ply fracture surfaces of a* [45/0/−45]_s *laminate.*

Laminate Type B

The $+60°$ plies of the $[\pm 60/0]_s$ laminate failed by a matrix crack extending from the notch-tip to the specimen edge. The $-60°$ plies failed by a combination of broken fibers at $90°$ (colinear with the notch) and matrix cracks at $-60°$. The $0°$ plies failed by broken fibers extending at combinations of $90°$ and $-60°$ directions which, to some extent, paralleled the damage in the adjacent $-60°$ plies.

The $0°$ plies of the $[0/\pm 60]_s$ laminate failed by broken fibers extending at $+60°$ and $90°$. The $+60°$ plies failed by a combined mode of matrix cracks and broken fibers at $90°$. The interior $-60°$ plies uncoupled from the $+60°$ plies and failed by matrix cracks.

The plies of the $[60/0/-60]_s$ laminate completely uncoupled. The $+60°$ plies and $-60°$ plies failed by matrix cracks, while the $0°$ ply failed by a pattern of broken fibers that was colinear with the notch.

Laminate Type C

The $30°$ plies of the $[\pm 30/90]_s$ laminate failed by a matrix crack which extended from the notch-tip to the specimen edge. The $-30°$ plies failed by broken fibers extending colinear with the notch. The $90°$ plies failed by a matrix crack also colinear with the notch.

Both the $+30°$ and $-30°$ plies of the $[30/90/-30]_s$ laminate failed by a combination of matrix cracks and broken fibers parallel to the notch. The $90°$ ply failed by a matrix crack.

The $90°$ plies of the $[90/\pm 30]_s$ laminate failed by matrix cracks. The interior $-30°$ plies also failed by matrix cracks. However, the $+30°$ plies failed by broken fibers extending at a $-60°$ direction. This direction of the fiber break pattern is an intermediate direction to the $90°$ and $-30°$ matrix cracks in the adjacent plies.

Laminate Type D

The $+45°$ plies of the $[\pm 45/0/90]_s$ laminate failed by combination of matrix cracks and broken fibers parallel to the notch. The $-45°$ ply failed similarly. The pattern of broken fibers in the $0°$ plies was parallel to the notch as was the matrix crack in the $90°$ plies.

The $0°$ plies of the $[0/90/\pm 45]_s$ laminate exhibited a pattern of broken fibers that was colinear with the notch and the matrix crack in the adjacent $90°$ ply. The $+45°$ plies failed by a combination of matrix cracks and broken fibers parallel to the notch. The $-45°$ ply failed by a matrix crack extending from the notch-tip to the specimen edge.

The $[0/\pm 45/90]_s$ laminate fractured in a manner that was self-similar with the machined notch. The $0°$, $+45°$, and $-45°$ plies failed by broken fibers

in a pattern more or less colinear with the notch. Only the failure of the 90°
ply was governed by matrix cracking which paralleled the notch.

Discussion of Fractographic Results

Notched strength, σ_N, was selected as a parameter to quantify and compare the notched laminate fracture behavior. Laminate notched strength is
defined as the value of far field uniform tensile stress at failure. This is given
by

$$\sigma_N = P/Wt \tag{1}$$

where

$P =$ load at failure,
$W =$ specimen width, and
$t =$ specimen thickness.

Four or more replicate tests were conducted to establish the notched strength
of each laminate. These values of notched strength are compared in the bar
chart of Fig. 10. The mean value, high value, low value, and the number of

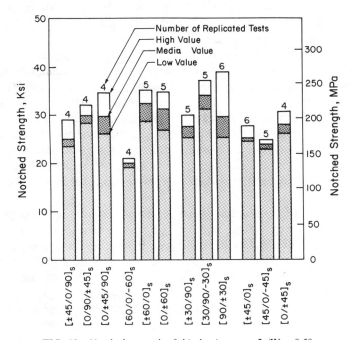

FIG. 10—*Notched strength of thin laminates as* 2a/W = 0.50.

replicate tests are shown for each laminate. These are widely varying values of notched strength. These differences in notched strength are due to the differences in the notch-tip damage zone and the role of the damage zone toward influencing the final laminate fracture. These effects are described in the following paragraphs.

Typically, when a zone of damage forms at a notch-tip, stress relief takes place and the notch-tip stress concentration is reduced. This effect tends to elevate the final fracture load. Axial splits in $0°$ plies are particularly effective in reducing the notch-tip stress-concentration. This is because axial splits have the effect of separating the $0°$ ply into a strip containing the notch and two outer strips without a notch. It follows, then, that larger damage zones typically provide greater stress relief and, hence, result in a more elevated value of notched strength. The Type D laminates are governed by this process. Axial splits did not form in the $0°$ plies of the weakest laminate, $[\pm 45/0/90]_s$. On the other hand, the two stronger Type D laminates exhibited axial splits and had more extensive zones of damage.

A second way in which the notch-tip damage zone influences the final fracture is by ply uncoupling which is brought about by extensive delaminations. When individual plies become uncoupled, they are free to fail by the fracture mode that provides the least resistance. This process is exhibited to some degree by all of the laminates of Types A, B, and C. In fact, for any specific laminate type, the value of notched strength is apparently directly related to the degree of coupling that is present at fracture. For example, the lowest value of notched strength belonged to the $[60/0/-60]_s$ laminate, which exhibited a completely uncoupled fracture.

The uncoupled fracture of the $[60/0/-60]_s$ laminate may at first appear contradictory with the earlier description of the damage evaluation where it was reported that no damage zone had formed prior to fracture. However, the sequence of fracture events clearly reveals the consistency in these two observations. The fracture of the $[60/0/-60]_s$ laminate was catastrophic. The formation of the delaminations which brought about the ply uncoupling was followed immediately by the complete fracture of the laminate. It was not possible to stop the test between the formation of the delaminations and laminate fracture. Therefore, all damage evaluations occurred prior to the formation of the delaminations.

One measure of ply coupling is the degree of fracture that takes place by fiber fracture. For example, if adjacent $\pm \theta$ plies remain coupled at fracture, at least one ply must fail by broken fibers. It was found that the percent of fracture by broken fibers was in direct correlation to the values of laminate notched strength. These results are tabulated in Table 2. It should be noted that not all differences in notched strength would be expected to be reflected by the percent of fiber fracture. For example, laminates with or without axial splits may have differences in notched strength without differences in the percent of fiber fracture.

TABLE 2—*Percent fracture by broken fibers.*

Laminate Stacking Sequence	Number of Plies That Failed by Broken Fibers	% Fracture by Broken Fibers Excluding 90° Plies	% Fracture by Broken Fibers Including 90° Plies	Notched Strength, ksi (MPa)
$[\pm 30/90]_s$	2	50	33	27.5 (189.6)
$[90/\pm 30]_s$	2	50	33	29.2 (201.3)
$[30/90/-30]_s$	4	100	66	33.8 (233.1)
$[60/0/-60]_s$	2	33	33	19.9 (137.2)
$[0/\pm 60]_s$	4	66	66	30.9 (213.1)
$[\pm 60/0]_s$	4	66	66	32.3 (222.7)
$[45/0/-45]_s$	2	33	33	23.7 (163.4)
$[\pm 45/0]_s$	2	33	33	25.4 (175.1)
$[0/\pm 45]_s$	2	33	33	27.7 (191.0)
$[\pm 45/0/90]_s$	3	50	38	24.9 (171.7)
$[0/90/\pm 45]_s$	4	66	50	29.9 (206.2)
$[0/\pm 45/90]_s$	6	100	75	30.4 (209.6)

Finally, it is not a simple matter to identify the influence of delaminations on notched strength. The mere presence of a delamination in the damage zone may also not be sufficient. The minor delaminations of Type D laminates provided stress relief, and the effect of ply uncoupling was confined to the local scale. These delaminations did not extend catastrophically at some critical load. On the other hand, the $[60/0/-60]_s$ laminate completely uncoupled at fracture. However, there are no delaminations evident in the radiograph, Fig. 4c, taken at 99% of P_{AVE}. This implies that otherwise very small delaminations extended catastrophically completely across the specimen ligaments once their critical load was achieved. The formation of the delaminations was apparently followed immediately by the fracture of each individual ply of the laminate.

Summary and Conclusions

A study of the fracture processes in T300/5208 notched laminated composites concentrated on characterizing the influences of laminate stacking sequence on the formation of notch-tip damage zones and the subsequent mode of fracture of each ply of the laminate. The results revealed that laminate notched strength varied considerably with both a specified set of ply fiber orientations as well as the order in which the specific plies were arranged in the laminate. The fractographic examinations identified differences in the laminate fracture process which are believed to be the primary contributor to the differences in notched strength. The twelve laminates, examined herein, exhibited varying degrees of notch sensitivity as a result of the ability of a laminate to develop "subcritical" damage in a manner that relieved the high stress concentration at the notch tip.

The laminate fracture processes observed in this study have been separated into two relatively distinct processes. The first fracture process is general stress relief. Damage formation at the notch tip prior to catastrophic fracture relieved some portion of the high stress concentrated at the notch tip. Laminates with more extensive damage zones achieved more stress relief and, therefore, were less notch-sensitive. A completely notch-insensitive laminate would fail at "net section ultimate." On the other hand, some laminates formed major delaminations which resulted in ply uncoupling. The uncoupled plies were free to fail in their preferred failure mode independently of the other plies. Once a weaker ply failed completely, the remaining plies were subjected to a much higher load. Laminate failure typically occurred shortly after the delamination damage event. Laminates that exhibited this fracture process were not afforded stress relief at the notch-tip and, therefore, failed at lower laminate stress levels. These laminates are highly notch-sensitive.

Finally, the percent of laminate fracture by broken fibers was in direct correlation with the laminate notched strength. Laminates which were not characterized by damage zones containing major delaminations exhibited higher percentages of failure by broken fibers and higher notched strength. Conversely, when delaminations resulted in widespread ply uncoupling, relatively few plies of a laminate failed by broken fibers producing a weaker laminate.

Acknowledgment

This work was funded by a grant from NASA Langley Research Center. C. C. Poe, Jr. was the technical monitor. The authors also wish to thank JoAnn McCoy, who assisted in the deply laboratory work.

References

[1] Harris, C. E. and Morris, D. H., "Fracture Behavior of Thick Laminated Graphite/Epoxy Composites," NASA Contractor Report 3784, NASA Langley Research Center, Hampton, VA., 1984.

[2] Poe, Jr., C. C. and Sova, J. A., "Fracture Toughness of Boron/Aluminum Laminates with Various Proportions of 0° and ±45° Plies," NASA Technical Paper 1707, NASA Langley Research Center, Hampton, VA, 1980.

[3] Garber, D. P., "Tensile Stress-Strain Behavior of Graphite/Epoxy Laminates," NASA Contractor Report 3592, NASA Langley Research Center, Hampton, VA, 1982.

[4] Freeman, S. M., "Characterization of Lamina and Interlaminar Damage in Graphite-Epoxy Composites by the Deply Technique," *Composite Materials: Testing and Design (Sixth Conference), ASTM STP 787*, I. M. Daniel, Ed., American Society for Testing and Materials, Philadelphia, 1982, pp. 50–62.

[5] Jamison, R. D., Schulte, K., Reifsnider, K. L., and Stinchomb, W. W., "Characterization and Analysis of Damage Mechanics in Tension-Tension Fatigue of Graphite/Epoxy Laminates," *Effects of Defects in Composite Materials, ASTM STP 836*, D. J. Wilkins, Ed., American Society for Testing and Materials, 1984, pp. 21–55.

DISCUSSION

N. Ranganathan[1] (*written discussion*)—You showed interesting aspects of the damage zones at the crack/notch tip as a function of the stacking sequence. Could you comment on the analogy which might exist between these damage zones and the plastic zones observed on metallic materials?

C. E. Harris and D. H. Morris (*authors' closure*)—The load-induced zones of damage that form at the notch tips of center-notched laminated composites is to some degree analogous to the crack-tip plastic zones that form in metals. While the microstructural damage events are entirely different, the influence of the damage zone on the notched strength of laminates is often similar to the influence of the plastic zones on the notch sensitivity of metals. From the macro or continuum fracture mechanics viewpoint, the zone of damage diminishes the strength of the notch-tip singularity (or stress concentration). As the damage forms in the high-stress regions, load or stress redistribution takes place. This alters the stress gradient in the vicinity of the notch. After the damage formation and stress redistribution, higher laminate loads are required to further produce critical stresses in the laminate and bring about complete fracture.

However, often this is an oversimplified view of the laminate fracture process. On the microscale, as a single damage event occurs such as the formation of a matrix crack, the crack produces localized stress relief, but also develops highly localized stress concentrations. These newly developed stress concentrations will then precipitate the formation of a new damage event such as broken fibers in the adjacent ply or an interlaminar delamination. As the laminate load increases, this sequence or progression of damage continues until complete laminate fracture occurs. One of the significant results of the research described in the subject paper is that the laminate stacking sequence affects the localized fracture process in the notch-tip damage zones, thereby altering the laminate notched strength.

While it is tempting to draw comparisons or analogies between the fracture mechanics of metals, it is the authors' opinion that this must be done with extreme caution. Continuum fracture mechanics concepts are useful for describing the notched behavior of some laminated composites. However, the micromechanics of damage in a laminated composite is quite different than is the case with homogeneous metals.

[1]E.N.S.M.A., Poihers 86034, Cedex, France.

Brian W. Smith[1] and Ray A. Grove[1]

Determination of Crack Propagation Directions in Graphite/Epoxy Structures

REFERENCE: Smith, B. W. and Grove, R. A., **"Determination of Crack Propagation Directions in Graphite/Epoxy Structures,"** *Fractography of Modern Engineering Materials: Composites and Metals, ASTM STP 948*, J. E. Masters and J. J. Au, Eds., American Society for Testing and Materials, Philadelphia, 1987, pp. 154–173.

ABSTRACT: This paper presents the results of an investigation into the fracture surface characteristics of delaminations in graphite/epoxy composite materials. Relationships between fracture surface features and imposed singular failure conditions are illustrated for tension (Mode 1) and shear (Mode 2) loading. Interlaminar tension fractures were generated utilizing a double cantilever beam geometry. Interlaminar shear fractures were generated utilizing an end-notch flexural geometry. In both cases controlled crack initiation and progression were produced between various cross-ply orientations. Interlaminar tension fracture features are illustrated showing a relatively flat surface morphology and river markings resulting from microplane coalescence. The direction of coalescence is shown to correspond to the direction of propagation. Fractures produced by interlaminar shear exhibit numerous epoxy platelets (hackles) oriented perpendicular to the direction of resolved tension at the crack tip. These platelets were found to occur predominantly on one fracture surface, where the direction of applied shear coincided with the direction of propagation. The overall relationship of each of these features with fiber orientation is reviewed and its value in postfailure identification of crack directions and load states presented.

KEYWORDS: fractography, composite materials, graphite/epoxy, fracture, crack propagation, failure analysis, delamination

When material failure occurs, the fracture surfaces produced generally contain a physical record of the events and conditions leading to fracture. The science of understanding and interpreting this physical record, "fractography," has been relatively well-established since about the mid-1950s. However, modern fractographic analyses have only recently been applied to composite material structures. In general, fractographic studies of other

---

[1]Supervisor—Metals and Finishes and materials engineer, respectively, The Boeing Commercial Airplane Co., Seattle, WA 98124.

materials (metals and unreinforced polymers) have made significant impacts in (1) understanding the microscopic mechanisms of cracking and (2) identifying the causes of component failure. Of these two areas, the latter is perhaps the most significant. Through fracture examinations, the origin, direction of crack growth, and load conditions involved in premature component fractures can generally be identified. In many cases, the definition of defects, damage conditions, or anomalous fracture modes by such studies may be sufficient to identify the cause of fracture. In those cases where such causes are not apparent, understanding the sequence of events leading to fracture on a microscopic scale is often invaluable in directing stress or materials characterization analyses. With the recent influx of graphite/epoxy materials into aircraft use, there is a particular need to be able to examine and interpret the fracture features of composite materials. Through an understanding of these features, the cause(s) of premature fractures occurring either in service or during component tests can be identified to provide valuable feedback to designers and engineers.

In this investigation the microscopic characteristics of interlaminar fractures in graphite/epoxy laminates were examined. This work was conducted as part of an Air Force–funded composites failure analysis program [1]. Long-term efforts within this Air Force program are aimed at developing an overall postfailure analysis capability for composite materials. In the current investigation, the principal objectives were to develop a fractographic capability to identify (1) the direction of crack propagation and (2) the relative load state (shear or tension) involved in interlaminar fractures. These two objectives were selected with the postmortem analysis of failed components specifically in mind. In general, experience with composite materials has shown delaminations to be one of the more dominant fracture modes commonly observed. As with most material systems in which several failure modes exist, delaminations may in some instances comprise the principal failure mode. In other cases delaminations may represent secondary fractures formed due to the extremely low interlaminar toughness of most composite materials. In either case, the ability to identify the origin, direction, and load states of such fractures is crucial to understanding their significance and in reconstructing the sequence of events leading to failure.

In this study, these objectives were addressed by generating interlaminar fractures under controlled load and crack propagation conditions. Two interlaminar load states were examined, pure Mode 1 tension and pure Mode 2 shear. In real structures, fractures are likely to occur by neither pure interlaminar shear nor tension. However, these two load states provide a logical framework for understanding the characteristics of interlaminar fractures under model conditions. Because of the interest in this study in identifying the direction of crack growth, fracture toughness test geometries were utilized to provide controlled initiation and crack growth conditions. The use of such toughness coupons loaded under singular load states represents a relatively standard approach commonly utilized in metal fractography. However, in composite materials, the cross-ply interface and the orientation of fibers with respect to the direction of crack propagation must also be considered. In this study, fractures were produced between 0/0, 0/90, and +45/−45° ply interfaces such that the effects of differing interfacial fiber orientations could be investigated.

Procedure

Materials

A wide variety of resin systems and fiber combinations exists for use in current graphite/epoxy composite applications. However, due to the high performance demands of aircraft structures, the number of materials commonly used is limited to those systems which exhibit good environmental behavior and ultimate strengths. These matrix systems are typically 176.6°C, curing tetraglycidyl diaminodiphenyl methane (TGDDM)–diaminodiphenyl sulphone (DDS) expoxies with BF3 amine catalysts. For the following investigation, Hercules 3501-6 with AS-4 fibers was selected for examination as a material representative of such systems. A resin content of 35% by weight on 0.145 kg/m² meter unidirectional tape was selected for examination because of its extensive use in commercial and military aircraft structures.

Test Procedure

In this study, delaminations were produced in which both the direction of cracking and load state at fracture were controlled. As discussed previously, fractures were produced under pure Mode 1 tension and pure Mode 2 shear. Mode 1 tension fractures were generated utilizing a double cantilever beam geometry, whereas Mode 2 shear fractures were produced using an end-notch flexure geometry. Both specimen geometries have been used by several investigators [2–7] in the measurement of the interlaminar toughness of composite materials. Figure 1 illustrates the double cantilever beam geometry. In this specimen, two beam halves are formed at one end by an implanted fluoronated ethylene propylene (FEB) insert. Interlaminar tension conditions are produced by deflecting these two beam halves in opposite directions. The end-notch flexural specimen geometry used to generate Mode 2 shear fractures is illustrated in Fig. 2. Similar to the double cantilever beam (DCB) specimen, two beam halves are formed at one end by an implanted FEP insert. In this case, conditions of interlaminar shear are imposed at the specimen midplane by cantilever deflection of both beam halves in the same direction. Fracture along the desired ply interfaces (0/0, 0/90, and +45/−45°) was produced by fabricating $(0)_{24}$, $(0/90)_{12s}$, and $(+45/−45°)_{12s}$ laminates. In each case the FEP insert was located at the

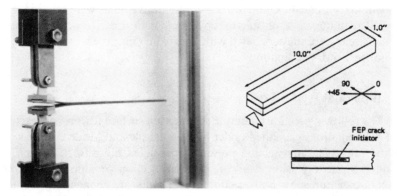

FIG. 1—*Double cantilever beam (DCB) specimen geometry and test configuration used to generate Mode 1 delaminations.*

approximate specimen midplane between the cross-plies of interest. Each laminate was cured in accordance with standard Boeing procedures at 176.6°C and 59 811 kg/m².

Testing was performed on a Mechanical Testing System (MTS) servohydraulic load frame under deflection-controlled loading. For both specimen geometries, increasing crosshead speeds were required to provide a relatively uniform rate of crack propagation. Mode 1 crack growth generally occurred in small stable increments, while Mode 2 crack growth exhibited a mixture of both rapid and slower, more stable growth. During the subject tests, macroscopic crack growth was visually observed to proceed from the implanted FEP initiator toward the opposite specimen end along the 0° direction. Following testing, fracture surfaces from five specimens of each type (Mode 1 and Mode 2) were examined using optical microscopy to verify that fracture had occurred between the intended ply interfaces. Based on these

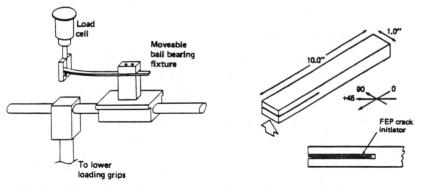

FIG. 2—*End-notch flexure (ENF) geometry and test configuration utilized to generate Mode 2 delaminations.*

preliminary optical examinations, representative areas were selected for detailed scanning electron microscopy (SEM). In order to prevent charging, these areas were sputter coated with approximately 20 nm of gold-palladium prior to SEM examination.

Experimental Work (Results)

The following sections describe the fracture surface features characteristic of delaminations generated under Mode 1 tensile and Mode 2 shear loading conditions. The first section presented considers those fractures generated under Mode 1 tension, while the second section presents Mode 2 shear fracture features. In both sections the results are sequentially organized based on the three different ply orientations (0/0, 0/90, and +45/−45°) between which fractures were generated.

Mode 1 Tension Fractures

Typically, interlaminar fractures in fiber-reinforced composite materials involve a combination of fiber/matrix separation and cohesive fracture of the surrounding matrix material. The extent to which either of these two features occur depends upon both the volume percent of reinforcing fibers and the proximity of the fracture plane to these fibers. In this study, delaminations produced under interlaminar tension typically exhibited a mixture of both fiber/matrix separation and cohesive resin fracture. In general, areas of fiber/matrix separation were found to be relatively smooth and featureless. For the most part, conditions of cohesive resin fracture dominated the overall fracture surface topography. As discussed in following paragraphs, such areas of fracture typically appeared flat and exhibited pronounced river markings and resin microflow features. The combination of these latter features generally appeared unique to interlaminar tension and was found to provide a means for identifying the direction of fracture.

0/0° Interface Fracture Surface

The fracture surface characteristics of delaminations produced under tension between adjacent 0° plies are illustrated in Figs. 3 and 4. One of the most distinctive features of this fracture is its relatively smooth, flat, planar topography. However, at higher magnifications the overall fracture surface can be seen to be composed of areas of cohesive resin fracture, fiber/matrix separation, and partially buried fibers. Of these three, regions of cohesive resin fracture constitute the predominant feature. These areas are roughly divided into longitudinal segments by locations of fiber/matrix separation and partially buried fibers. As illustrated at higher magnification, partially buried fibers occur when the fracture plane intersects underlying fibers at a shallow

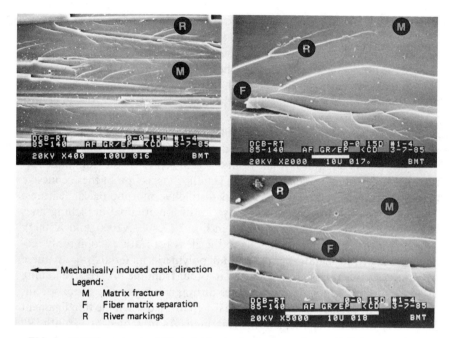

Mechanically induced crack direction
Legend:
M   Matrix fracture
F   Fiber matrix separation
R   River markings

FIG. 3—*SEM fractographs of Mode 1 delamination between adjacent 0° plies. (Figure reduced 35% for printing purposes.)*

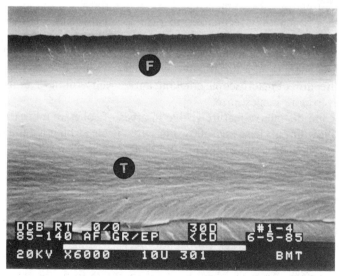

FIG. 4—*Micrograph illustrating adhesive areas of (F) fiber/matrix separation and (T) textured microflow. (Figure reduced 25% for printing purposes.)*

angle. In such cases, a general progression from cohesive matrix fracture to interfacial separation occurs. In this transition region, residual resin can generally be seen adhering to the fiber surface. However, as illustrated in Fig. 4, once the fiber intersects the surface the cleanly replicated impression of the AS-4 fiber becomes apparent. This condition indicates that fiber separation under interlaminar tension occurs adhesively, except where the fiber just intersects the fracture surface. Because of this adhesive characteristic, areas of fiber/matrix separation were generally found to be devoid of any specific morphological features related to the direction of crack propagation.

As just discussed, areas of flat resin fracture represent the dominant feature of Mode 1 tension delaminations. Detailed inspection of these cohesive matrix fracture regions reveals several distinctive morphological features. The most easily distinguishable are a series of longitudinal branching lines which form a riverlike pattern, shown in Fig. 3 (X400, X2000, and X5000). These river markings are analogous to the cleavage features commonly recognized in brittle metals, ceramics, and polymeric materials [8]. In these materials, such features are recognized to occur from the progressive joining of adjacent microscopic fracture planes during crack growth. More specifically, each line segment represents a local step formed when the thin ligament separating these displaced planes is fractured. As presented by Griffith [9], the amount of strain energy involved in fracture is proportional to the area of created fracture surface and the amount of material plastic deformation. A large number of locally displaced fracture planes represents a higher energy condition than a single continuous fracture surface. Consequently, there is a large driving force for the number of planes involved in fracture to decrease during fracture. In metals and other material systems, the joining of these planes is generally recognized to produce coalescence of each of the ligaments, resulting in a riverlike pattern. As a result, the direction of coalescence of these ligaments indicates the local direction of crack growth. This argument appears to be valid for the river marks visible in Fig. 3, since the general direction of macroscopic cracking agrees well with the average direction of river mark coalescence.

The second distinctive feature visible in areas of cohesive matrix fracture are microflow lines. These microflow lines are visible at high magnifications as a fine-grained structure or texture on the fractured resin surface (Fig. 4). Detailed inspection of the overall flow of this grained structure reveals a herringbone-shaped pattern with its open end oriented in the direction of induced crack propagation. This pattern is similar in appearance to chevron patterns commonly encountered in the fracture of metallic structures. The appearance of chevrons in metals is associated with microscopic deformation in the direction of local crack propagation. The herringbone pattern arises from the inherent tendency of a propagating crack to take the shortest path to a free surface. As such, chevrons tend to rotate from the direction of overall crack growth towards adjacent cracks or free surfaces. Based on this interpretation, the localized direction of crack propagation can be determined by examining the direction and orientation of herringbone-shaped patterns. As illustrated in Fig. 4, the induced crack propagation direction coincides with the direction of expanding and radiating microflow texture.

Cross-Ply Interfaces (0/90, and +45/−45°) Fracture Surfaces

The fracture topography typical of delaminations produced between cross-plied orientations under tension is illustrated in Figs. 5 and 6. For both cross-ply orientations examined (0/90 and +45/−45°), matrix fracture and regions of fiber/matrix separation remained the dominant fracture features. In general, cross-ply orientations produced significant differences in the extent of fiber/matrix separation and size and shape of these fractured matrix areas. As illustrated in Fig. 5, fracture between 0 and 90° ply orientations produced considerably more exposed resin fracture than in the adjacent 0° ply case just discussed. Once again, the overall flatness of these areas appears to be characteristic of Mode 1 tension. In general, examination of exposed fiber areas reveals both partially buried fibers as well as regions of adhesive fiber/matrix separation. As before, the degree of adhesive separation appears to depend upon the amount of fiber exposure, with more exposure favoring adhesive separation.

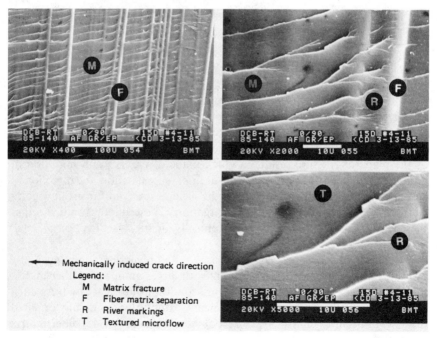

Mechanically induced crack direction
Legend:
M   Matrix fracture
F   Fiber matrix separation
R   River markings
T   Textured microflow

FIG. 5—*Fracture topography typical of Mode 1 delamination between 0 and 90° plies. (Figure reduced 35% for printing purposes.)*

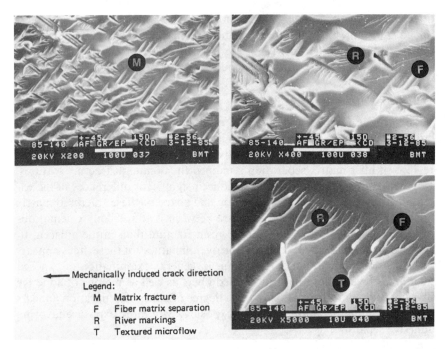

Mechanically induced crack direction
Legend:
M   Matrix fracture
F   Fiber matrix separation
R   River markings
T   Textured microflow

FIG. 6—*Fractographs of Mode 1 delamination between* +45 *and* −45° *plies.* (*Figure reduced 35% for printing purposes.*)

As illustrated at X400, locations of 0 and 90° fiber/matrix separation tend to divide the overall fracture surface into roughly rectangular areas of matrix fracture. Examination of these areas reveals both distinct river marks as well as subtle conditions of resin microflow. In this fracture, it is particularly interesting to note that river mark branching appears to be associated with areas of partially buried or exposed fibers. This suggests that local disruption of the plane of fracture results in crack plane divergence and initiation of multiple crack microplanes. As visible at high magnification, the direction of coalescence of these microplanes and the general texture of microflow coincide with the direction of crack propagation. These findings indicate that these features can be used to fractographically identify the direction of crack propagation.

Fractures produced between +45 and −45° ply orientations generally exhibited a rougher overall topography with larger amounts of fiber/matrix separation than either of the two fracture conditions just discussed. As illustrated in Fig. 6, the fracture surface typically exhibits longitudinally oriented rows of fan-shaped matrix fracture bounded by areas of 45° fiber/matrix separation. In general, these fan-shaped areas are oriented with a slight tilt to the overall fracture surface. Further detailed examination reveals river markings associated with areas of fiber/matrix separation as well as subtle conditions of resin microflow. As illustrated, the direction of river mark coalescence and microflow progression often deviates from the direction of induced crack propagation. This condition can be attributed to the local rotation of crack propagation towards adjacent microscopic fracture zones. However, as illustrated at X400, these local variations average together so that the overall direction of river mark branching and resin microflow correspond well with the direction of induced fracture. This finding indicates that the cumulative behavior of the fracture surface must be considered when utilizing microscopic features to identify the direction of crack propagation.

Mode 2 Shear Fractures

Interlaminar fractures produced under Mode 2 shear appeared distinctly different from delaminations produced under Mode 1 tension. In both cases, two principal zones of fracture were observed: (1) areas of resin matrix fracture and (2) regions of fiber/matrix separation. However, fractures produced under interlaminar shear were found to exhibit larger amounts of fiber/matrix separation and smaller, finely spaced areas of cohesive matrix fracture. In contrast to Mode 1, these finely spaced areas of matrix fracture exhibited a relatively rough topography. Detailed inspection of these areas revealed numerous inclined platelets (hackles) of fractured epoxy oriented normal to the direction of resolved tension. Under certain instances, the orientation of these hackles was found to positively correlate with the direction of induced cracking.

0°/0° Interface Fracture Surface

The fracture characteristics of delaminations produced between adjacent 0° plies under interlaminar shear are illustrated in Fig. 7. At low magnification (X400), the overall fracture surface topography is noticeably rougher than that typical of Mode 1 tension (see Fig. 3). At this magnification, regions of fiber/matrix separation interspersed with narrow rows of cohesive resin fracture are visible. In general, a larger degree of fiber/matrix separation appears evident under Mode 2 shear as compared to Mode 1 tension. Under Mode 2, flexural loads, carried by the fiber, are transferred into the interlaminar matrix region. As a result, increased conditions of fiber/matrix separation appear reasonable inasmuch as higher interfacial stresses exist. As illustrated in Fig. 7, this condition is reflected in the existence of relative zones of cohesive resin fracture confined to small narrow rows by areas of extensive fiber/matrix separation. In general, detailed inspection of these areas of fiber separation revealed the lightly crenulated impression of the AS-4 fiber, indicating adhesive interfacial separation.

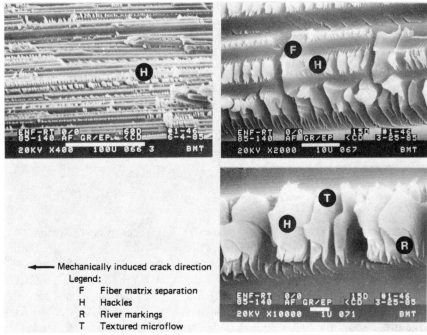

Mechanically induced crack direction
Legend:
F   Fiber matrix separation
H   Hackles
R   River markings
T   Textured microflow

FIG. 7—*Fracture topography typical of Mode 2 delamination between adjacent 0° plies. (Figure reduced 35% for printing purposes.)*

Upon detailed examination, the narrow rows of cohesive resin fracture located between areas of fiber separation were found to exhibit numerous inclined platelets. As illustrated, these platelets are almost exclusively oriented cross-wise to the 0° fiber orientation. In a variety of other investigations, these platelets have often been referred to as hackles. Other investigators [10] have suggested that hackles are a result of flexural loading associated with local bending of the fracture surface just behind the crack tip. However, the presence of hackles exclusively under conditions of Mode 2, rather than Mode 1, suggests interlaminar shear as the primary load source behind their formation. The appearance of these hackles suggests that their formation occurs by the coalescence of numerous microcracks inclined at an angle to the plane of applied shear. This model appears to be supported by the observation of S-shaped cracks located at the midplane (highest shear loading) of short beam shear specimens in previous Boeing studies [11] as illustrated in Fig. 8. In general, the hackles structures visible in Fig. 7 and those microcracks visible in Fig. 8 tend to be oriented at approximately 40 to 60° to the plane of applied shear. This orientation is approximately normal to the resolved tension component of applied shear. This finding suggests that hackle and microcrack formations occur under locally resolved

FIG. 8—*Microstructure of cracks found in short beam shear specimen tested at 132°C. (Figure reduced 25% for printing purposes.)*

tension conditions. Based on these observations, the sign of applied inter-laminar shear (clockwise or counterclockwise, that is, + or −) can be determined by examining the tilt of hackles with respect to the plane of fracture.

Further inspection of the fracture surface just discussed also reveals areas of scalloped, or concave-shaped, resin fracture (Fig. 9). Detailed examination of the size and shape of these scalloped features suggests that their formation results from the separation of hackles from the fracture surface. This conclusion appears substantiated by the existence of nearly separated hackles and the formation of underlying scalloped areas as shown in Fig. 9.

Separation of hackles can produce two possible relationships between hackle tilt and the direction of crack propagation, depending upon which side of the shear plane hackles is retained (Fig. 10). In Mechanism A, separation occurs such that hackles are retained on the side in which the direction of crack propagation coincides with the local shear component direction. This first condition produces hackles tilted in the direction of crack propagation, as well as normal to the direction of resolved tension. Conversely, in Mechanism B, separation occurs such that hackles are retained on the side in which the direction of crack propagation opposes the local shear component direction. In this latter condition, the tilt of the hackles opposes the direction of crack propagation.

In order to identify the dominant mechanism of hackle separation, both sides of each fracture surface were examined. As shown in Fig. 11, both sides

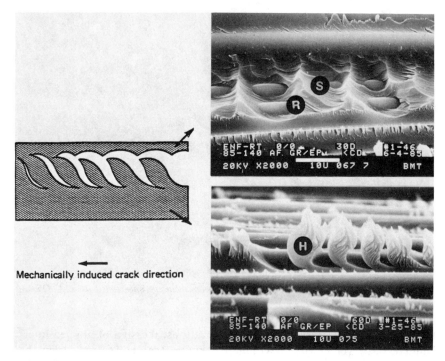

FIG. 9—*SEM micrographs illustrating scalloped resin fracture areas and their development. The top micrograph illustrates scallops and (R) river markings, while the micrograph on the bottom illustrates (H) hackle separation and underlying scallop formation. (Figure reduced 30% for printing purposes.)*

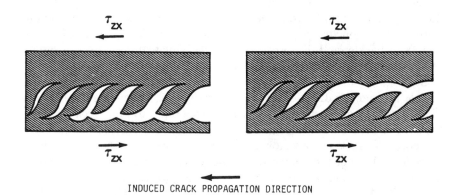

FIG. 10—*Schematic illustration of possible hackle separation mechanisms. Mechanism A illustrates hackle formation coincident with the direction of crack propagation, whereas Mechanism B illustrates the formation of hackles opposite to the direction of crack propagation.*

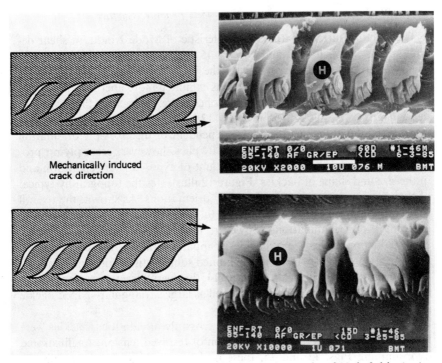

Mechanically induced crack direction

FIG. 11—*Illustration of opposing (H) hackle tilts on mating sides of Mode 2 delamination between adjacent 0° plies. (Figure reduced 33% for printing purposes.)*

of fracture between adjacent 0° plies exhibit pronounced hackle formations. The existence of hackles on both surfaces indicates that hackle separation occurs by both Mechanisms A and B. This finding indicates that hackle tilt is not a viable means of fractographically determining the direction of crack propagation for shear fractures between adjacent 0° plies.

Several finer morphological features are apparent on both hackle and scalloped areas. Both hackles (Fig. 7) and scalloped areas (Fig. 9) exhibited textured resin microflow and branched river markings analogous to those identified under Mode 1. For the most part, these features appear to emanate from areas of fiber/matrix separation and rotate to align with the direction of hackle tilting, either coincident with or opposite to the direction of imposed crack propagation (Mechanism A or B). Based on their microscale, these features probably reflect the local direction of crack propagation involved in hackle formation and separation. The emanation of these features from areas of fiber/matrix separation suggests that hackle formation initiates along the fiber interface and progresses either coincident with or opposite to the direction of overall propagation.

Cross-Ply Interfaces (0/90 and +45/−45°) Fracture Surface

The fracture surface features characteristic of Mode 2 induced shear delaminations between 0/90 and +45/−45° oriented plies are illustrated in Figs. 12 through 15. Consistent with the just-cited findings, delaminations produced between cross-ply orientations exhibited more fiber/matrix separation and smaller, narrower rows of cohesive resin fracture than Mode 1 delaminations. As illustrated in Figs. 12 and 14, areas of cohesive resin fracture exhibited inclined hackles independent of the cross-ply orientation examined. In comparison to adjacent 0° plies, however, cross-plying produced some slight alteration in the amount of exposed fiber separation and in the size and shape of hackles. Figure 12 illustrates the topography typical of that produced between 0 and 90° ply orientations. As shown, the overall topography appears somewhat similar to the morphology characteristic of adjacent 0° ply delaminations. However, in comparing Figs. 7 and 12, some slight reduction in the size of hackles is apparent. In contrast, delamination between +45/−45° ply interfaces exhibited some distinct differences in fracture morphology. As illustrated in Fig. 14, the overall fracture surface exhibited more cohesive matrix fracture as well as larger, triangular-shaped hackle structures.

With respect to both of the just-cited cross-ply orientations, hackles were found tilted normal to the local direction of resolved tension. As illustrated by examining Figs. 13 and 15, these hackles were found to occur almost exclusively on one side of the fracture surface, with scallops on the adjoining side. With respect to Fig. 10, these hackles are tilted coincident with the direction of crack propagation, indicating separation by Mechanism A. In contrast to the adjacent 0° ply condition, this observation indicates that a positive correlation exists between hackle tilt and the direction of crack propagation for cross-ply interfacial conditions. However, further studies

 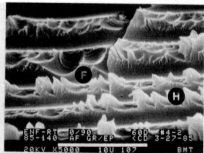

Mechanically induced crack direction

FIG. 12—*Fracture morphology of Mode 2 delamination between 0 and 90° plies illustrating areas of (F) fiber/matrix separation and (H) hackles. (Figure reduced 35% for printing purposes.)*

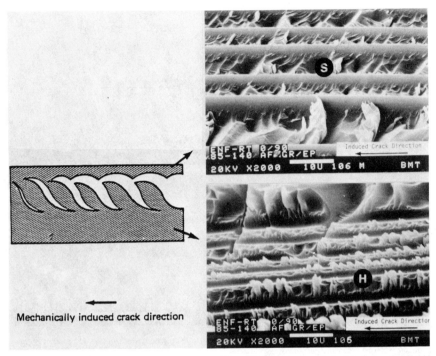

FIG. 13—*Topography of mating fracture surfaces produced under Mode 2 between 0 and 90°ply orientations. (H) Hackle separation occurs by Mechanism A with hackles tilted in the direction of crack propagation. (Figure reduced 30% for printing purposes.)*

need to be performed to determine if the mechanism of hackle separation is dependent upon conditions such as specimen geometry, lay-up, and degree of mixed mode loading.

Further inspection of both hackles and scalloped fracture areas reveals several finer morphological features in addition to the gross topography just discussed. Detailed examination of hackles and scalloped areas revealed conditions of both textured resin microflow as well as branched river mark features. Both the river marks and resin microflow features appear for the most part to emanate from regions of fiber/matrix separation. In the case of delamination between adjacent 0° plies, discussed previously, these features were found to progress both towards and away from the macroscopic direction of crack propagation. However, in contrast to this case, microflow and river markings for both cross-ply orientations were found to progress only in the direction of crack propagation. Because of their microscopic nature, these features probably reflect the microscopic direction of cracking involved in fracture. Consequently, hackle formation between cross-ply orientations appears to occur by the initation and growth of S-shaped microcracks in the

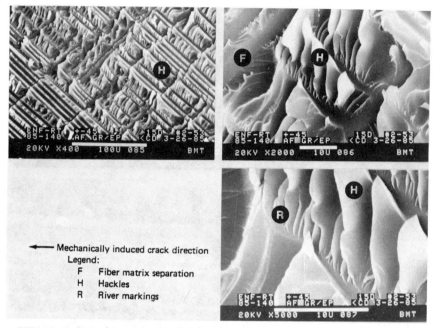

FIG. 14—*Fractographs of Mode 2 shear delamination between +45 and −45° plies. (Figure reduced 35% for printing purposes.)*

direction of macroscopic crack growth. This situation supports the interpretation of hackle separation by Mechanism A (see Fig. 10).

Discussion

These findings have considered the characteristics of interlaminar fractures generated under model conditions. While real world delamination fractures usually occur due to a mixture of tension and shear loads, the results presented do suggest several fractographic trends which may be used to identify the direction and load state of more complex fractures. In examining fractures from in-service or test components, topographies which are relatively flat can probably be associated with Mode 1 loading. The determination of the fracture direction for these types of delamination surfaces should be relatively straightforward, using either the cumulative direction of river mark coalescence or direction of resin microflow. On the other hand, fractures which exhibit a relatively rough topography are somewhat more difficult to analyze. Based on the just-cited findings, these fractures are likely to have involved a substantial amount of Mode 2 loading. In such cases, the tests performed within this paper suggest that microscopic crack growth can occur either coincident with or opposite to the direction of overall separa-

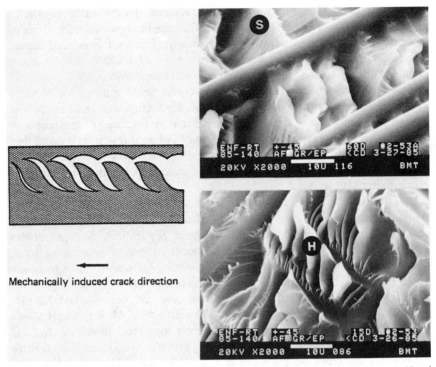

Mechanically induced crack direction

FIG. 15—*Fractographs of mating fracture surfaces produced under Mode 2 between +45 and −45 degree plies. Coincident with Mechanism A, (H) hackles are retained on one side, tilted in the direction of crack growth. (Figure reduced 30% for printing purposes.)*

tion, depending on which mechanism (A or B) predominates. In this study Mechanism A was observed to predominate for the 0/90 and +45/−45° interfacial fracture conditions. However, given the complexity of component fractures it is doubtful that a single mechanism can be assumed. Consequently, it is unlikely that a single direction of crack growth can be defined for shear fractures. The symmetry and shape of hackles do, however, align parallel with the direction of propagation. This may provide a relatively easy means for determining the direction parallel to which crack growth occurred.

Conclusions

In this study, the fracture surface characteristics of delaminations produced under Mode 1 tension and Mode 2 shear were investigated for a variety of interfacial ply orientations (0/0, 0/90, and +45/−45°). Conclusions of this investigation were:

1. Mode 1 tension fractures could be differentiated from Mode 2 shear fractures based upon their relative fracture surface topographies. Fractures

produced under Mode 1 tension were characterized by the existence of relatively large areas of flat cohesive matrix fracture interspersed with areas of fiber/matrix adhesive separation. Conversely, fractures produced under Mode 2 shear exhibited relatively narrow rows of hackled resin fracture separated by extensive amounts of fiber/matrix separation.

2. The direction of crack propagation could be determined for Mode 1 tension fractures by the examination of river markings and resin microflow features present on flat areas of cohesive matrix fracture. Examination of these features with respect to the direction of imposed fracture revealed that river mark branch coalescence and expanding microflow coincide with the direction of crack propagation in the same manner as for metallic and polymeric materials.

3. Detailed examination of cohesive resin fracture areas generated under Mode 2 shear revealed epoxy platelets inclined approximately normal to the resolved tensile component of applied shear. Separation of these hackles from the fracture surface was shown to occur by one of two mechanisms, yielding hackles aligned either coincident with (Mechanism A) or opposite to (Mechanism B) the direction of crack propagation.

4. The direction of crack propagation and tilt of observed hackles were found to correlate positively for 0/90 and +45/−45° interfacial Mode 2 fractures, which occurred predominantly by Mechanism A. In this case, the tilt of observed hackles may provide an identifiable feature by which local crack propagation can be determined. However, given the complexity of actual component fractures, this correlation may not be usable.

5. Fractures produced between adjacent 0° plies typically occurred by a mixture of both Mechanisms A and B. Therefore, the direction of crack propagation could not be determined via the examination of hackle tilt orientation for this interface condition. However, further studies should be performed to determine if the mechanism of hackle separation is dependent upon specimen test geometry.

6. In general, both hackles and their regions of separation (scallops) exhibited both river markings and resin microflow conditions similar to that noted under Mode 1 fracture. These features were found to initiate at regions of fiber/matrix separation and to progress in the direction of hackle tilt.

Acknowledgments

The authors wish to acknowledge the contributions of R. E. Smith, S. D. Pannell, F. J. Fechek, and Dr. A. G. Miller for their assistance in these investigations.

References

[1] "Failure Analysis for Composite Structure Materials," Contract F33615-84-C-5010, Air Force Systems Command, Aeronautical Systems Division/PMR RC, Wright Patterson Air Force Base, 1984.

[2] Miller, A. G., Hertzberg, P. E., and Rantala, V. W., *Proceedings*, Twelfth National SAMPE Technical Conference, Seattle, WA, October 1980, Society for the Advancement of Material and Process Engineering, New York, p. 279.

[3] Devitt, D. F., Schapery, R. A., and Bradely, W. L., "A Method for Determining the Mode I Delamination Fracture Toughness of Elastic and Viscoelastic Composite Materials," *Journal of Composite Materials*, Vol. 14, 1980, pp. 270–285.

[4] Jurf, R. A. and Pipes, R. B., "Interlaminar Fracture of Composite Materials," *Journal of Composite Materials*, Vol. 16, 1982, pp. 386–394.

[5] Vanderkley, P. S., "Mode I-Mode II Delamination Fracture Toughness of Unidirectional Graphite/Epoxy Composite," Report MM 3724-81-15, Texas A&M University, College Station, TX, December 1981.

[6] Wilkins, D. J., Eisenmann, J. R., Camin, R. A., Margolis, W. S., and Bemson, R. A., "Characterizing Delamination Growth in Graphite-Epoxy," in *Damage in Composite Materials, ASTM STP 775*, 1982, pp. 168–183.

[7] Ramkumar, R. L., "Performance of a Quantitative Study of Instability-Related Delamination Growth," NASA CR 166046, March 1983.

[8] *Failure Analysis And Prevention*, Vol. 10, American Society for Metals, Metals Park, OH, 1975.

[9] Griffith, A. A., *Philosophical Transactions, Royal Society of London*, Vol. 221A, 1920, pp. 163–198.

[10] Morris, G. E., "Determining Fracture Directions and Fracture Origins on Failed Graphite/Epoxy Surfaces," *Nondestructive Evaluation and Flaw Criticality for Composite Materials, ASTM STP 696*, 1979, pp. 274–297.

[11] Miller, A. G. and Wingert, L. E., "Fracture Surface Characterization of Commercial Graphite/Epoxy Systems," *Nondestructive Evaluation and Flaw Criticality for Composite Materials, ASTM STP 696*, pp. 223–273.

Composites III: Particulate Composites

Golam M. Newaz[1]

Microstructural Aspects of Crack Propagation in Filled Polymers

REFERENCE: Newaz, G. M., **"Microstructural Aspects of Crack Propagation in Filled Polymers,"** *Fractography of Modern Engineering Materials: Composites and Metals, ASTM STP 948*, J. E. Masters and J. J. Au, Eds., American Society for Testing and Materials, Philadelphia, 1987, pp. 177–188.

ABSTRACT: A study was undertaken to evaluate the fracture behavior of sand and clay-filled polyester composites. A two-pronged effort was made (1) to characterize the fracture toughness of sand-polyester and sand-clay-polyester composites and (2) to identify the mechanisms which account for fracture resistance in these composites.

Separate laminates of sand-resin and sand-clay-resin were made using a polyester resin, sand (300-μm mesh), and clay (0.5- to 3-μm mesh) particulates. The volume fraction of the constituents were 30/70 and 23/7/70 in the laminates, respectively. Three-point bend notch tests were conducted to record the load-deflection response of the composites. Fracture energy was calculated based on linear elastic fracture mechanics. Results show that the fracture energy of the sand-polyester composite was about 25% greater than that of the sand-clay-polyester composite. Fractographic study was undertaken using a scanning electron microscope to determine the mechanisms which contribute to fracture resistance in the two composites. This evaluation was considered critical to evaluate the role of the sand and clay fillers on the overall fracture behavior.

In the sand-filled specimens, fracture steps were observed behind the particles, which are characteristic of the pinning mechanisms. These features arise from the crack front breaking away from the pinning position, that is, from the rigid sand particles. The toughening mechanism in this composite is essentially attributed to the crack-pining phenomenon and subsequent interaction of various crack fronts. In this case, the particles act as obstacles to the moving crack front, and additional energy is required to break away from the pinned positions, contributing to higher fracture energy. This increase in line energy of the crack front is analogous to the line-tension effect due to dislocation pinning in metals. However, in the sand-clay-polyester composite, the pinning mechanism appears to be absent. Rather, the fracture surface is rough and is reminiscent of quasicleavage-type damage. Fracture energy in this case may be related to the size and density of cavities. The toughening mechanism in this type of hybrid particulate-filled composite is not well understood and requires further investigation.

KEYWORDS: fracture energy, particulate composite, crack-pinning mechanism, fracture step, crack propagation

[1]Principal research scientist, Mechanics Section, Battelle Memorial Institute, Columbus, OH 43201; formerly, Composites and Textile Applications Laboratory, Owens-Corning Fiberglas Corp., Technical Center, Granville, OH 43023.

A new generation of cost-effective composites, namely, particulate-filled composites, are receiving considerable attention. Particulate-reinforced polymers have found application in many areas requiring improved composite properties such as (1) compressive strength, (2) fracture toughness, (3) stiffness, (4) dimensional stability, (5) impact strength, and (6) creep characteristics. A particulate composite may be designed to offer one or more of these properties. An important consideration in the use of particulate composite is their cost-effectiveness. Quite often, to achieve one or more of the just-cited properties, usually the more costly matrix material can be replaced by less costly particulates, ensuring material cost advantages. Discussions on various aspects of filler properties and their usage in various polymers are given by Katz and Milewski [1] and Nielsen [2].

In dealing with the issue of fracture behavior of rigid particulate-filled composites in this study, two low-cost polymer composites were selected. Sand particulates are used to provide stiffness in particulate composites. This is a cost-effective and efficient way of gaining stiffness. Clay particulates are normally used in polymer composites to improve flow properties during processing and to provide uniform laminate properties. However, their influence on fracture behavior of particulate composites is not well known. The primary motivations for this study were (1) to characterize the fracture energy of the two composites and (2) to evaluate the crack propagation characteristics, that is, to study the dominant mechanisms responsible for overall toughness in the filled polymers. An understanding of the fracture processes is considered essential to develop tough filled composites.

The subject of fracture energy as it relates to filled composites has received much attention in the past 15 years. Studies by Lange and Radford [3], Broutman and Sahu [4], Evans [5], Lange [6,7], Maxwell et al. [8], and Newaz [9] have addressed several aspects of fracture in polymers filled with rigid fillers. Maxwell et al. [8] have also referred to the work of Spanoudakis[2] in this area. It is generally recognized that the dominant mechanism contributing to fracture energy in filled polymers with rigid particulates is the crack pinning phenomenon. The rigid particulates act as obstacles and cause the crack front to bow out between the fillers. This fracture process is associated with additional energy absorption. This increase in line energy of the crack front is analogous to the line-tension effect due to dislocation pinning in metals. An attempt is made in this study to further understand this fracture process. Also, a hybrid particulate composite, sand-clay-polyester composite, is studied to evaluate its fracture behavior. Hybrid particulate composites have not been studied as thoroughly as the single particulate-filled composites.

[2]J. Spanoudakis, Ph.D thesis, Queen Mary College, University of London, 1980.

Approach

In this section, a brief discussion is forwarded on fracture energy, G_c, which is used to assess the resistance to crack propagation in materials.

G_c can be indirectly evaluated, first by determining the critical stress intensity factor, K_c. For a notched specimen subjected to bending, the critical stress intensity factor is given by

$$K_c = \frac{PS}{BW^{3/2}} \cdot \frac{Y_1}{Y_2} \tag{1}$$

where

$Y_1 = 3(a/W)^{1/2}[1.99 - a/W(1 - a/W)(2.15 - 3.93a/W - 2.7(a/W)^2]$
$Y_2 = 2(1 + 2a/W)(1 - a/W)^{3/2}$
P = maximum load,
S = span length,
B = specimen thickness,
W = specimen depth, and
a = crack length.

These parameters are illustrated in Fig. 1.

Equation 1 is a wide-range stress intensity factor proposed by Srawley [10] for flexure specimens. It may be mentioned that many other expressions are available. However, they are restricted to certain a/W ratios.

From the load deflection response of notched specimens under bending, the maximum load can be estimated, and the critical stress intensity factor, K_c, can be calculated according to Eq 1. Subsequently, fracture energy can be evaluated for a plane stress condition using

$$G_c = K_c^2/E \tag{2}$$

where E = modulus of the material.

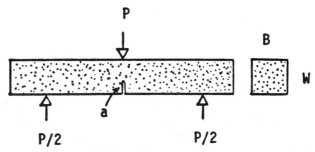

FIG. 1—*Three-point loading of specimen for fracture energy evaluation.*

Materials and Testing

Separate laminates of sand-resin and sand-clay-resin were made using sand particulates having a nominal diameter of 300 μm, clay platelets having a dimension of 0.5 to 3 μm, and an orthophthalic polyester resin. All particulates were untreated prior to mixing. The resin was diluted with styrene. The resin-to-styrene ratio was 100:32. The promoter, accelerator, and initiator levels used were 0.20%, cobalt octate (12%), 0.05% dimethylaniline (DMA), and 2.00% methyl ethly ketone peroxide (MEKP) based on resin weight.

To reduce separation of materials in laminates, the mixture of resin and particulates with initiator was continually mixed (low speed) for a period of 6 to 6.5 min before pouring it into casting cells. The remaining 1.5 to 2.0 min prior to the system's gelations were used to pour the casting cells (40.64 by 40.64-cm glass plates, with rubber at edge) to the required levels. Gelation took place 10 to 35 s after the pouring cycle.

Both castings were gelled at room temperature and cooled prior to a 121°C (250°F) postcure cycle for 2 h. Specimens were cooled slowly (still in casting cells) to room temperature after their postcure cycle. The volume fractions of the sand-resin and sand-clay-resin laminates were 30/70 and 23/7/70, respectively.

The tension and fracture toughness tests were conducted with dogbone, and three-point bending specimens (Fig. 2) were cut from these laminates. The notches in fracture toughness specimens were cut using a circular diamond saw. For each material, a total of six specimens were tested.

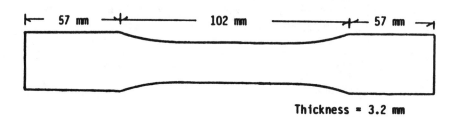

Thickness = 3.2 mm

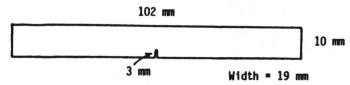

FIG. 2—*Geometry and dimension of tension and fracture toughness specimens.*

Scanning electron microscopy (SEM) was used for both types of specimens to determine the fracture surface characteristics.

Results and Discussion

The tensile stress-strain data of the two composites are given in Table 1. It is clear that the presence of clay tends to embrittle the filled composite. This is evidenced by the fact that the modulus is higher for the sand-clay-polyester composite and the strain to failure is lower compared to the sand-polyester composite. In a related study, Newaz [9] investigated acoustic emission from these composites under applied tensile loading. The microcracking behavior and subsequent fracture as monitored using acoustic emission also confirmed the relative brittleness of the sand-clay-polyester composite.

In assessing propagation characteristics, evaluation of fracture energy or G_c provides an indication of resistance of the composite to crack growth from preexisting cracks. First K_c-values were determined using Eq 1 from load-deflection curves (Fig. 3). G_c-values were then evaluated using Eq 2. For the sand-polyester and sand-clay-polyester composites, fracture energy values are given in Table 2. The neat unsaturated polyester resin fracture energy is $56 \text{ N} \cdot \text{m/m}^2$. Both the composites offer substantially improved energy over the neat resin. However, between the two composites, the sand-polyester composite exhibits greater fracture resistance based on the average G_c-values. We had already noted that the sand-clay-polyester composite was more brittle compared to the sand-polyester composite. From an energy standpoint, both the composites can be classified as semibrittle.

TABLE 1—*Tensile properties of filled polymers.*

	Sand-Polyester	Sand-Clay-Polyester
Modulus, GPa	6.70, 0.49[a]	8.02, 0.53
Strain to inelasticity, %	0.32, 0.04	0.26, 0.04
Proportional limit, MPa	19.00, 0.30	18.10, 0.50
Failure strain, %	0.75, 0.09	0.49, 0.05
Strength, MPa	26.40, 0.47	25.20, 1.24

[a]Standard deviation.

TABLE 2—*Fracture energy of filled polymers.*

	Fracture Energy, G_c, $(\text{N} \cdot \text{m/m}^2)$
Sand-polyester	325, 20[a]
Sand-clay-polyester	260, 12

[a]Standard deviation.

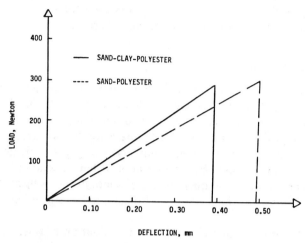

FIG. 3—*Load-deflection response of notched specimens under three-point bending.*

In dealing with particulate composites, it is essential to determine what controls the fracture energy of such systems. The possible parameters or mechanisms are as follows:

1. Increased surface area of fillers.
2. Energy absorption by dispersed phase.
3. Interaction between crack front and dispersed phase.
4. Size and density of defects.

The first mechanism is often claimed to have the most influence on fracture energy. Generally, particulate composites with smaller particles tend to have higher fracture energy than the ones with larger particles. Total surface area difference between such composites is considered to be the source of this difference in fracture energy. In a pioneering work, Lange and Radford [3] have shown that, although surface area difference has some influence, this is not the most important parameter which controls fracture energy in rigid particulate composites. The work of Maxwell et al. [8] also confirms this. The surface area argument can be easily disputed since fracture energy data as a function of particulate volume fraction shows a maxima for a given particle dimensions as shown in Fig. 4. The implication is that if surface area was the dominant parameter, then fracture energy would continue to increase as a function of particulate volume fraction. The second item as just mentioned is not a possibility in the case of rigid particulate composites, for example, sand-filled composites. However, for rubber-type particulates, energy absorption by the dispersed phase could be an important contributing factor.

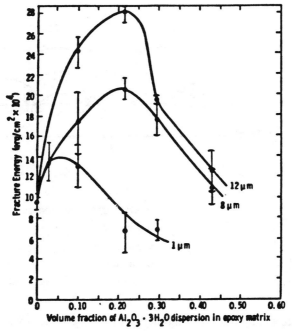

FIG. 4—*Fracture energy versus volume fraction of alumina trihydrate-filled epoxy composites showing presence of maxima* [3].

Lange and Radford [3], Lang [6,7], and Maxwell et al. [8] have clearly demonstrated that with rigid particulates, the dominant mechanism of energy absorption is due to the interaction of the crack front and dispersed phase. The mechanism is illustrated in Fig. 5. Lange proposed that the increase in fracture energy over the matrix fracture energy is due to the energy required to move the crack between particles as given by the following equation

$$G_c = 2(\gamma_0 + T/d) \tag{3}$$

where

G_c = fracture energy of filled composite,
γ_0 = matrix fracture energy,
T = line energy per unit length of crack front, and
d = interparticle spacing.

In our own studies, the presence of this mechanism for sand-filled particulate composites has been confirmed as shown in Fig. 6. In this figure, fracture steps are clearly seen on the fracture surface of the matrix material.

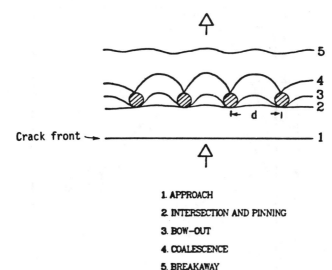

1. APPROACH

2. INTERSECTION AND PINNING

3. BOW–OUT

4. COALESCENCE

5. BREAKAWAY

FIG. 5—*Various stages of crack-pinning mechanism in rigid particulate-filled composites.*

The steps are a consequence of momentary pinning of the crack front at rigid sand particulates as it moves through the composite. The influence of inter-particle spacing on crack pinning is discussed at length by Evans [5]. The steps are formed perpendicular to the moving crack front. The pattern of these steps, commonly referred to as "river" patterns, can be used to trace the shape of the crack front at any position during fracture. In studying the

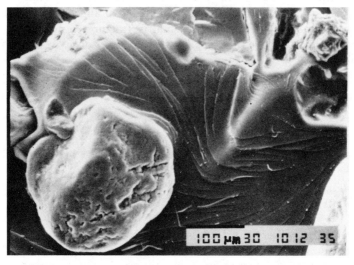

FIG. 6—*Scanning electron photomicrograph showing fracture steps due to crack-pinning mech-anism in sand-polyester composite.*

fracture energy problem, previous workers have identified the pinning or crack-front interaction mechanism quite well. Lange and Radford [3] pointed out that beyond the maxima in the fracture energy our findings clearly show that in this regime defects within the composites, that is, voids and poor particle-matrix bonding due to processing, may also play a siginficant role on overall fracture energy [11,12]. Composites with higher volume fraction of particulates beyond maxima have lower fracture energy, and crack-pinning steps are not evident. This is due to the fact that defect population increases with increasing particulate volume fraction [11]. Thus, the spectrum of fracture energy versus volume fraction can be divided into two regions. The pinning mechanism controls the fracture energy up to the maxima as shown in Fig. 7. And beyond it the fracture energy is to a large extent controlled by defects in the composite.

In the sand-clay-polyester composite, the pinning mechanism appears to be absent (Fig. 8). Rather, the fracture surface is rough with numerous crests and crevices. A high magnification (X20 000) SEM fractograph of this composite did not show any fracture steps (Fig. 9). The toughening mechanism in this type of hybrid particulate-filled composite is not well understood and requires further investigation. The difference in fracture energy may in part be due to the greater distance of rigid sand particulates in sand-clay-polyester composite provided clay particulates are considered not to influence the fracture.

Studies to determine the mechanisms of fracture in hybrid particulate systems have been initiated [9,13]. Preliminary results show that clay particles embrittle the polyester matrix, and the clay-polyester pseudomatrix is expected to have lower fracture energy than the original polyester matrix. This embrittling effect is also substantiated by the composites data in Table 1.

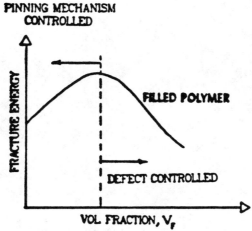

FIG. 7—*Controlling mechanisms influencing fracture in rigid particulate-filled polymers in relation to the maxima in fracture energy versus volume fraction of filler curve.*

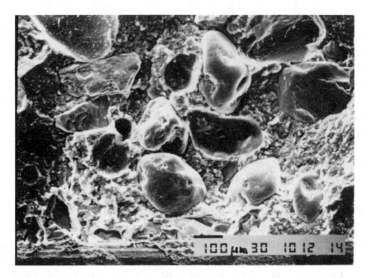

FIG. 8—*Scanning electron photomicrograph showing fracture surface without the presence of fracture steps in sand-clay-polyester composite.*

The clay phase is considered to blend with the matrix to form a pseudomatrix phase. Certainly the clay phase takes part in the fracture process as do the other constituents. However, the role of clay phase is not clearly understood although studies in this area have been initiated [*13*]. Furthermore, since the amount of sand particulates is lower in a sand-clay-polyester composite than in a sand-polyester composite for the same particle volume frac-

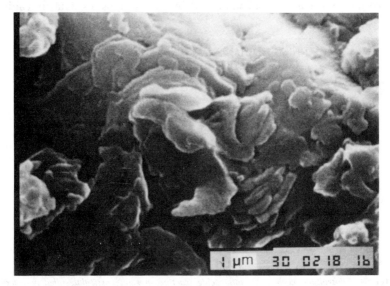

FIG. 9—*Absence of crack-pinning mechanism in sand-clay-polyester composite (X20 000).*

tion, the sand interparticle distance, d, is greater. This will lower the fracture energy. Therefore, based on Eq 3, the sand-clay-polyester is expected to have lower G_c. However, much work is needed to clearly determine the complex role of clay particulates in altering the fracture behavior of hybrid particulate composites.

Summary

1. The average fracture energy of the sand-polyester composite is greater than the average fracture energy of the sand-clay-polyester composite. However, considering the scatter in modulus values, the difference in fracture energy is not significant. The difference in fracture energy is entirely due to the modulus values according to Eq 2.

2. The toughening mechanism in two-phase rigid particulate composites such as a sand-filled polyester composite is primarily due to a crack pinning mechanism and subsequent interaction of various crack fronts.

SEM fractography has clearly confirmed the presence of fracture steps indicative of the crack-pinning mechanism in the sand-polyester composite.

3. The toughening mechanism in sand-clay-polyester is not due to the crack-pinning process. The fracture energy is likely to depend, physically, on the size and density of the cavities on the fracture surface. The fracture process in hybrid particulate composites, for example, sand-clay-polyester, requries further investigation for the determination of the physical processes which control the energy release rate.

Acknowledgment

The author acknowledges the assistance of Terry Beaver of Materials Engineering and Rodney Holman of Materials Science Laboratories, Fiberglas Technical Center, Granville, Ohio for specimen preparation.

References

[1] Katz, H. S. and Milewski, J. V., *Handbook of Fillers and Reinforcements for Plastics*, Van Nostrand Reinhold Co., New York, 1978.
[2] Nielsen, L. E., *Mechanical Properties of Polymers and Composites*, Vols. 1 and 2, Marcel Dekker, Inc., New York, 1974.
[3] Lange, F. F. and Radford, K. C., *Journal of Material Science*, Vol. 6, 1971, pp. 1197–1203.
[4] Broutman, L. J. and Sahu, S., *Materials Science and Engineering*, Vol. 8, 1971, p. 98.
[5] Evans, A. G., *Philosophical Magazine*, 1972, pp. 1327–1344.
[6] Lange, F. F., *Philosophical Magazine*, 1970, pp. 983–992.
[7] Lange, F. F. in *Fracture and Fatigue*, Vol. 5, L. J. Broutman and R. H. Krock, Eds., Academic Press, New York, 1974, pp. 1–44.
[8] Maxwell, D., Young, R. J., and Kinloch, A. J., *Journal of Materials Science Letters*, Vol. 3, 1984, pp. 9–12.
[9] Newaz, G. M., "Fracture in Particulate Composites," *Journal of Materials Science Letters*, Vol. 5, 1986, pp. 71–72.

[10] Srawley, J. E., *International Journal of Fracture*, Vol. 12, 1976, pp. 475–476.

[11] Newaz, G. M., unpublished data, Fiberglas Technical Center, Granville, Ohio, 1985.

[12] Joneja, S. K. and Newaz, G. M., "Fracture Toughness and Impact Characteristics of a Hybrid System: Glass-Fiber/Sand/Polyester," *Effects of Defects in Composite Materials, ASTM STP 836*, D. J. Wilkins, Ed., American Society for Testing and Materials, Philadelphia, 1984, pp. 3–20.

[13] Newaz, G. M., "Tensile Behavior of Clay-Filled Polyester Composites," *Journal of Polymer Composites*, Vol. 7, No. 3, 1986, pp. 176–181.

David H. Allen,[1] Charles E. Harris,[1] Eric W. Nottorf,[2] and Graeme G. Wren[2]

A Fractographic Study of Damage Mechanisms in Short-Fiber Metal Matrix Composites

REFERENCE: Allen, D. H., Harris, C. E., Nottorf, E. W., and Wren, G. G., "A Fractographic Study of Damage Mechanisms in Short-Fiber Metal Matrix Composites," *Fractography of Modern Engineering Materials: Composites and Metals, ASTM STP 948,* J. E. Masters and J. J. Au, Eds., American Society for Testing and Materials, Philadelphia, 1987, pp. 189–213.

ABSTRACT: A general constitutive theory is formulated for a short-fiber metal matrix composite. The constitutive theory is based on continuum mechanics with constraints provided by fracture mechanics and thermodynamics. The basic premise of the constitutive theory is that load-induced microstructural damage associated with the inclusion of the fibers in the metal matrix results in a loss of material stiffness.

The concept of the damage-dependent constitutive theory was evaluated by an experimental investigation of the microstructural damage in an Aluminum 6016-T6 matrix with silicon carbide (SiC) whiskers. The primary objective of the investigation was the identification of microstructural damage such as voids or cracks that were associated with the presence of the SiC whiskers.

The results of the experimental investigation include measured reductions in the elastic modulus of tension specimens which were loaded beyond yield and unloaded. Also, there was an associated change in the extent of the microstructural damage. These results lead to the conclusion that the effect of microstructural damage must be included in the constitutive relationship in order to fully describe the behavior of a metal matrix composite.

KEYWORDS: composite material, metal matrix, damage, fatigue, scanning electron microscopy, nondestructive evaluation, continuum mechanics, thermodynamics, internal state variables, SiC, Aluminum 6061-T6

In recent years, metal matrix composites have received increasing attention as a viable material for structural applications. Along with the normal advantages associated with tailorability of directional strength, these composites can be used at elevated temperatures that would adversely affect the

[1]Associate professor and assistant professor, respectively, Aerospace Engineering Dept. Texas A & M University, College Station, TX 77843.
[2]Research assistant, Aerospace Engineering Dept., Texas A & M University, College Station, TX 77843.

integrity of conventional polymeric matrix composite systems. Metal matrices also offer higher strength and stiffness than polymeric matrices over a broad range of temperatures. Further advantages of metal matrix composites are that machining is possible by using conventional techniques and equipment, most metal matrices are weldable, and short-fiber or particulate composites can be extruded or rolled to various desired shapes.

A metal matrix composite consists of a metal matrix that is reinforced with short fiber, particulate, or continuous fiber material. The matrix is generally aluminum or titanium, but reinforcements vary from silicon carbide (SiC) particles to boron, graphite, or aramid fibers. Production techniques for the material include powder metallurgy for particulate or short-fiber composites and diffusion bonding, vapor deposition, and plasma spraying for continuous fiber composites. Along with polymeric composites, the continuous fibers in the metal matrix composites can be oriented in different directions to tailor directional strength and stiffness to satisfy particular design requirements.

To effectively utilize the capabilities of metal matrix composites, a model that simulates the behavior of the material must be developed. Since composites develop load-induced microstructural fracture during their operational life, this damage must be incorporated into the model in order to construct a valid constitutive model. Both theoretical and experimental efforts are required to develop the model.

This paper presents a model that will predict the thermomechanical constitutive behavior of randomly oriented short-fiber, quasi-isotropic metal matrix composites under both monotonic and cyclic fatigue loading conditions. The model incorporates various effects such as damage, inelastic strain, and viscoelasticity in the constitutive equations with constraints obtained from continuum mechanics and thermodynamics with internal state variables.

The concept of the damage-dependent constitutive model requires experimental verification. Also, the full development of the model requires an experimentally generated data base. Therefore, this paper also presents the results of an experimental program aimed at addressing the just-cited two issues. The material system selected for study is a 6061-T6 aluminum matrix with chopped silicon carbide whiskers. The objective of the experimental program is to determine if there is a load-induced reduction in the elastic modulus of tension specimens along with an associated observable physical phenomenon such as a change in the state of microstructural damage.

Previous Research

A vast amount of research has been reported on composite materials. Most of the work involves experimental studies on the behavioral response of materials. To a lesser extent, general constitutive models have been developed. General information pertaining to metal matrix composites is presented by Vinson and Chow [1] and Renton [2]. These authors discuss fabrication techniques for both continuous fiber and particulate reinforced metal matrix composites as well as selected applications and properties of the material. Fabrication techniques include powder metallurgy, liquid metal infiltration, diffusion bonding, electroforming, rolling, extrusion, pneumatic impaction, and plasma spray. There have been a number of investigations [3–19] that have studied the physical behavior of metal matrix composites. Divecha et al. [7] have studied aluminum with short whisker (SiC) reinforcement. The authors found that the aluminum (Al)/SiC composite properties varied significantly with the amount of SiC whiskers used. Maclean and Misra [8] have examined the applications of Al/SiC composites for aerospace use. They found that although the strength and stiffness of the material were higher than that of conventional aluminum, fracture toughness decreased.

In order to observe damage in metal matrix composites, various experimental techniques have been investigated. Pipes et al. [10] have observed the acoustic emission response of various laminates of boron (B)/Al composites. Both mechanical and acoustic emission responses of the composites are presented. Various nondestructive examining techniques such as ultrasonic C-scan, radiography, acoustic emissions, stiffness change, edge replication, and eddy current were investigated by Ulman and Henneke [11] to evaluate large-scale imperfections and damage. X-ray radiography and ultrasonic C-scan were found to be excellent methods to observe both initial imperfections and damage due to loading.

Numerous investigative groups [12–19] have reported the observation of damage in metal matrix composites. Several different material systems have been studied, including B/Al, BSiC/titanium (Ti), and Al/SiC. It has been found that complex damage states develop in both continuous fiber laminated composites and chopped fiber composites.

A number of theoretical papers have also been published on constitutive equations for metal matrix composites [20–26]. However, to date these efforts have not included a damage parameter. Rate-dependent viscoplastic constitutive models have been developed for neat metals [27–39]. These models have not included a damage parameter applicable to composites. Theoretical models have been developed for predicting constitutive properties of cracked bodies [40–50]. However, to date these models have not been applied to metal matrix composites. A review of the current literature indicates that no constitutive model has yet been developed which is applicable to metal matrix composites that undergo large-scale matrix yielding and load-induced damage.

Model Development

This section presents a short review of a constitutive theory for composite media with damage which is described in detail in Refs 49–52. The model is presented herein because it provides the motivation for the experimental research program undertaken as part of this research.

The constitutive framework is based on a continuum mechanics approach with constraints on the relations provided by thermodynamics and fracture mechanics. The general model is applicable to materials with damage (such as voids, cracks, etc.) and includes inelastic effects such as plasticity [52]. The model is constructed within the framework of continuum mechanics and thermodynamics. The governing conservation laws are integrated over a small local volume element which is assumed to have a statistically homogeneous damage state, as shown in Fig. 1. The Helmholtz free energy can be expressed as

$$h^{TOT} = h^{EP} + u_L^c \tag{1}$$

where

h^{TOT} = the total Helmholtz free energy,
h^{EP} = the Helmholtz free energy due to the elastic-plastic response in the absence of damage, and
u_L^c = the energy due to damage.

It is therefore hypothesized that

$$h^{EP} = h^{EP}(\varepsilon_{ij}, \varepsilon_{ij}^I, \Delta T) \tag{2}$$

where

ε_{ij} = the total strain tensor,
ε_{ij}^I = the inelastic strain tensor, and
ΔT = the temperature difference from the reference temperature.

Furthermore

$$u_L^c = u_L^c(\varepsilon_{ij}, \varepsilon_{ij}^I, \Delta T, \alpha_{ij}) \tag{3}$$

where α_{ij} is the internal state variable representing damage, which is defined by

$$\alpha_{ij} = \frac{1}{V_L} \int_S u_i n_j dS \tag{4}$$

where

S = the surface area of cracks in the local volume V_L, and
u_i and n_j = the crack opening displacements and normals, as described in Fig. 2.

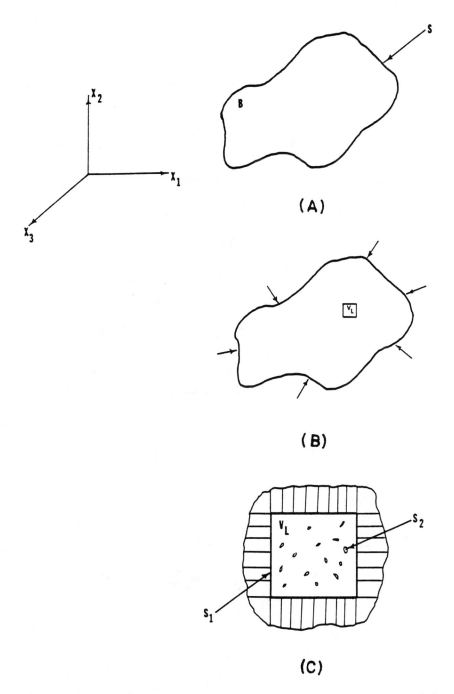

FIG. 1—*Material Body B*: (A) *virgin "undamaged" state*; (B) *body with tractions applied*; *and* (C) *local control volume with damage*.

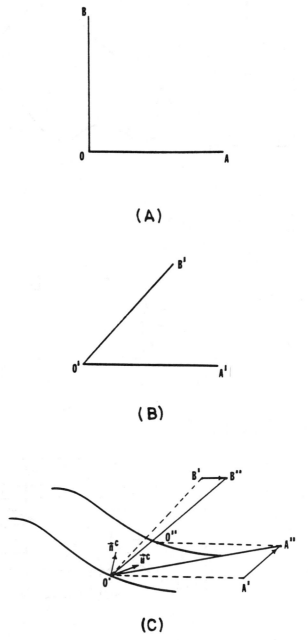

FIG. 2—*Kinematics of the damage process*: (A) *Point O before deformation*; (B) *Point O after deformation, prior to fracture*; and (C) *Point O after fracture.*

Constraints imposed by the second law of thermodynamics give the following result [51]

$$\sigma_{ij} = \rho \frac{\partial h^{TOT}}{\partial \varepsilon_{ij}} \tag{5}$$

Therefore, expanding Eqs 2 and 3 in Taylor series expansions in terms of their arguments, substituting into Eq 5, and truncating higher order terms results in

$$\sigma_{ij} = \sigma_{ij}^R + C_{ijkl}(\varepsilon_{kl} - \varepsilon_{kl}^I - \alpha_{kl} - \varepsilon_{kl}^T) \tag{6}$$

where

σ_{ij}^R = the residual stress tensor,
ε_{ij}^I = the inelastic strain tensor,
ε_{ij}^T = the thermal strain tensor, and
C_{ijkl} = the linear elastic modulus tensor.

For the uniaxial case in which there is negligible temperature change, the following form results

$$\sigma = \sigma^R + E_L(\varepsilon - \varepsilon^I - \alpha) \tag{7}$$

where E_L is the initial loading elastic modulus. Now define the initial unloading modulus E_U such that (see Fig. 3)

$$E_U = \frac{\partial \sigma}{\partial \varepsilon} = E_L\left(1 - \frac{\partial \varepsilon^I}{\partial \varepsilon} - \frac{\partial \alpha}{\partial \varepsilon}\right) \tag{8}$$

It is assumed that at relatively low homologous temperatures the inelastic strain remains constant when unloading, so that

$$\frac{\partial \varepsilon^I}{\partial \varepsilon} = 0 \tag{9}$$

Assuming linearly elastic unloading of the matrix, the change in damage is proportional to the change in strain

$$\frac{\partial \alpha}{\partial \varepsilon} = \text{constant (upon unloading)} = \beta \tag{10}$$

Therefore for unloading

$$E_U = E_L(1 - \beta) \tag{11}$$

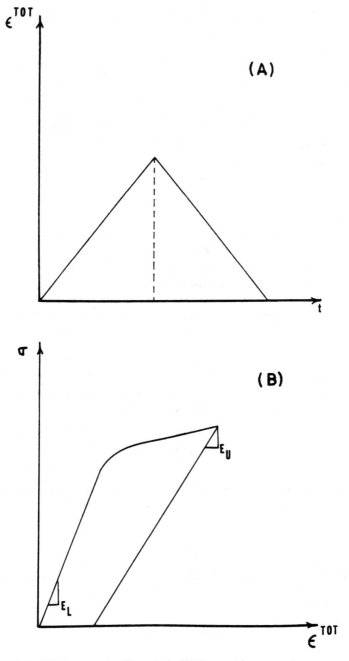

FIG. 3—(A) *Input strain*; (B) *typical Al/SiC material stress-strain response.*

The model just described provides the motivation for the experimental research. It is hypothesized that the two parameters that are defined in this model (α_{ij} and β) can be determined by experimental methods. By determining the change between the initial loading and unloading moduli of the composite in a uniaxial mechanical test, β (defined in Eq 11), can be found. It is also observed that α_{ij} (defined in Eq 4) can be determined by evaluating the amount of surface area in the composite. It is therefore desirable to determine if a cause-and-effect relationship exists between the microstructural damage (α_{ij}) and the stiffness loss (β).

Experimental Program

As stated before, the primary objective of the experimental effort is to provide support and documentation for the constitutive model for metal matrix composites developed in the previous section. This initial experimental study will provide an indication of the type of damage that may be present in the "virgin" material and an indication of any additional damage that is created from mechanical loading of the specimen. The term *damage* includes cracks, voids, and debonding of particles and matrix material. All of these modes of damage cause new surface area to be created in the material. One of the goals of this effort is to identify and quantify the amount of damage in the composite. To accomplish this objective, it is necessary to develop an experimental evaluation technique and mechanical test program. The procedure must be capable of inducing and then examining microstructural damage in the composite.

Material and Specimen Fabrication

The material used in this study was obtained from ARCO Metals Silag Operation in Greer, South Carolina. The composition of the material is 6061 aluminum with a 20% volume fraction of F-9 grade silicon carbide whiskers. The SiC whiskers average 2 μm in diameter and 20 μm long. The constituents were cast into a billet form by a powder metallurgy process and then rolled into a plate. The composite has a T-6 temper. Prior to specimen machining, the plate was examined by C-scan to observe whether or not macroscopic voids or density changes could be detected. No defects were observed except for surface imperfections. Specimens were machined both parallel and perpendicular to the principal rolling direction as shown in Fig. 4. Tension test coupons were machined in accordance with ASTM Methods of Tension Testing of Metallic Materials (E 8-85) to the dimensions shown in Fig. 4.

Mechanical Testing

An Instron Model 1125 screw-driven test system with 50.8-mm (2-in.) wedge action friction grips was used for mechanical testing. Longitudinal

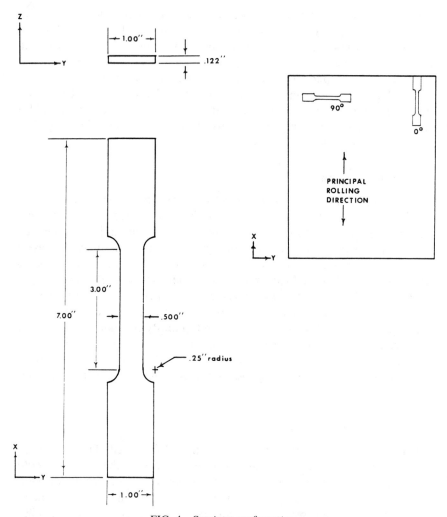

FIG. 4—*Specimen configuration.*

and transverse displacement data were obtained by the use of an extensometer. Load, time, and longitudinal and transverse displacememt data were digitized and saved for subsequent data reduction. Also, an analog plot of load versus displacement was recorded and monitored during the actual tests. Several specimens were tested to failure to obtain the yield point and ultimate strengths of the material. Other tests involved loading the specimens past their yield points and incrementally unloading and reloading to obtain unloading moduli at various load levels. All tests were performed at a crosshead speed of 0.02 mm/s (0.05 in./min). Both the loading and unloading moduli were determined for each specimen and loading sequence. The moduli were computed by a least squares curve fit to the digital data. Either six or ten sequential data points were utilized in the calculations. Also, the unloading moduli were computed and compared at several load levels.

Microstructural Damage Evaluation

Once the specimens were mechanically loaded, it was necessary to examine the microstructure for damage. It was desirable to use a nondestructive technique in the evaluation process since the most convenient way to observe the initiation and growth of damage is to study the behavior of a particular region of one specimen. However, the photographic resolution of non-destructive evaluation (NDE) techniques such as ultrasonics or X-ray radiography was not adequate to distinguish the microstructural damage of interest, which could be as small as one tenth of a micrometre. Therefore, it was decided that scanning electron microscopy (SEM) would be the best method of microstructural evaluation. Resolution of the photomicrographs obtained from SEM is better than one tenth of a micrometre. The disadvantage of using SEM is that the technique is destructive by nature. This implies that growth of damage in one particular specimen under various conditions cannot be measured.

The preparation of the internal surfaces of the tension specimens followed a procedure that was established after evaluating the application of several methods [53,54] to the Al/SiC specimens. The specimens were sectioned into pieces about 12.5 by 12.5 mm (0.5 by 0.5 in.) for mounting in the scanning electron microscope (SEM). Since there was uncertainty about the degree of anisotropy of the fiber orientation in the plane of the rolled plate, two sections were cut from each coupon so that two orthogonal views of each coupon could be examined. The sectioned pieces were then mounted in a conductive mounting material (Konductomet I) manufactured by Buehler. After mounting, the specimens were polished with diamond paste, chemically etched, and cleaned in an ultrasonic cleaner. It should be noted that specimen surface preparation requires great care because of the vast difference between the hardness of SiC and Al. Unless care is taken and an appropriate grade and hardness of abrasive compound is used, an uneven terrain may be created by the removal of the softer aluminum matrix leaving exposed SiC whiskers.

The final step in specimen preparation was to vacuum deposit a thin (about 300 Å) coating of gold to the viewing surfaces. This is necessary to achieve a good image on the SEM. When the voltage is applied to the specimens by the SEM, free electrons are released from the specimens. It is these electrons that are received by a sensor which forms the specimen image. Light molecular weight compounds have less free electrons, thus the image formed is not as sharp as an image of a compound with a large molecular weight. Since gold is an element with a large molecular weight, a thin coating on the specimens provides an electron source without losing surface detail.

The specimens were examined in a Jeol JSM-25 II SEM. After choosing the optimum acceleration voltage for the type of specimens, a series of SEM photographs were taken of the surfaces of orthogonal edges of the specimens under virgin, loaded, and failed conditions. These photographs included magnifications ranging from X2000 to X10 000. Along with the photographs of these polished surfaces, additional photographs of fracture surfaces were taken to observe the effect of the whisker reinforcement.

The photomicrographs were qualitatively and quantitatively evaluated for microstructural damage associated with the SiC particles. A scheme for determining the actual surface area of the damage was based on standard quantitative metallographic techniques [55]. The evaluation method consisted of covering the photograph with a transparent grid of 20 by 20 divisions per inch. A count was made of the intersections of the grid that covered damage. The percent of damage surface area was then obtained by dividing the intersections covering the damage by the total number of intersections and multiplying by 100.

Results

As noted before, the results presented herein were obtained for the primary purpose of evaluating the concept of the damage-dependent constitutive relationships. No attempt was made to construct a complete data base to support the full development of the constitutive model. A comparison of the typical stress-strain response of the Al 6061-T6, Al-SiC with 0° orientation, and Al-SiC with 90° orientation is given in Fig. 5. As can be seen, the effect of adding the SiC particles is to increase the elastic modulus and to increase the yield and ultimate strengths. These typical stress-strain curves were very useful for determining the intermediate strain levels used to evaluate the unloading moduli.

Several tension specimens were monotonically loaded past yield, followed by continued loading with periodic unloading to obtain unloading moduli at different maximum load levels. This test scheme allowed for the determination of the initial loading modulus and several unloading moduli at different load levels. Typical material mechanical responses are shown in Figs. 6 and 7, for the 0° and 90° orientation, respectively. Values of initial loading moduli are compared to numerous unloading moduli for several specimens in Figs. 8 and 9 for the 0° orientation and 90° orientation, respectively. The moduli were calculated using a least squares curve fitting program. It is obvious that the unloading moduli decrease with increasing strain for both specimen orientations.

Since the value of the modulus was very dependent on the location of the data points on the unloading curve, a consistent method of data reduc-

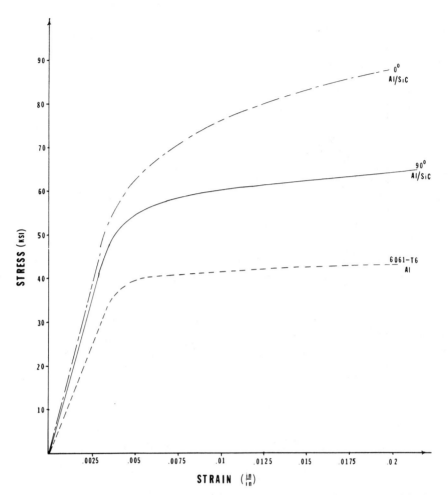

FIG. 5—*Typical stress-strain curves for Al 6061-T6 and Al/SiC parallel (0°) and perpendicular (90°) to the principal rolling direction.*

tion was adhered to during all modulus calculations. Three different sets of data points were selected on each unloading curve for each test. The unloading modulus was then calculated from each set of data points. The three sets of data points used were:

1. The first six data points following load reversal.
2. The middle ten data points on the unloading curve.
3. All the data points on the unloading curve.

Three values of the initial loading modulus were also calculated using the same technique. The mean and standard deviations of these data point sets

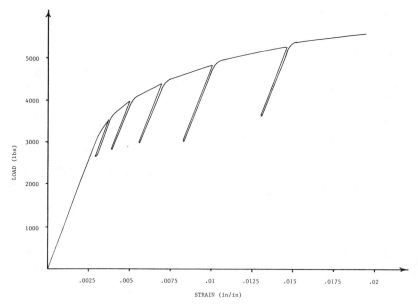

FIG. 6—*Typical load-unload curve for the 0° specimen orientation.*

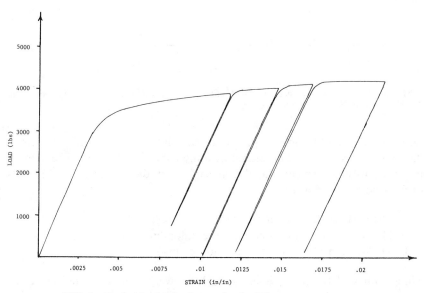

FIG. 7—*Typical load-unload curve for the 90° specimen orientation.*

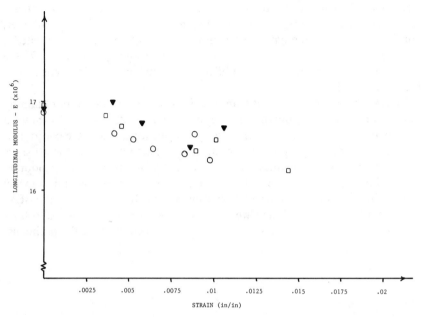

FIG. 8—*Comparison of initial loading and subsequent unloading moduli for the 0° specimen orientation.*

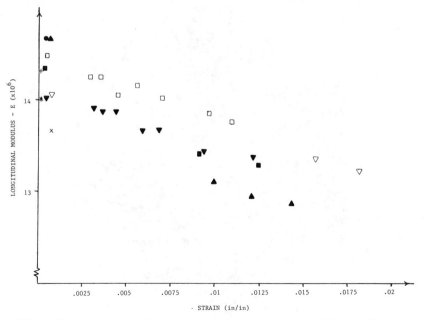

FIG. 9—*Comparison of initial loading and subsequent unloading moduli for the 90° specimen orientation.*

were then calculated for the initial loading curve and each subsequent unloading curve. By comparing sets of unloading moduli calculated in this manner, consistent trends in the data were identified. The 90° specimens exhibited a clear reduction in unloading modulus with increasing applied strain. On the other hand, the unloading moduli of the 0° specimens were less sensitive to increasing strain.

Photomicrographs obtained from the SEM were very useful in obtaining information about the microstructure of the composite. Figure 10 shows a composite view of three mutually orthogonal sections, where the coordinate axes are the same as in Fig. 4. The orientation and actual shape of the whiskers should be observed. The question of whether the SiC whiskers are randomly oriented in the plane that the Al/SiC billet was rolled was important in view of the differences in the 0° and 90° specimen data. It is clear from the SEM photographs that the whiskers are oriented somewhat in the principal rolling direction (x), which coincides with the loading direction of the 0° specimens.

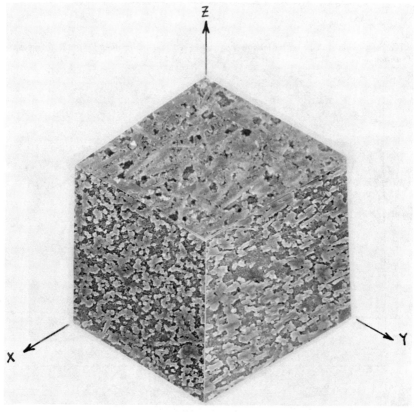

FIG. 10—*SEM photograph of orthogonal sections showing the SiC particle orientation.*

It is necessary to look at the microstructure of both virgin and failed specimens to determine if there are load-induced changes in the state of microstructural damage. The damage state was determined from five photomicrographs of randomly selected edge and side views of each loading state for a single coupon. Figures 11 and 12 are representative photomicrographs from virgin and loaded specimens for the 0° and 90° orientations, respectively. Upon close examination of the photomicrographs, it was observed that there were numerous voids and crack-like features of the microstructural surfaces of both the virgin and loaded specimens. Therefore, the fractographic examinations attempted to identify only changes in the surface features between the virgin and loaded specimens. The most extensive change (see Fig. 12) in the microstructure was due to cracks in the aluminum matrix extending from the SiC particles. There were no other extensively observed load-induced microstructural damage modes. Table 1 lists the amount of damage that was measured in each group of photomicrographs for specimens loaded in both the 0° and 90° orientations. It is apparent from Figs. 11 and 12 as well as from Table 1 that there is little change in the microstructure of damage for specimens loaded in the 0° direction, while specimens loaded in the 90° direction exhibit significant load-induced microstructural damage.

The most dramatic photomicrographs of the microstructural response of the Al/SiC material were fracture surface photos. One of the most important considerations in the determination of crack onset in composite materials is the bond between the reinforcing particles and the matrix material. Figure 13 is a low magnification (X150) photograph of the fracture surface. When the fracture surface is viewed under a higher magnification (X3000, see Fig. 14) two major observations can be made. First, the many cusps in the matrix show that, on the microstructural level, the material is very ductile. The second observation is that the SiC whiskers form a very strong bond with the aluminum. This is due to the observation that the whiskers remain in the matrix, although they appear to be bonded only on a small portion of their surface area (see Fig. 14). The result of this strong bond is that when stress concentrations exist at the interfaces of the aluminum matrix and reinforcing whiskers, the matrix may yield and inelastically deform before interfacial cracks initiate.

As stated earlier, specimens with the 0° orientation (see Fig. 8) indicated relatively little load-induced reduction in modulus. This tends to suggest that there is no mechanism present to produce a reduction in the stiffness of the specimen. This is supported by a comparison of the photomicrographs of a virgin specimen and a loaded specimen (see Fig. 11), which shows no apparent difference in microstructural damage state. Furthermore, a specimen subjected to 1000 cycles with the maximum strain exceeding yield also did not exhibit an increase in the microstructural damage. On the other hand, the specimens with the 90° orientation (see Fig. 9) exhibited a significant

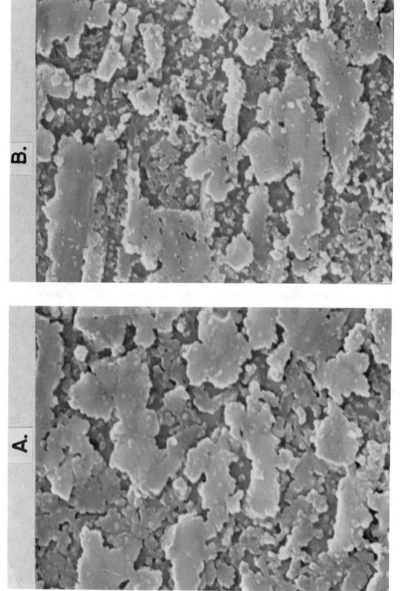

FIG. 11—SEM photographs of 0° specimens at X10 000: (A) virgin specimen; (B) sectioned view of failed specimen.

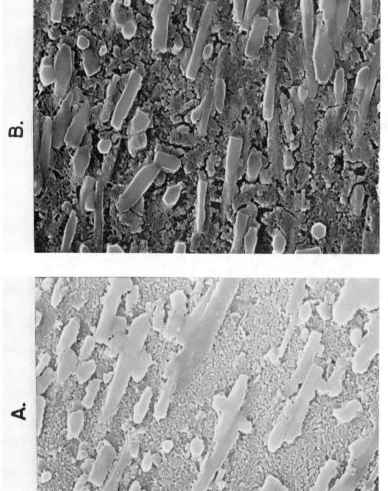

FIG. 12—*SEM photographs of 90° specimens at X7000:* (A) *virgin specimen;* (B) *loaded specimen at 94% ε_{ult}.*

TABLE 1—*Percent of damage (crack) surface areas.*

Specimen Orientation	Specimen Load History	Percent Crack Surface Area[a]	Standard Deviation
0° (end view)	Virgin	11.17	2.29
0° (end view)	Monotonic failed	9.48	1.55
0° (end view)	1000 cycle fatigue	9.31	1.16
0° (side view)	Virgin	9.53	1.55
0° (side view)	Monotonic failed	8.62	1.38
0° (side view)	1000 cycle fatigue	8.37	1.50
90° (end view)	Virgin	1.20	0.90
90° (end view)	Monotonic	14.7	2.30
90° (end view)	1000 cycle fatigue	31.2	3.10

[a]Both large voids and cracks were counted as damage in the specimens with the 0° orientation.

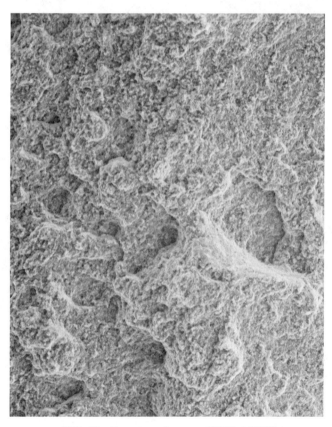

FIG. 13—*Fracture surface area (SEM at X150).*

FIG. 14—*Fracture surface area* (*SEM at X3000*).

reduction in modulus with increasing strain level. This trend does suggest the presence of a mechanism that produces a reduction in the stiffness of the specimen. This contention is supported by the comparison of photomicrographs of a virgin specimen and a loaded specimen (see Fig. 12), which reveals an obvious and considerable increase in the microstructural damage associated with the SiC particles. Finally, a fatigue loaded 90° specimen also exhibited both a substantial increase in microstructural damage along with a significant reduction in the elastic modulus. A similar observation has been made by Johnson in continuous fiber metal matrix composites [15,16,45].

Conclusion

A constitutive model of the behavior of short-fiber reinforced metal matrix composites has been formulated. The constitutive model was developed from the continuum mechanics viewpoint with constraints imposed by thermodynamic considerations. The model utilizes an internal state variable to characterize the development of load-induced microstructural damage associated with the inclusion of the fiber particles in the metal matrix. The internal state variable is defined by the kinematic behavior of the microcracks and enters the stress-strain relationships as a strain-like quantity.

It was shown in the development of the constitutive model that the damage parameter, α_{ij}, can be used to predict stiffness losses. Therefore, it was the objective of this research to measure stiffness loss and the associated microstructural damage as a function of strain level in order to qualitatively assess the applicability of the model to the Al-SiC metal matrix composite. This experimental objective was carried out by determining the initial loading and subsequent unloading moduli of tension specimens oriented parallel (0°) and perpendicular (90°) to the principal rolling direction of a plate fabricated from 6061-T6 aluminum with silicon carbide whiskers. In addition, SEM was utilized to characterize and quantify load-induced changes in the microstructural damage associated with the silicon carbide particles.

The results showed that the Al-SiC plate was anisotropic with approximately 15 to 20% difference in the moduli of the specimens oriented in the 0° and 90° directions. Also, the SEM photomicrographs indicated that the SiC whiskers were oriented more or less parallel to the principal rolling direction.

There was very little difference between the initial loading and subsequent unloading moduli of the 0° specimens. Also, there was no apparent load-induced change in the state of microstructural damage in these specimens. On the other hand, there was a significant reduction in the moduli of specimens oriented in the 90° direction. Furthermore, the photomicrographs revealed very obvious and significant load-induced changes in the state of microstructural damage in the 90° specimens.

The results illustrate a clear cause and effect between the increase in load-induced microstructural damage and a decrease in the elastic modulus of the Al-SiC metal matrix composite. It is concluded from these results that the constitutive behavior of a short-fiber reinforced metal matrix composite can only be modelled by an appropriate treatment of the microstructural damage associated with the fiber particles. The constitutive model proposed herein accounts for this microstructural damage through a load history dependent internal state variable. Although the results must be considered qualitative at this time, it appears that the proposed model may be acceptable for predicting stiffness loss as a function of microstructural damage.

Acknowledgment

This research was sponsored by the Air Force Office of Scientific Research under Contract No. F49620-83-C-0067.

References

[1] Vinson, J. R. and Chow, T., *Composite Materials and Their Use in Structures*, John Wiley & Sons, New York, 1975, Chap. 2.5.

[2] Renton, W. J., *Hybrid and Select Metal-Matrix Composites: A State of the Art Review*, American Institute of Aeronautics and Astronautics, New York, 1977, Chap. 1.

[3] Hancock, J. R. and Shaw, G. G., "Effect of Filament-Matrix Interdiffusion on the Fatigue Resistance of Boron-Aluminum Composites," in *Composite Materials: Testing and Design (Third Conference)*, *ASTM STP 546*, American Society for Testing and Materials, Philadelphia, 1974, pp. 497–506.

[4] Nielsen, L. E., "Mechanical Properties of Particulate-Filled Systems," *Journal of Composite Materials*, Vol. 1, 1967, pp. 100–119.

[5] Martin, J. W., *Micromechanics in Particle-Hardened Alloys*, Cambridge University Press, London, 1980.

[6] Herakovich, C. T., Davis, J. G., and Dexter, H. B., "Reinforcement of Metals with Advanced Filamentary Composites," in *Composite Materials: Testing and Design (Third Conference)*, *ASTM STP 546*, American Society for Testing and Materials, Philadelphia, 1974, pp. 52.

[7] Divecha, A. P., Fishman, S. G., and Karmarkar, S. D., "Silicon Carbide Reinforced Aluminum—A Formable Composite," *Journal of Metals*, Sept. 1981, pp. 12–17.

[8] Maclean, B. J. and Misra, M. S., "SiC Reinforced Alloys for Aeorspace Applications," Martin Marietta Aeorspace, Denver, CO, 1983.

[9] Parry, C. W., Ed., "Aluminum Metal Matrix Composites Outperform Unreinforced Alloys," *Aluminum Developments Digest*, Winter 1984–85, p. 3.

[10] Pipes, R. B., Ballintyn, N. J., and Carlyle, J. M., "Acoustic Emission Response Characteristics of Metal Matrix Composites," in *Composite Materials: Testing and Design (Fourth Conference)*, *ASTM STP 617*, American Society for Testing and Materials, Philadelphia, 1977, pp. 153–169.

[11] Ulman, D. A. and Henneke II, E. G., "Nondestructive Evaluation of Damage in FP/Aluminum Composites," in *Composite Materials: Testing and Design (Sixth Conference)*, *ASTM STP 787*, I. M. Daniel, Ed., American Society for Testing and Materials, Philadelphia, 1982, pp. 323–342.

[12] Hoover, W. R., "Crack Initiation in B-Al Composites," *Journal of Composite Materials*, Vol. 10, April 1976, p. 106.

[13] Awerbuch, J. and Hahn, H. T., "Hard Object Damage of Metal Matrix Composites," *Journal of Composite Materials*, Vol. 10, July 1976, pp. 231–257.

[14] Awerbuch, J. and Hahn, H. T., "K-Calibration of Unidirectional Metal Matrix Composites," *Journal of Composite Materials*, Vol. 12, July 1978, pp. 222–237.

[15] Johnson, W. S., "Mechanisms of Fatigue Damage in Boron/Aluminum Composites," in *Damage in Composite Materials*, *ASTM STP 775*, K. L. Reifsnider, Ed., American Society for Testing and Materials, Philadelphia, 1982, pp. 83–102.

[16] Dvorak, G. J. and Johnson, W. S., "Fatigue of Metal Matrix Composites," *International Journal of Fracture*, Vol. 16, 1980, pp. 585–607.

[17] Hancock, J. P. and Swanson, G. D., "Toughness of Filamentary Boron/Aluminum Composites," in *Composite Materials: Testing and Design (Second Conference)*, *ASTM STP 497*, American Society for Testing and Materials, Philadelphia, 1972, pp. 299–310.

[18] Sun, C. T. and Prewo, K. M., "The Fracture Toughness of Boron Aluminum Composites," *Journal of Composite Materials*, Vol. 11, April 1977, p. 164.

[19] Awerbuch, J. and Hahn, H. T., "Crack-Tip Damage and Fracture Toughness of Boron/Aluminum Composites," *Journal of Composite Materials*, Vol. 13, April 1979, pp. 82–107.

[20] Stüwe, H. P., "The Work Necessary to Form a Ductile Fracture Surface," *Engineering Fracture Mechanics*, Vol. 13, 1980, pp. 231–236.

[21] Aboudi, J., "A Continuum Theory for Fiber-Reinforced Elastic-Viscoplastic Composites," *International Journal Engineering Science*, Vol. 20, No. 5, 1982, pp. 605–621.

[22] Aboudi, J., "Effective Constitutive Equations for Fiber-Reinforced Viscoplastic Composites Exhibiting Anisotropic Hardening," *International Journal of Engineering Science*, Vol. 21, No. 9, 1983, pp. 1081–1096.

[23] Bodner, S. R and Partom, Y., "Constitutive Equations for Elastic-Viscoplastic Strain-Hardening Materials," *Journal of Applied Mechanics*, Vol. 42, No. 2, 1975, pp. 385–389.

[24] Bodner, S. R., "Representation of Time Dependent Mechanical Behavior of Rene 95 by Constitutive Equations," AFML-TX-79–4116, Air Force Materials Laboratory, Wright-Patterson Air Force Base, Dayton, OH, 1979.

[25] Bodner, S. R., Partom, I., and Partom, Y., "Uniaxial Cyclic Loading of Elastic-Viscoplastic Materials," *Journal of Applied Mechanics*, 1979, pp. 805–810.

[26] Stouffer, D. C. and Bodner, S. R., "A Relationship Between Theory and Experiment for a State Variable Constitutive Equation," AFWAL TR-80-4194, Air Force Materials Laboratory, Wright-Patterson Air Force Base, Dayton, OH, 1981.

[27] Krieg, R. D., Swearengen, J. C., and Rohde, R. W., "A Physically-Based Internal Variable for Rate-Dependent Plasticity," *Proceedings ASME/CSME PVP Conference*, 1978, American Society of Mechanical Engineers, New York, pp. 25–27.

[28] Miller, A. K., "An Inelastic Constitutive Model for Monotonic, Cyclic, and Creep Deformation: Part I—Equations Development and Analytic Procedures and Part II—Application to Type 304 Stainless Steel," *ASME Journal of Engineering Materials and Technology*, 98-H, 1976, p. 97.

[29] Miller, A. K., "Modelling of Cyclic Plasticity with Unified Constitutive Equations: Improvements in Simulating Normal and Anomalous Bauschinger Effects," *Journal of Engineering Materials and Technology*, Vol. 102, April 1980, pp. 215–220.

[30] Hart, E. W., "Constitutive Relations for the Nonelastic Deformation of Metals," *ASME Journal of Engineering Materials and Technology*, 98-H, 1976, p. 193.

[31] Walker, K. P., "Representation of Hastelloy-X Behavior at Elevated Temperatures With a Functional Theory of Viscoplasticity," presented at the ASME Pressure Vessels Conference, San Francisco, 12 Aug. 1980, American Society of Mechanical Engineers, New York.

[32] Krempl, E., "On the Interaction of Rate and History Dependence in Structural Metals," *Acta Mechanica*, Vol. 22, 1975, pp. 53–90.

[33] Cernocky, E. P. and Krempl, E., "A Theory of Viscoplasticity Based on Infinitesimal Total Strain," *Acta Mechanica*, Vol. 36, 1980, pp. 263–289.

[34] Cernocky, E. P. and Krempl, E., "A Theory of Thermoviscoplasticity Based on Infinitesimal Total Strain," *International Journal of Solids and Structures*, Vol. 16, 1980, pp. 723–741.

[35] Liu, M. C., Krempl, E., and Nairn, D. C., Transactions of ASME, Series H, and *Journal of Engineering Materials and Technology*, Vol. 98, 1976, pp. 322–329.

[36] Cernocky, E. P. and Krempl, E., "A Nonlinear Uniaxial Integral Constitutive Equation Incorporating Rate Effects, Creep, and Relaxation," *International Journal of Nonlinear Mechanics*, Vol. 14, 1979, pp. 183–203.

[37] Liu, M. C. and Krempl, E., "A Uniaxial Viscoplastic Model Based on Total Strain and Overstress," *Journal of the Mechanics and Physics of Solids*, Vol. 27, 1979, pp. 377–391.

[38] Zienkiewicz, O. C. and Cormeau, I. C., "Visco-Plasticity—Plasticity and Creep in Solids—A Unified Numerical Approach," *International Journal for Numerical Methods in Engineering*, Vol. 8, 1974, pp. 821–845.

[39] Allen, D. H. and Haisler, W. E., "A Theory for Analysis of Thermoplastic Materials," *Computers and Structures*, Vol. 13, 1981, pp. 124–135.

[40] Horii, H. and Nemat-Nasser, S., "Overall Moduli of Solids with Microcracks: Load-Induced Anisotropy," *Journal of the Mechanics and Physics of Solids*, Vol. 31, No. 2, 1983, pp. 155–171.

[41] Eshelby, J. D., "The Determination of the Elastic Field of an Ellipsoidal Inclusion, and Related Problems," *Proceedings of the Royal Society*, A241, 1957, pp. 376–396.

[42] Margolin, L. G., "Elastic Moduli of a Cracked Body," *International Journal of Fracture*, Vol. 22, 1980, pp. 65–79.

[43] Laws, N., Dvorak, G. J., and Hejazi, M., "Stiffness Changes in Unidirectional Composites Caused by Crack Systems," *Mechanics and Materials*, Vol. 2, 1983, pp. 123–127.

[44] Min, B. K. and Crossman, F. W., "Analysis of Creep for Metal Matrix Composites," *Journal of Composite Materials*, Vol. 16, May 1982, pp. 188–203.

[45] Johnson, W. S., "Modelling Stiffness Loss in Boron/Aluminum Laminates Below the Fatigue Limit," in *Long-Term Behavior of Composites, ASTM STP 813*, T. K. O'Brien, Ed., American Society for Testing and Materials, Philadelphia, 1983, pp. 160–176.

[46] Schapery, R. A., "On Viscoelastic Deformation and Failure Behavior of Composite Materials with Distributed Flaws," in *Advances in Aerospace Structures and Materials*, AD-01, American Society of Mechanical Engineers, New York, 1981, pp. 5–20.

[47] Schapery, R. A., "Models for Damage Growth and Fracture in Nonlinear Viscoelastic Particulate Composites," in *Proceedings*, Ninth U. S. National Congress of Applied Mechanics, American Society of Testing and Materials, Philadelphia, 1982.

[48] Talreja, R., "A Continuum Mechanics Characterization of Damage in Composite Materials," *Proceedings of the Royal Society London*, 1984.

[49] Allen, D. H., Groves, S. E., and Harris, C. E., "A Thermomechanical Constitutive Theory for Elastic Composites with Distributed Damage Part I: Theoretical Development," MM-5023-85-17, Mechanics and Materials Center, Texas A & M University, College Station, TX, Oct. 1985.

[50] Allen, D. H., Harris, C. E., and Groves, S. E., "A Thermomechanical Constitutive Theory for Elastic Composites with Distributed Damage Part II: Model Applications," MM-5023-85-15, Mechanics and Materials Center, Texas A & M University, College Station, TX, Oct. 1985.

[51] Allen, D. H., "Thermodynamic Constraints on the Constitution of a Class of Thermoviscoplastic Solids," Report No. MM-12415-82-10, Mechanics and Materials Center, Texas A & M, College Station, TX, Dec. 1982.

[52] Wren, G. and Allen, D. H, "Development of a Theoretical Framework for Constitutive Equations for Metal Matrix Composites with Damage," Report No. MM-4875-85-9, Mechanics and Materials Center, Texas A & M, College Station, TX, June 1985.

[53] Samuels, L. E., *Metallographic Polishing by Mechanical Methods*, American Society for Metals, Metals Park, OH, 1982.

[54] McCall, J. L. and Mueller, W. H., *Metallographic Specimen Preparation*, Plenum Press, N.Y., 1974.

[55] Rostoker, W. and Dvorak, J. R., *Interpretation of Metallographic Structures*, Academic Press, N.Y., 1965, Chap. 5.

Composites IV: Environmental Effects

Scott M. Milkovich,[1] *George F. Sykes, Jr.,*[2] *and Carl T. Herakovich*[3]

Fracture Surfaces of Irradiated Composites

REFERENCE: Milkovich, S. M., Sykes, G. F., Jr., and Herakovich, C. T., **"Fracture Surfaces of Irradiated Composites,"** *Fractography of Modern Engineering Materials: Composites and Metals, ASTM STP 948,* J. E. Masters and J. J. Au, Eds., American Society for Testing and Materials, Philadelphia, 1987, pp. 217–237.

ABSTRACT: Electron microscopy was used to analyze the fracture surfaces of T300/934 graphite/epoxy, unidirectional, off-axis tensile coupons which were subjected to 1.0 MeV electron radiation at a rate of 5.0×10^7 rad/h for a total dose of 1.0×10^{10} rad. Fracture surfaces from irradiated and nonirradi.ted specimens tested at 116 K ($-250°$F), room temperature, and 394 K ($+250°$F) were analyzed to assess the influence of radiation and temperature on the mode of failure and variations in constituent material as a function of environmental exposure. Micrographs of fracture surfaces indicate that irradiated specimens are more brittle than nonirradiated specimens at low temperatures. However, at elevated temperatures the irradiated specimens exhibit significantly more plasticity than nonirradiated specimens. The increased plasticity in irradiated specimens tested at elevated temperature is much more evident in 10° off-axis specimens. This is the result of the high shear stresses in these specimens. The same high degree of plasticity is not observed in the 90° specimens which fail at a much lower ultimate strain. Little difference in fracture surfaces and material behavior for irradiated and nonirradiated specimens is noted for room temperature specimens. The analysis of the photomicrographs is shown to correspond well with mechanical behavior of the specimen.

KEYWORDS: composites, electron radiation, microscopy, fracture, graphite-epoxy

The use of fiber-reinforced composite materials as efficient high-performance structural materials has greatly increased in recent years. The main advantage of these materials is their superior strength-to-weight and stiffness-to-weight ratios. The success of space exploration and exploitation using composite materials rests in the ability of the composites to withstand the hostile space environment. In an almost perfect vacuum, the cold of space

[1]Senior member, Technical Staff, Northrop Defense System Division, Rolling Meadows, IL 60008.

[2]Research scientist, NASA Langley Research Center, Hampton, VA 23665.

[3]Professor of engineering science and mechanics, Virginia Polytechnic Institute and State University, Blacksburg, VA 24061.

coupled with radiant solar heating effects can lead to a wide range of operating temperatures. In addition, a space structure will be subjected to ultraviolet, electron, and proton irradiation. Ultraviolet is electromagnetic radiation produced by the sun, while electron and proton radiation are present from trapped particles in the earth's Van Allen radiation belts. Ultraviolet and proton radiation will affect only the surface of a space structure, but electron radiation will be highly penetrating. Since composite materials will be used in future space structures with life spans of 20 to 30 years [1], a key materials technology need is the ability to understand how fiber-reinforced materials will behave under such harsh conditions for long periods of time.

The material chosen for this study is the "NASA space-approved" T300/934 graphite-epoxy. Radiation exposure consisted of 1.0 MeV electrons at a rate of 5.0×10^7 rad/h for a total dose of 1.0×10^{10} rad. This total dose simulates a "worst-case" exposure of 30 years in a geostationary use orbit [2,3]. Great care has been taken to assure that the specimens did not overheat during their accelerated irradiation exposure. Testing and characterization covers the temperature range of 116 to 394 K (-250 to $+250°F$), which represents the extremes that may be encountered in a space environment [2].

Procedure

The facility used for specimen radiation exposures was the Space Materials Durability Laboratory located at the NASA Langley Research Center in Hampton, Virginia. This facility uses a Radiation Dynamics Corp. Dynamatron to produce a uniform 10-in.-diameter electron beam, and a water-cooled backplate to keep the composite samples from exceeding 303 K (85°F) during exposure. Thermocouples monitored temperature during exposure, and a Faraday cup measured electron dose and dose rate.

The dimensions of the specimens used in this study were dictated by limitations of the radiation facility. The specimen thickness was limited to four plys to insure uniform radiation exposure through the thickness. Unidirectional tension specimens of $[0]_4$, $[10]_4$, $[45]_4$, and $[90]_4$ laminates were cut from four-ply composite panels and then exposed to electron radiation. A specimen size of 6 by 0.5 in. was chosen to optimize the number of specimens that could be placed in the electron beam without overly compromising the aspect ratio of the tensile coupons. Fiber glass tabs were used for load introduction (leaving a gage length of 3.5 in.), and strain was measured with wide-temperature-range strain gages.

The $[0]_4$ laminates were used to measure the longitudinal strength, X_T, longitudinal stiffness, E_1, and major Poisson's ratio v_{12}, and the $[90]_4$ laminates were used to measure the transverse strength, Y_T, and transverse modulus, E_2. The shear strength, S, was calculated from the $[10]_4$ laminate [4], and the shear modulus, G_{12}, from the $[45]_4$ laminate [5]. Work by Pindera and Herakovich has shown that shear coupling in the $[10]_4$ laminate leads to less than satisfactory results for the shear modulus [6]. They recommend that the $[10]_4$ laminate be used for determining shear strength only and that the 45° laminate be used to measure shear modulus [7].

Mechanical tests were performed on both the nonirradiated and irradiated material at 116 K ($-250°$F), room temperature 300 K (80°F), and 394 K ($+250°$F). All tests were performed on a tension testing machine fitted with an environmental chamber that employs resistance elements for heating and liquid nitrogen for cooling. Three specimens were tested for each test condition. Additional details for the specimens and test program can be found in Ref 8.

A scanning electron microscope (SEM) was used to examine the fracture surfaces of the laminates that failed during the mechanical testing phase of this study. SEM inspection was made on the matrix-dominated fracture surfaces of the $[90]_4$ laminate, which fails primarily in tension, and also on the $[10]_4$ laminate, which should fail mostly by shear [4]. Inspection of the $[0]_4$ laminates was impossible since these laminates were almost completely destroyed at failure. The complete destruction of the $[0]_4$ laminates may have been due to the thinness of the specimens. Inspection of the $[45]_4$ fracture surface was not deemed appropriate since this laminate fails in a mixed tension and shear mode. All fracture surfaces were coated with gold-palladium prior to SEM inspection.

Results

The actual experimental results and averages for elastic and strength properties as functions of temperature and irradiation exposure are presented in Tables 1–4. Comparisons of average principal values, with percent changes, are presented in Tables 5–7.

Stress-Strain Curves

The curves presented in Fig. 1 show typical stress-strain behavior for both the nonirradiated and irradiated $[0]_4$ laminates at all test temperatures. The $[0]_4$ laminate exhibits a stiffening behavior at high strains, as expected, at all temperatures [7]. Very little temperature-dependence is noted (Tables 5 and 6), although electron irradiation tends to increase slightly the modulus at all test temperatures (2 to 4%, Table 7).

Axial stress-strain behavior for the other laminates is roughly linear at low and room temperatures for both the nonirradiated and irradiated cases (Figs. 2 to 4). At 116 K ($-250°$F), the stress-strain curves are essentially linear; as the test temperature is increased, the elastic modulus and proportional limit decrease and the degree of nonlinearity increases. Significant nonlinearity is noted in the behavior of the nonirradiated laminates at the

TABLE 1—*Test results for [0]₄ T300/934 graphite-epoxy.*

Specimen	Temperature	X_T, ksi (MPa)		E_1, msi^a (GPa)		v_{12}
		NONIRRADIATED				
1	−250°F (116 K)	143.8	(992)	18.70	(129)	0.3326
2		142.8	(985)	18.29	(126)	0.2999
3		136.2	(939)	18.88	(130)	0.3061
Avg		140.9	(972)	18.62	(128)	0.3129
1	Room temperature	221.2	(1525)	18.74	(129)	0.3161
2		220.4	(1520)	19.32	(133)	0.3140
3		224.6	(1549)	18.58	(128)	0.3129
Avg		222.1	(1531)	18.88	(130)	0.3143
1	+250°F (394 K)	210.4	(1451)	19.14	(132)	0.3634
2		195.7	(1349)	19.05	(131)	0.3350
3		174.9	(1206)	18.90	(130)	0.3356
Avg		193.7	(1336)	19.03	(131)	0.3447
		1.0×10^{10} RAD				
1	−250°F (116 K)	149.3	(1029)	19.45	(134)	0.3829
2		134.6	(928)	18.93	(131)	0.3497
3		98.0	(676)	19.11	(132)	0.3721
Avg		127.3	(878)	19.16	(132)	0.3682
1	Room temperature	221.7	(1529)	19.37	(134)	0.2948
2		243.8	(1681)	19.52	(135)	0.2941
3		202.2	(1394)	19.03	(131)	0.2603
Avg		222.6	(1535)	19.31	(133)	0.2831
1	+250°F (394 K)	184.5	(1272)	19.77	(136)	0.3768
2		150.3	(1036)*	19.23	(133)	0.3830
3		149.7	(1032)*	20.29	(140)	0.4311
Avg		161.5	(1114)	19.76	(136)	0.3970

^{a}Msi = million pounds per square inch.
*Tab slippage.

TABLE 2—*Test results for* [10]₄ *T300/934 graphite-epoxy.*

Specimen	Temperature	$\sigma_{ult}(10°)$, ksi (MPa)		S, ksi (MPa)		$E_x(10°)$, msi (GPa)	
		NONIRRADIATED					
1	−250°F (116 K)	53.84	(371)	9.21	(63.5)	12.52	(86.3)
2		38.68	(267)	6.61	(45.6)	13.53	(93.3)
3		36.32	(250)	6.21	(42.8)	13.20	(91.0)
Avg		42.95	(296)	7.34	(50.6)	13.08	(90.2)
1	Room temperature	64.15	(442)	10.97	(75.6)	10.93	(75.4)
2		57.79	(398)	9.88	(68.1)	10.61	(73.2)
3		52.11	(359)	8.91	(61.4)	10.63	(73.3)
Avg		58.02	(400)	9.92	(68.4)	10.72	(73.9)
1	+250°F (394 K)	31.83	(219)	5.44	(37.5)	8.19	(56.5)
2		38.22	(264)	6.54	(45.1)	8.52	(58.7)
3		34.75	(240)	5.94	(41.0)	8.52	(58.7)
Avg		34.93	(241)	5.97	(41.2)	8.41	(58.0)
		1.0×10^{10} RAD					
1	−250°F (116 K)	43.57	(300)	7.45	(51.4)	12.60	(86.9)
2		41.18	(284)	7.04	(48.5)	11.01	(75.9)
3		42.42	(292)	7.25	(50.0)	12.75	(87.9)
Avg		42.39	(292)	7.25	(50.0)	12.12	(83.6)
1	Room temperature	56.21	(388)	9.61	(66.3)	9.98	(68.8)
2		58.26	(402)	9.96	(68.7)	10.16	(70.1)
3		47.87	(330)	8.19	(56.5)	10.60	(73.1)
Avg		54.11	(373)	9.25	(63.8)	10.25	(70.7)
1	+250°F (394 K)	22.63	(156)	3.87	(26.7)	5.37	(37.0)
2		26.02	(179)	4.45	(30.7)	6.44	(44.4)
3		22.49	(155)	3.85	(26.5)	6.84	(47.2)
Avg		23.71	(163)	4.06	(28.0)	6.22	(42.9)

TABLE 3—*Test results for [45]$_4$ T300/934 graphite-epoxy.*

Specimen	Temperature	$\sigma_{ult}(45°)$, ksi (MPa)		$E_x(45°)$, msi (GPa)		G_{12}, msi (GPa)	
		NONIRRADIATED					
1	−250°F (116 K)	14.50	(100.0)	2.892	(19.9)	1.225	(8.4)
2		13.33	(91.9)	2.725	(18.8)	1.110	(7.7)
3		9.67	(66.7)	2.825	(19.5)	1.178	(8.1)
Avg		12.50	(86.2)	2.814	(19.4)	1.170	(8.1)
1	Room temperature	15.06	(103.8)	1.823	(12.6)	0.692	(4.8)
2		13.75	(94.8)	1.806	(12.5)	0.682	(4.7)
3		13.83	(95.4)	1.819	(12.5)	0.689	(4.8)
Avg		14.21	(98.0)	1.816	(12.5)	0.688	(4.8)
1	+250°F (394 K)	11.59	(79.9)	1.381	(9.5)	0.482	(3.3)
2		11.16	(76.9)	1.617	(11.1)	0.605	(4.2)
3		8.01	(55.2)	1.618	(11.2)	0.606	(4.2)
Avg		10.25	(70.7)	1.539	(10.6)	0.563	(3.9)
		1.0×10^{10} RAD					
1	−250°F (116 K)	9.36	(64.5)	2.978	(20.5)	1.165	(8.0)
2		11.68	(80.5)	2.963	(20.4)	1.156	(8.0)
3		9.25	(63.8)	2.785	(19.2)	1.051	(7.2)
Avg		10.10	(69.6)	2.909	(20.1)	1.123	(7.7)
1	Room temperature	12.95	(89.3)	2.087	(14.4)	0.810	(5.6)
2		14.71	(101.4)	2.029	(14.0)	0.776	(5.4)
3		15.77	(108.7)	1.974	(13.6)	0.744	(5.1)
Avg		14.48	(99.8)	2.030	(14.0)	0.777	(5.4)
1	+250°F (394 K)	6.28	(43.3)	1.174	(8.1)	0.407	(2.8)
2		7.84	(54.1)*	1.326	(9.1)	0.484	(3.3)
3		5.39	(37.2)	0.956	(6.6)	0.309	(2.1)
Avg		6.50	(44.8)	1.152	(7.9)	0.397	(2.7)

*Tab slippage.

TABLE 4—*Test results for* [90]₄ *T300/934 graphite-expoxy.*

Specimen	Temperature	Y_T, ksi (MPa)		E_2, msi (GPa)		v_{21}
		NONIRRADIATED				
1	−250°F (116 K)	4.47	(30.8)	1.824	(12.6)	
2		4.19	(28.9)	1.854	(12.8)	0.0307
3		5.02	(34.6)	1.808	(12.5)	
Avg		4.56	(31.4)	1.829	(12.6)	
1	Room temperature	10.34	(71.3)	1.410	(9.7)	
2		10.85	(74.8)	1.369	(9.4)	0.0187
3		6.92	(47.7)	1.350	(9.3)	
Avg		9.37	(64.6)	1.376	(9.5)	
1	+250°F (394)	7.70	(53.1)	1.227	(8.5)	
2		8.68	(59.8)	1.199	(8.3)	0.0225
3		3.91	(27.0)	1.296	(8.9)	
Avg		6.76	(46.6)	1.241	(8.6)	
		10×10^{10} RAD				
1	−250°F (116 K)	2.18	(15.0)	1.939	(13.4)	
2		2.80	(19.3)	2.146	(14.8)	0.0408
3		3.44	(23.7)	2.284	(15.7)	
Avg		2.81	(19.4)	2.123	(14.6)	
1	Room temperature	8.73	(60.2)	1.545	(10.7)	
2		6.52	(45.0)	1.534	(10.6)	0.0222
3		5.68	(39.2)	1.465	(10.1)	
Avg		6.98	(48.1)	1.515	(10.4)	
1	+250°F (394 K)	6.22	(42.9)	1.142	(7.9)	
2		5.75	(39.6)	1.037	(7.2)	0.0214
3		5.67	(39.1)	1.014	(7.0)	
Avg		5.88	(40.5)	1.064	(7.3)	

TABLE 5—*Average test results for nonirradiated T300/934.*

Temperature	Material Property	Nonirradiated, ksi/msi (MPa/GPa)		Change from Room Temperature, %
−250°F (116 K)	X_T	140.9	(972)	−36.6
	Y_T	4.56	(31)	−51.3
	S	7.34	(51)	−26.0
	E_1	18.62	(128)	1.4
	E_2	1.829	(12.6)	+32.9
	v_{12}	0.3129		−0.4
	G_{12}	1.170	(8.1)	+70.1
Room temperature	X_T	222.1	(1531)	...
	Y_T	9.37	(65)	...
	S	9.92	(68)	...
	E_1	18.88	130	...
	E_2	1.376	(9.5)	...
	v_{12}	0.3143		...
	G_{12}	0.688	(4.7)	...
+250°F (394 K)	X_T	193.7	(1336)	−12.8
	Y_T	6.76	(47)	−27.9
	S	5.97	(41)	−39.8
	E_1	19.03	(131)	+1.0
	E_2	1.241	(8.6)	−9.8
	v_{12}	0.3447		+9.7
	G_{12}	0.563	(3.9)	−18.2

TABLE 6—*Average test results for irradiated T300/934.*

Temperature	Material Property	1.0×10^{10} rad, ksi/msi (MPa/GPa)		Change from Room Temperature, %
−250°F (116 K)	X_T	127.3	(878)	−42.8
	Y_T	2.81	(19)	−59.7
	S	7.25	(50)	−21.6
	E_1	19.16	(132)	−0.8
	E_2	2.123	(14.6)	+40.1
	v_{12}	0.3682		+30.1
	G_{12}	1.123	(7.7)	+44.5
Room temperature	X_T	222.6	(1535)	. . .
	Y_T	6.98	(48)	. . .
	S	9.25	(64)	. . .
	E_1	19.31	(133)	. . .
	E_2	1.515	(10.4)	. . .
	v_{12}	0.2831		. . .
	G_{12}	0.777	(5.4)	. . .
+250°F (394 K)	X_T	161.5	(1114)	−27.4
	Y_T	5.88	(41)	−15.8
	S	4.06	(28)	−56.1
	E_1	19.76	(136)	+2.3
	E_2	1.064	(7.3)	−29.8
	v_{12}	0.3970		+40.2
	G_{12}	0.397	(2.7)	−48.9

TABLE 7—*Comparison of irradiated and nonirradiated T300/934.*

Temperature	Material Property	Nonirradiated, ksi/msi (MPa/GPa)		1.0×10^{10} rad, ksi/msi (MPa/GPa)		Percent Change
$-250°F$ (116 K)	X_T	140.9	(972)	127.3	(878)	9.7
	Y_T	4.56	(31)	2.81	(19)	-38.4
	S	7.34	(51)	7.25	(50)	0.0
	E_1	18.62	(128)	19.16	(132)	$+2.9$
	E_2	1.829	(12.6)	2.123	(14.6)	$+16.1$
	v_{12}	0.3129		0.3682		$+17.7$
	G_{12}	1.170	(8.1)	1.123	(7.7)	-4.0
Room temperature	X_T	222.1	(1531)	222.6	(1535)	0.0
	Y_T	9.37	(65)	6.98	(48)	-25.5
	S	9.92	(68)	9.25	(64)	-6.8
	E_1	18.88	(130)	19.31	(133)	$+2.2$
	E_2	1.376	(9.5)	1.515	(10.4)	$+10.1$
	v_{12}	0.3143		0.2831		-11.0
	G_{12}	0.688	(4.7)	0.777	(5.4)	$+12.9$
$+250°F$ (394 K)	X_T	193.7	(1336)	161.5	(1114)	-16.6
	Y_T	6.76	(47)	5.88	(41)	-13.0
	S	5.97	(41)	4.06	(28)	-32.0
	E_1	19.03	(131)	19.76	(136)	$+3.8$
	E_2	1.241	(8.6)	1.064	(7.3)	-14.3
	v_{12}	0.3447		0.3970		$+15.2$
	G_{12}	0.563	(3.9)	0.397	(2.7)	-29.5

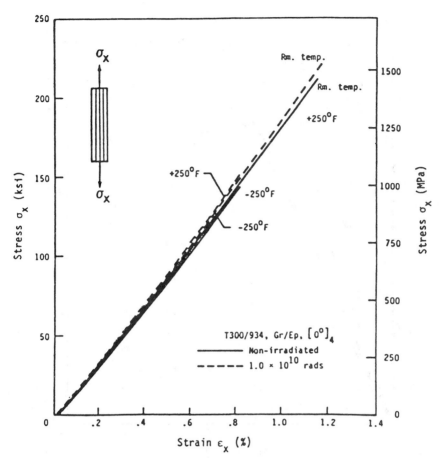

FIG. 1—*Nonirradiated stress-strain curves compared to irradiated stress-strain curves for the* [0]₄ *laminate as a function of temperature.*

elevated temperatures, especially in those laminates exhibiting large shear stresses (Figs. 2 and 3). This behavior is expected in view of the influence of temperature on the response of the epoxy matrix material [8]. However, very low proportional limits and extreme nonlinearity are observed in the irradiated material at 394 K (+250°F) (Figs. 2 and 3). This behavior is greatest for the [10]₄ and [45]₄ laminates, which are subjected to high shear stresses and which exhibit large strains prior to failure.

Radiation greatly increases the plasticity of the epoxy matrix material at elevated temperatures, but not significantly at lower temperatures. The plasticity is attributed to the epoxy because it is known that fibers exhibit linear behavior, and further because this nonlinearity is more pronounced in matrix-dominated behavior with high shear.

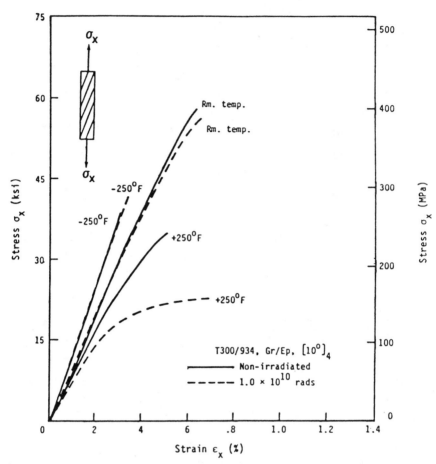

FIG. 2—*Nonirradiated stress-strain curves compared to irradiated stress-strain curves for the* [10]₄ *laminate as a function of temperature.*

Fracture Surfaces of Irradiated Composites

SEM micrographs of representative fracture surfaces of the irradiated laminates and the nonirradiated laminates are presented in Figs. 5 through 10. Figures 5 through 7 were taken from the [10]₄ laminates, and Figs. 8 through 10 were taken from the [90]₄ laminates. Figures 5 and 8 were taken at a magnification of X375, and all the others were taken at X3400. In all figures, the column of fracture surfaces on the left are from nonirradiated laminates, and the column of fracture surfaces on the right are from irradiated laminates. Specimens in the top row were tested at 116 K (−250°F). Those in the middle row were tested at room temperature, 300 K (80°F). Those in the bottom row were tested at 394 K (+250°F).

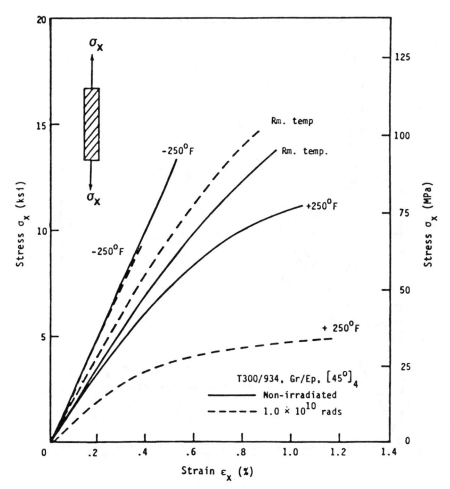

FIG. 3—*Nonirradiated stress-strain curves compared to irradiated stress-strain curves for the* [45]₄ *laminate as a function of temperature.*

Little difference over the fracture surfaces of the composite can be noted at X375 (Figs. 5 and 8). At this magnification, the failure surfaces are, for the most part, very uniform. The small differences that can be identified appear to be in the failure patterns of the epoxy matrix.

The microphotographs presented in Figs. 6 and 9 were taken at X3400 and focused primarily on the matrix. Differences in matrix failure patterns are immediately apparent for each of the exposure conditions. At 116 K (−250°F), both the nonirradiated and the irradiated laminates exhibit brittle failure in their epoxy matrices. Brittle cleavage planes are noted for both radiation exposure conditions. The cleavage planes in the irradiated laminate are smaller and more numerous than those observed in the nonirradiated

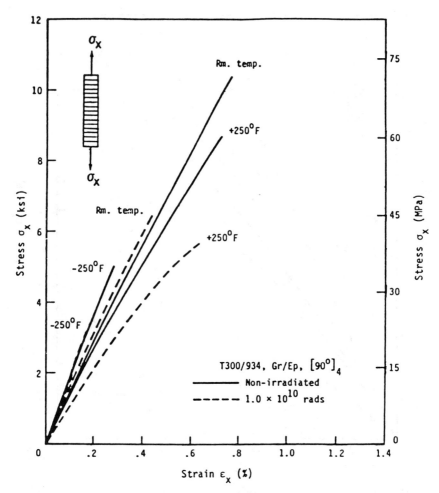

FIG. 4—*Nonirradiated stress-strain curves compared to irradiated stress-strain curves for the* [90]₄ *laminate as a function of temperature.*

laminate. This is an indication that the irradiated laminate exhibits a more brittle fracture than the nonirradiated composite. Indeed, this is the conclusion that was drawn by observing the stress-strain curves presented earlier. From these stress-stain curves, it can be seen that at 116 K (−250°F), the irradiated composite produces a more brittle failure and a higher modulus of elasticity than the nonirradiated material. In general, at low temperatures, radiation tends to embrittle the epoxy resin matrix, resulting in a stiffer matrix and a more brittle failure than that exhibited by the nonirradiated laminate.

The photographs in the middle row of Figs. 6 and 9 were taken from

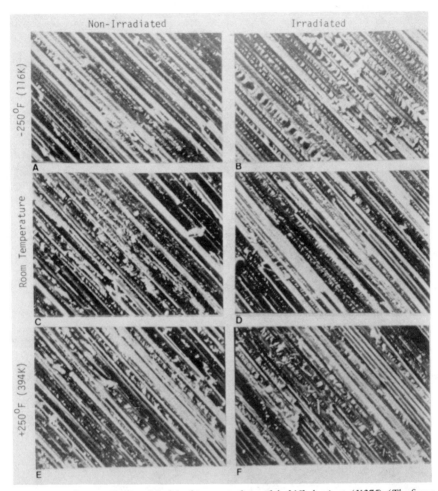

FIG. 5—*Electron micrographs of the fracture surfaces of the* [10]₄ *laminate* (X375). (*The figure was reduced 34% for printing purposes.*)

specimens that had been tested at room temperature. At this temperature, the failure patterns exhibited in the matrix are plastic. Instead of failing in brittle cleavage planes, the epoxy stretched and plasticity deformed before failing. At room temperature, only slight differences are observed between the nonirradiated and irradiated laminates. The irradiated material exhibits plastic deformation, but it appears "rougher" than the nonirradiated material. This is consistent with mechanical tests which indicate little difference between nonirradiated and irradiated laminates at room temperature.

At 394 K (+250°F), the plasticity observed in the epoxy matrix, before failure, is much greater than that observed at the lower temperatures. This increased plasticity is evident in both the stress-strain curves (Figs. 2–4) and

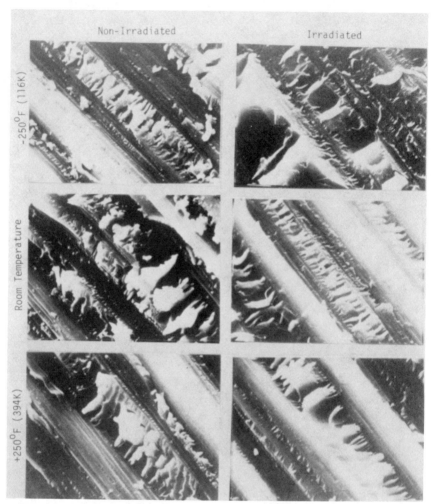

FIG. 6—*Electron micrographs of the fracture surfaces of the* [10]$_4$ *laminate concentrating on the matrix* (X3400). (*The figure was reduced 34% for printing purposes.*)

the photomicrographs (Figs. 6, 8 and 9). For the [10]$_4$ laminate (Fig. 6), the irradiated laminate exhibits extreme amounts of plasticity prior to failure. However, this same type of behavior is not noted in the irradiated [90]$_4$ laminate (Fig. 9). This difference in failure mode is undoubtedly due to the difference in the local stress state at failure. The [10]$_4$ laminate fails primarily in shear, exhibiting very large strains prior to failure. The [90]$_4$ laminates fails in a triaxial stress state (at the micro level) at relatively low strains. At high temperatures, the irradiated composite produces a more plastic failure and a lower modulus than the nonirradiated material. In general, at 394 K (+250°F), radiation tends to greatly increase the plastic behavior of the

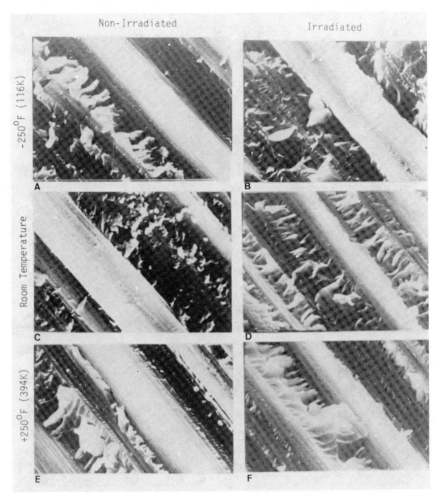

FIG. 7—*Electron micrographs of the fracture surfaces of the* [10]₄ *laminate concentrating on the fibers* (X3400). (*The figure was reduced 27% for printing purposes.*)

epoxy resin matrix, resulting in a more pliable matrix and a more plastic failure mode than that exhibited by the nonirradiated laminate.

The photographs presented in Figs. 7 and 10 were also taken at X3400, but focused on the fibers. Inspection of the fibers, in all photographs, shows little difference as a function of radiation exposure condition. In all cases, the fibers remain relatively inert. Therefore, it can be concluded that the majority of radiation and temperature dependence of the fracture surfaces of this graphite-epoxy composite is due to the matrix and not the fibers. The epoxy matrix is degraded by radiation and influenced by temperature while the fibers remain relatively inert.

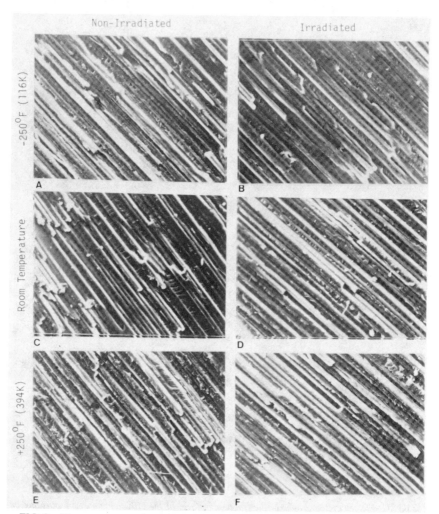

FIG. 8—*Electron micrographs of the fracture surfaces of the* [90]$_4$ *laminate* (X375). (*The figure was reduced 33% for printing purposes.*)

Summary

Inspection of the fracture surfaces of the nonirradiated and irradiated laminates shows differences in failure due to radiation and temperature exposure. At low temperatures, where mechanical results indicate brittle behavior, brittle cleavage planes are observed in the failed epoxy matrix. The cleavage planes in the irradiated laminate are smaller and more numerous than those observed in the nonirradiated laminate, indicating a more brittle fracture. Mechanical results, especially the stress-strain curves, exhibit significant nonlinearity at elevated temperatures. Inspection of the epoxy

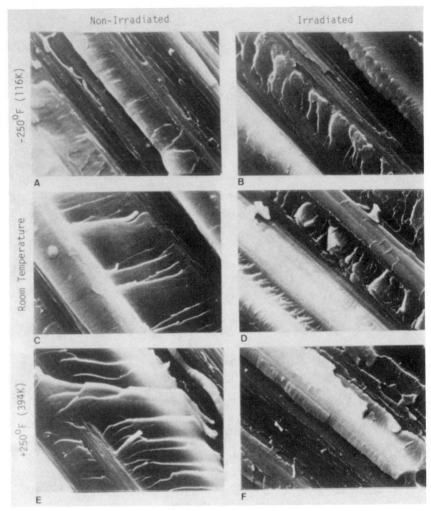

FIG. 9—*Electron micrographs of the fracture surfaces of the [90]₄ laminate concentrating on the matrix (X3400). (The figure was reduced 27% for printing purposes.)*

matrix at high magnifications shows that large amounts of plastic deformation are present in those specimens tested at the elevated temperature.

The experimental results produced in other investigations indicate that the radiation-induced degradation is due primarily to changes in the epoxy resin [8,9]. Microscopic examination of the laminates in the current study shows that the matrix fails by different modes that are dependent on the radiation and temperature exposure. Microscopic examination of the fibers, on the other hand, reveals that they remain inert.

Microscopic examination of the fracture surface of the laminates tested

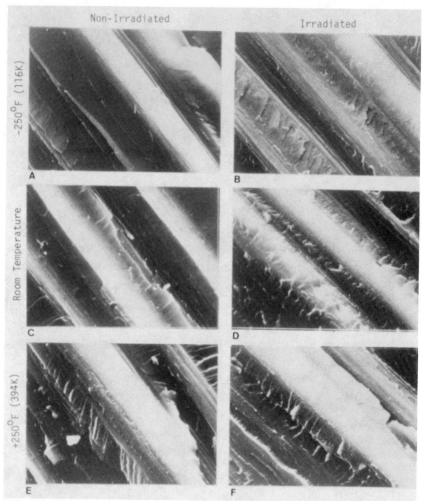

FIG. 10—*Electron micrographs of the fracture surfaces of the* [90]₄ *laminate concentrating on the fibers* (*X3400*). (*The figure was reduced 31% for printing purposes.*)

supports the differences in mechanical behavior that were measured. At low temperature, radiation tends to embrittle the epoxy resin. At elevated temperature, irradiation produces a more plastic matrix, which is most evident for shear response.

Acknowledgment

This work was supported by the NASA-Virginia Tech Composites Program under NASA Grant NAG-1-343.

References

[1] Kelley, H. M., Rummler, D. R., and Jackson, L. R., "Research in Structures and Materials for Future Space Transportation Systems—An Overview," *Journal of Spacecraft and Rockets*, Vol. 20, No. 1, Jan.-Feb. 1983, pp. 89–96.

[2] Tenney, D. R., Sykes, G. F., and Bowles, D. E., "Space Environmental Effects on Materials," presented at Conference on Environmental Effects on Materials for Space Applications, Toronto, Canada, 22–24 Sept. 1982, Advisories Group for Aerospace Research and Development, Neuilly sur Seine, France.

[3] Garrett, H. B., "Review of the Near-Earth Spacecraft Environment," *SPIE*, Vol. 216, Optics in Adverse Environments, 1980, pp. 109–115.

[4] Chamis, C. C. and Sinclair, J. H., "Ten-deg Off-axis Test for Shear Properties in Fiber Composites," *Experimental Mechanics*, Vol. 17, No. 9, Sept. 1977, pp. 339–346.

[5] Jones, R. M., *Mechanics of Composite Materials*, Scripta Book Co., Washington, DC, 1975.

[6] Pindera, M. J. and Herakovich, C. T., "An Endochronic Theory for Transversely Isotropic Fiberous Composites," VPI-E-81-27, Virginia Polytechnic Institute and State University, Blacksburg, VA, Oct. 1981.

[7] Pindera, M. J. and Herakovich, C. T., "An Elastic Potential for the Nonlinear Response of Unidirectional Graphite Composites," *Journal of Applied Mechanics*, Vol. 51, No. 3, Sept. 1984, pp. 546–550.

[8] Milkovich, S. M., "Space Radiation Effects on Graphite-Epoxy Composite Materials," Ph.D. dissertation, Virginia Polytechnic Institute and State University, Blacksburg, VA, May 1984.

[9] Sykes, G. F., Milkovich, S. M., and Herakovich, C. T., "Simulated Space Radiation Effects on a Graphite Epoxy Composite," presented at the ACS Spring National Meeting in Miami Beach, Florida, 28 April to 3 May 1985, American Chemical Society, Washington, DC.

John E. Masters[1]

Characterization of Impact Damage Development in Graphite/Epoxy Laminates

REFERENCE: Masters, J. E., "**Characterization of Impact Damage Development in Graphite/Epoxy Laminates,**" *Fractography of Modern Engineering Materials: Composites and Metals, ASTM STP 948*, J. E. Masters and J. J. Au, Eds., American Society for Testing and Materials, Philadelphia, 1987, pp. 238–258.

ABSTRACT: Experimental investigations indicate that the damage tolerance of graphite/epoxy systems, as measured by the laminate's residual compression strength after low-velocity impact, is directly proportional to the neat resins strain to failure. Results also indicate that laminate damage tolerance is significantly increased when thin, discrete layers of tough, ductile resin are "interlayered" or placed between the graphite/epoxy lamina. The physical mechanics which lead to this improved impact resistance are examined in this investigation. Interlayered and noninterlayered 36-ply pseudoisotropic $[(\pm45/0/90/0/90)_2/\pm45/0/90/\pm45]_s$ laminates were subjected to low-velocity, drop-weight impacts. Incident impact energy levels ranged from 1110 to 8890 N · m/m of laminate thickness (that is 250 to 2000 in. · lb/in. (Note: Original measurements were in English, which appear in parentheses in this paper.) The resulting impact-induced damage was characterized by a combination of destructive and non-destructive test methods, including ultrasonic C-scan, light microscopy, and scanning electron microscopy. The differences in the amounts and in the distributions of impact-induced damage in the interlayered and noninterlayered laminates will be discussed in detail. The ultrasonic C-scan data indicate that interlayering significantly reduced the size of internal delamination at all impact energy levels. Microscopic investigation of sectioned, polished impact specimens indicates that the pattern of damage development was also altered through interleafing. Although transverse cracks developed in the interlayered laminates at intermediate impact levels, they do not form delaminations at the ply interfaces as they do in the uninterleafed laminates. Damage at these levels is limited, for the most part, to a system of transverse cracks. While interlayered laminates do delaminate at the high-impact energy levels, the number and size of the delaminations are reduced.

KEYWORDS: impact damage, interleaf, delamination, transverse cracks, hackles, resin toughness, damage patterns, Mode II strain energy release rate, residual compression strength

---

[1]Senior research engineer, American Cyanamid Co., Stamford, CT 06904.

Damage tolerance is a prime area of concern in aircraft design. Structural aircraft components are susceptible to foreign-object impact during fabrication, assembly, ground handling, takeoff, in flight, and, finally, during landing operations. The damage containment capability of the structure is, therefore, a vitally important consideration. As Greszczuk [1] noted, however, the advantages inherent in first-generation graphite/epoxy material systems were overshadowed by their poor resistance to impact loading in applications where low velocity impact was a design consideration.

Internal damage, mainly in the form of interply delamination, as well as transverse cracking, developed in the highly cross-linked, first-generation resin systems at relatively low-impact energy levels. Internal damage induced by low-velocity, hard object impact can significantly reduce a composite structure's residual tensile and compression strengths, even when external structural damage is not visually observed [2,3]. Moreover, the reduction in laminate residual strength is accelerated with increasing impact energy [4].

The role of the matrix resin in laminate impact damage resistance has been the subject of several investigations. Palmer [5] tested 23 resin systems with widely varying tensile strength, stiffness, and elongation characteristics in an attempt to define desirable "tough" resin traits. He concluded that of the three properties just listed, resin elongation appears to be the major variable in controlling laminate impact resistance. Similarly, experimental investigations conducted at American Cyanamid [6] indicate that the damage tolerance of graphite/epoxy systems, as measured by the laminate residual compression strength after low-velocity impact, is directly proportional to the neat resin's flexural strain to failure.

One approach to improved damage tolerance, therefore, is to develop advanced matrix resin systems with increased strain to failure. The effectiveness of this approach is seen in Fig. 1, which plots neat resin flexural strain versus laminate residual compression strength. As the figure illustrates, advanced epoxy systems with strains to failure in the 8% range, for example, Cycom 1806 and Cycom 1808, have compression strengths after impact in the 220 to 255 MPa (32 to 37 ksi) range. A typical first-generation epoxy such as Hercules 3502 with a 2.5% strain to failure, on the other hand, has a residual compression strength of 152 MPa (22 ksi).

A second approach to improve laminate damage tolerance is to selectively toughen the ply interfaces which delaminate under impact. This concept, which is known as interleafing [7], is illustrated schematically in Fig. 2. This physical construction technique takes standard prepreg tape (that is, tape with a 60% fiber volume fraction) containing an improved, toughened matrix resin and adds to it a thin layer of a second resin. There are two key factors to this approach: first, the interlayer resin used must have a large ultimate shear strain; second, it must remain a discrete layer after the laminate has been cured. Experimental results indicate that this method can significantly increase impact resistance. The average residual compression

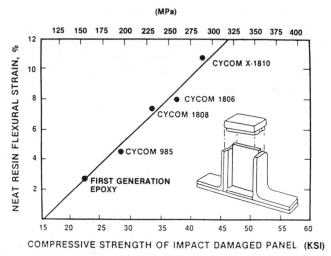

FIG. 1—*Neat resin flexural strain versus laminate residual compression strength.*

strength of Cycom 1808 laminates increased by 190 to 410 MPa (27.6 to 59.6 ksi) when it was interlayered with a developmental interleaf system [8].

This report will detail the results of an investigation of the damage induced in graphite/epoxy laminates by low-velocity impact. It will define the size and the distribution of damage induced in the laminates at various impact levels. Three graphite/epoxy prepreg systems will be examined: Cycom 907, Cycom 1808, and Hercules 3502. In addition, results will be presented for Cycom 1808 laminates which have been interlayered with thin layers of thermoplastic film. The report will begin by describing the material systems examined and the experimental procedures employed. This will be followed by a detailed description of the resultant impact-induced damage.

Materials and Experimental Procedures

Impact damage development in three diverse graphite/epoxy material systems was examined in this study. Hercules 3502 is an example of a first-generation epoxy system which has a long history of in-flight services. This highly cross-linked system has excellent hot/wet structural properties (that is, high glass transition temperature) but very limited damage tolerance. American Cyanamid's Cycom 907, on the other hand, has excellent impact resistance but low hot/wet structural performance. It is a rubber-modified resin system which has been used as a model material in several damage tolerance studies [9,10]. Those two resins, therefore, represent relative extremes in material response.

American Cyanamid's Cycom 1808 was also investigated in this study. It is a single-phase, high strain-to-failure epoxy system with good 121 to 132°C

Cured Laminate — Cross Section

Prepreg Form

Release Paper

Graphite Fibers in Improved Matrix Resin

Discrete Layer of Higher Strain Resin for Added "Toughness"

FIG. 2—*Interleafed composite construction.*

(250 to 270°F) hot/wet performance. As Fig. 1 illustrated, Cycom 1808 is representative of advanced resin systems with improved impact resistance. Cycom 1808 prepreg tape was also interlayered with an interleaf system suitable for 132°C (270°F) wet service and investigated in this study.

Unsized AS4 graphite fiber was used in all three prepreg systems. These grade 145 g/m^2 (fiber areal weight) prepreg tapes had a 34 weight % resin content. The total resin content in the interleafed system (that is, the total of the interleaf plus the matrix resin) was only 40% by weight.

The size and distribution of impact-induced damage were monitored in 36-ply [($\pm45/0/90/0/90_2$)/$\pm45/0/90/\pm45$]$_s$, pseudoisotropic laminates fabricated from the material systems just described. In these evaluations, 100-mm (4.0-in.)-wide, 480-mm (19.0-in.)-long test panels (Fig. 3) were clamped in a steel test fixture and subjected to low-velocity, drop-weight impact. The test fixture, shown in Fig. 4, has a 50-mm (2.0-in.)-diameter circular cutout and sits on a concrete floor. This is a severe test in that very little energy can be absorbed by deflection of the panel. These specimens were impacted at seven locations along their length. The incident impact energy level was increased after each impact to induce damage which ranged from the barely visible surface damage level to the severe damage level. Typical incident energies were: 1110, 2220, 3560, 4890, 6225, 7560, and 8890 N · m/m (250, 500, 800, 1100, 1400, 1700 and 2000 in. · lb/in.) of laminate thickness.

The specimens were ultrasonically C-scanned after being impacted to determine the size of the impact-induced delamination. They were then sectioned through the impact site, polished, and microscopically examined to determine the through-the-thickness delamination and transverse crack distributions. These microscopic inspections of the sectioned specimens were

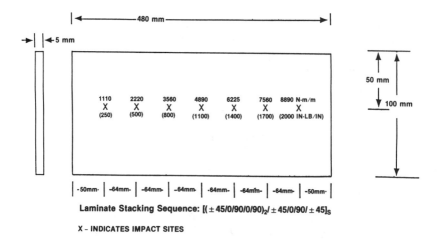

FIG. 3—*Specimen geometry.*

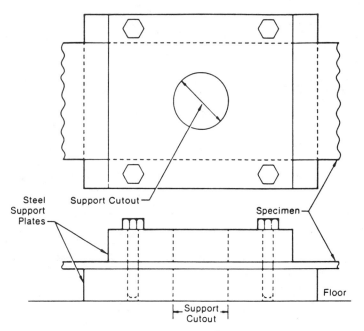

FIG. 4—*Impact test fixture.*

also used to verify the accuracy of the delamination measurements made from the C-scans.

Experimental Results

Baseline Material Systems

Figure 5 plots the size of internal delamination as measured by ultrasonic C-scan versus incident impact energy for the three baseline (uninterlayered) epoxy systems: AS4/3502, AS4/1808, and AS4/907. As expected, AS4/907 demonstrated the greatest impact resistance of these three epoxy systems. Ultrasonic inspection detected no internal damage at the 1110 N · m/m (250 in. · lb/in.) impact level. Further, less damage was induced in this material under 8890 N · m/m (2000 in. · lb/in.) loading than developed in the other systems at 2220 N · m/m (500 in. · lb/in.). By comparison, delamination in the AS4/3502 laminate extended over the entire unsupported specimen length under the initial 1110 N · m/m (250 in. · lb/in.) impact. Due to the clamping action of the test fixture, impact-induced delamination was arrested at the higher impact levels. As indicated in the figure, the cutout region covered 2025 mm² (3.14 in.²). Ultrasonic inspection indicated that the delamination actually extended slightly beyond this area at higher impact

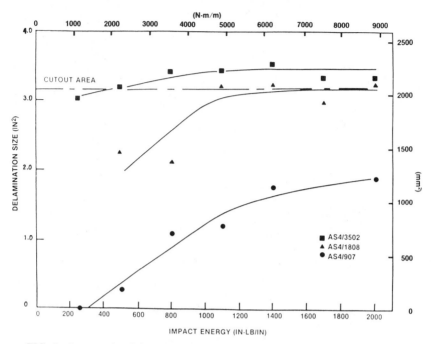

FIG. 5—*Impact-induced damage size versus impact energy for baseline epoxy systems.*

energies. The AS4/1808 specimen demonstrated an intermediate damage tolerance. Its performance at low impact levels was superior to AS4/3502 but not as good as the AS4/907 test standard's. An impact energy of 4890 N · m/m (1100 in · lb/in.) was required to extend the internal delamination to the clamped section.

The relative impact resistance of these systems as measured by the size of impact-induced delaminations is also reflected in their patterns of internal damage. Figure 6 shows polished cross sections of the AS4/3502 laminates impacted at 1110 and 3560 N · m/m (250 and 800 in. · lb/in.). While no damage is apparent directly below the point of contract in the 1110 N · m/m (250 in · lb/in.) specimen, a series of transverse cracks are evident in the laminate on either side of the impact site. Though not connected, these cracks define a truncated triangular region which extends over the upper two-thirds of the specimen. At this energy level, no transverse cracks or delaminations are apparent within this region; but many transverse cracks and several large delaminations have developed in the laminate outside of the triangular region.

A key feature is the association of the transverse cracks and interfacial delaminations. The delaminations appear to begin at the intersection of the transverse crack and the lamina interfaces. This observation has been con-

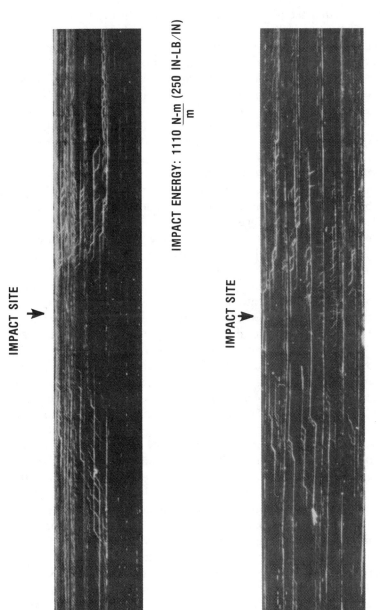

FIG. 6—*Internal damage in AS4/3502 laminate.*

firmed by Guynn and O'Brien [*11*], who employed a staining and deplying technique to examine internal impact damage in graphite/epoxy laminates. They concluded that the damage sequence starts with ply crack formation followed by delamination.

Damage in the 3560 N · m/m (800 in. · lb/in.) laminate is distributed throughout the laminate thickness. The number of delaminations was increased, and several have developed in the region below the point of contact. Back surface damage is also apparent in the laminate. As the impact energy level increases, the number of transverse cracks and delaminations increases. However, due to the clamping action of the test fixture, these delaminations cannot propagate far beyond a 25.4 mm (1.0 in.) radius from the impact site. Figure 7 shows the cross sections of AS4/3502 laminates impacted at 6225 and 8890 N · m/m (1400 and 2000 in. · lb/in.). As the figure indicates, virtually every ply has delaminated at the point of contact under the 8890 N · m/m (200 in. · lb/in.) impact. A number of transverse cracks are also linked together to form a continuous, jagged-fracture surface which extends from the top impact surface through the laminate thickness to the back surface.

As the C-scan data previously demonstrated, the AS4/1808 laminates are more damage tolerant than the AS4/3502 laminates. Unlike the 3502 laminates, no damage was evident in these specimens under 1110 N · m/m (250 in. · lb/in.) impacts. The pattern of damage development in these laminates, is, however, similar to the AS4/3502 laminates. Figure 8 shows photomicrographs of laminates impacted at 3560 and 8890 N · m/m (800 and 2000 in. · lb/in.). The truncated triangular damage pattern is again evident in the region below the point of impact in the 3560 N · m/m (800 in. · lb/in.) specimen. Delamination and transverse cracking are again distributed through the thickness, but, unlike the AS4/3502 laminate, no back surface damage developed under this load. The number of delaminations and transverse cracks again increases with increased impact energy, as is evidenced in the cross section of the laminate impacted at 8890 N · m/m (2000 in. · lb/in.). Although the delamination length has increased and extends to the clamped section, damage in this material is not nearly as severe as that seen in the AS4/3502 laminate at the same energy level. There are far fewer delaminations and transverse cracks in the AS4/1808 laminate.

The AS4/Cycom 1808 specimens were sectioned again through the damaged region so that the delaminated ply surfaces could be examined in the scanning electron microscope. Figure 9 contains photomicrographs taken of the surface of a 90° and a 0° ply which delaminated under a 7560 N · m/m (1700 in. · lb/in.) impact. These surfaces exhibit the hackle patterns commonly observed in Mode II shear specimens [*12,13*]. Although the raised epoxy platelets are more evident on the surface of the 90° ply which retained more of the resin in this region, they are also evident in the resin left behind between fibers on the 0° ply surface. These hackle patterns were also evident on the surfaces of −45° and +45° plies taken from the same 7560 N · m/m (1700 in. · lb/in.) impact specimen (Fig. 10). One important difference in

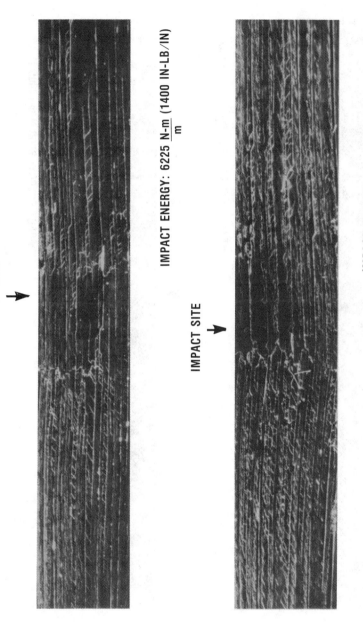

IMPACT SITE →

IMPACT ENERGY: 6225 $\frac{N-m}{m}$ (1400 IN-LB/IN)

IMPACT SITE →

IMPACT ENERGY: 8890 $\frac{N-m}{m}$ (2000 IN-LB/IN)

FIG. 7—*Internal damage in AS4/3502 laminate* (continued).

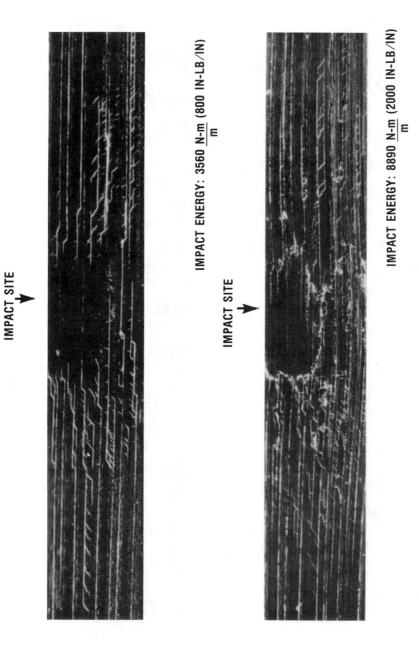

IMPACT SITE

IMPACT ENERGY: 3560 $\frac{\text{N-m}}{\text{m}}$ (800 IN-LB/IN)

IMPACT SITE

IMPACT ENERGY: 8890 $\frac{\text{N-m}}{\text{m}}$ (2000 IN-LB/IN)

FIG. 8—Internal damage in AS4/1808 laminate.

90° PLY SURFACE

0° PLY SURFACE

FIG. 9—*SEM photomicrographs of delaminated 90°/0° AS4/1808 lamina.*

−45° PLY SURFACE

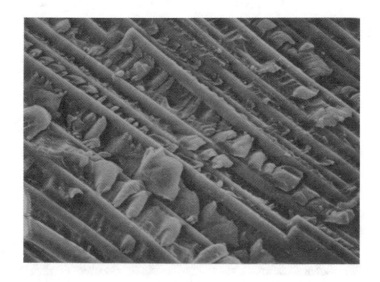

+45° PLY SURFACE

FIG. 10—*SEM photomicrographs of delaminated* − 45°/45° *AS4/1808 lamina.*

this case, however, is that the pattern at this interface is aligned with the 45° fibers. The hackles had been aligned with the 0° fibers in the previous case. This indicates that, depending on the orientation of the fibers at the interface, delaminations grow in different directions under impact. Guynn and O'Brien made a similar observation in their study.

Interlayered Cycom 1808 Laminate

The addition of the interleaf to the AS4/Cycom 1808 system greatly improved the materials impact resistance. Figure 11 plots the size of the impact-induced delamination versus impact energy for baseline and interlayered laminates. As the figure demonstrates, no internal delamination was detected at the initial 1110 N · m/m (250 in. · lb/in.). impact level. Moreover, the size of the damage induced in the interlayered specimen, even at the highest impact energy level, does not approach the size of the delaminations induced in the uninterlayered baseline laminate. In fact, less damage developed in the interlayered AS4/Cycom 1808 laminate than in the AS4/Cycom 907 laminate.

The pattern of damage development noted earlier is significantly altered

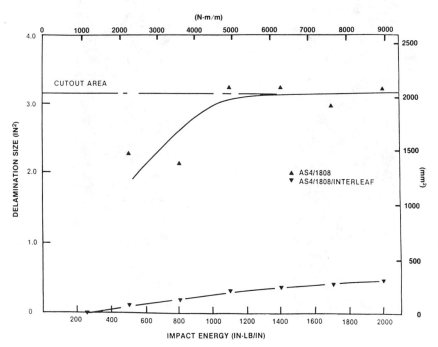

FIG. 11—*Impact-induced damage size versus impact energy for interlayered and baseline AS4/ 1808 laminates.*

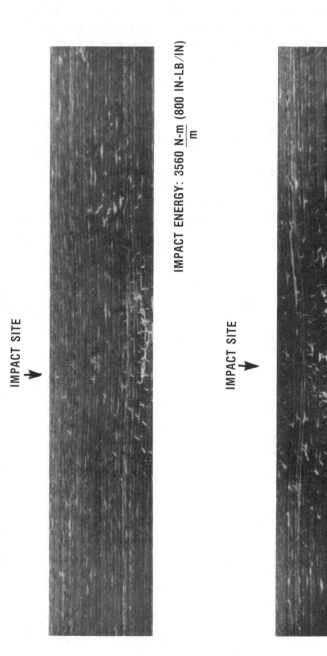

IMPACT SITE ➤

IMPACT ENERGY: 3560 $\frac{\text{N-m}}{\text{m}}$ (800 IN-LB/IN)

IMPACT SITE ➤

IMPACT ENERGY: 6225 $\frac{\text{N-m}}{\text{m}}$ (1400 IN-LB/IN)

FIG. 12—*Internal damage in interlayered AS4/1808 laminates.*

when interlayers are added to the laminate. Figure 12 contains photomicrographs of interlayered AS4/1808 laminates which have been impacted at 3560 and 6225 N · m/m (800 and 1400 in. · lb/in.). The photomicrographs demonstrate that while transverse cracks developed in these specimens, they do not initially form the characteristic triangular pattern through a cross section of laminate thickness as they did in the uninterleafed laminates. Instead, transverse cracks first appear in the lower plies (that is, near the back surface) in the interleafed laminates. With increased impact loads, they begin to appear in the upper portions of the laminate, forming a triangular crack zone analogous to those found in uninterleafed laminates. More significantly, very few delaminations formed at the laminae interfaces. Delamination development in the 6225 N · m/m (1400 in. · lb/in.) impact specimen was restricted to two or three minor delaminations near the back surface. On the other hand, the transverse cracks are distributed throughout the laminate thickness as in the baseline laminates. The number of delaminations was not significantly increased under the 8890 N · m/m (2000 in. · lb/in.) impact as is shown in Fig. 13. This figure shows the cross section of both the interlayered and noninterlayered 1808 laminates impacted at 8890 N · m/m (2000 in. · lb/in.). The significant reduction in internal damage development with interleafing, particularly in the number and extent of the delaminations, is evident in the figure.

The basic mechanism at work in the interlayered laminates is the suppression of delamination at the ply interfaces. Although the transverse cracks form in the prepreg material under impact, the toughened interfaces do not delaminate. Figure 14 shows 200X photomicrographs (reduced 20% for printing purposes) of typical transverse cracks which had developed in interlayered and baseline AS4/1808 laminates under 3560 N · m/m (800 in. · lb/in.) impact. As the figure demonstrates, while delamination developed in the baseline laminate, the interleafed system gives no evidence of delamination. In effect, damage is limited to a system of transverse cracks which develop in the interleafed laminates at intermediate impact levels as is evidenced in Fig. 12. When delaminations are induced in the interlayered laminates under higher impact loads, they are fewer and much smaller than the delaminations that develop in uninterleafed laminates at comparable energy levels.

Summary of Experimental Results

The evidence of the shear failure at the ply interfaces of the AS4/1808 lamina, Figs. 9 and 10, support the work of Bostaph and Elber [14]. They developed a fracture mechanics analysis based on large deflection plate mechanics to describe the progress of delamination damage in composite plates struck by a hard spherical object. This analysis indicates that the extent of impact-induced delamination is a function of the critical shear strain energy release rate of the matrix.

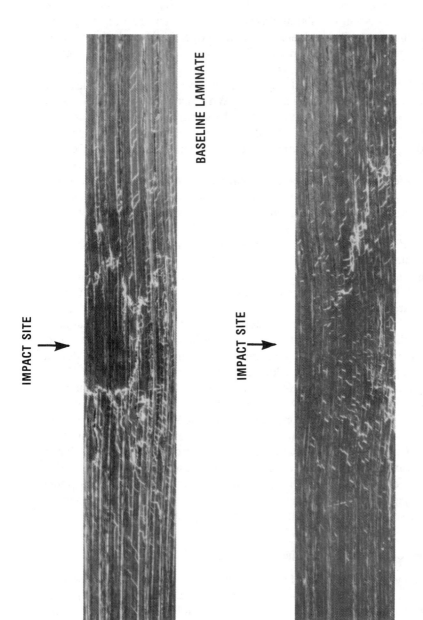

FIG. 13—*Internal damage in interlayered and baseline AS4/1808 laminates at 8890 N · m/m (2000 in. · lb/in.).*

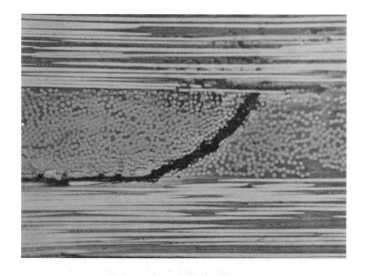

BASELINE LAMINATE (200 X)

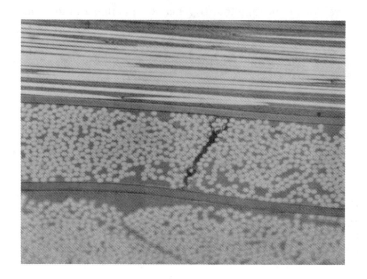

INTERLAYERED LAMINATE (200 X)

FIG. 14—*Transverse cracks in interlayered and baseline AS4/1808 laminates at 1110 N · m/m (800 in. · lb/in.).*

Further evidence for this theory is found when the material's Mode I and Mode II strain energy release rates are plotted versus their impact resistance (as measured by the residual compression strength after impact). Delamination resistance data developed at Cyanamid [8] indicate that the thin interlayers had only a nominal effect on the baseline material's Mode I strain energy release rate. The Mode II delamination resistance, on the other hand, is significantly increased with interleafing. The inteleafed laminates' G_{IIC}s were two-to-three times greater than their baseline counterparts.

Figure 15 plots the residual compression strengths after impact versus the Mode I strain energy release rates of these interlayered and baseline epoxy systems. The general trend of increased damage tolerance with increased G_{IC}, as observed by several investigators, hold for the baseline systems. This correlation is, however, not accurate when the interlayered laminate results are considered. As noted earlier, the Mode I strain energy release rate of these specimens is only slightly higher than the baseline laminate's; the residual compression strength, however, is twice as large as the baseline system's. G_{IC} does not, in this case, appear to be a good indicator of damage tolerance.

A good correlation is seen, however, when G_{IIC} is plotted versus residual compression strength. Figure 16 plots these data for the interlayered and baseline systems just discussed. In this case, both sets of data fall along the same general line, indicating that G_{IIC} may be a better overall indicator of a material's impact resistance.

In addition to the Cycom 1808-Interleaf A results presented in Figs. 15 and 16, the author has noted [8] that this correlation of G_{IIC} and residual

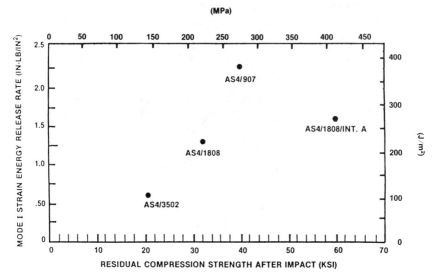

FIG. 15—*Mode I strain energy release rate versus compression strength after impact.*

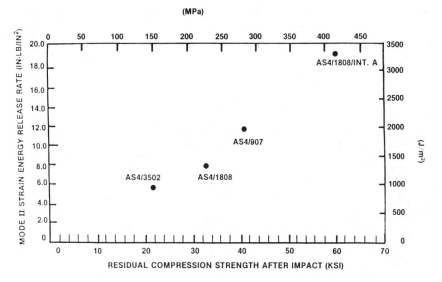

FIG. 16—*Mode II strain energy release rate versus compression strength after impact.*

compression strength after impact is observed for several interleaf graphite/epoxy material systems.

Conclusions

The results of this investigation of impact-induced damage in graphite/epoxy laminates led to the following conclusions:

1. Internal damage, in the form of transverse cracks and delaminations, develops in the matrix material at the point of contact. The transverse cracks form a truncated pyramid pattern; delaminations appear to begin at the intersection of these transverse cracks and the lamina interfaces.

2. The materials' shear strain energy release rate may control this delamination propagation.

3. The use of tougher matrix resin systems increases laminate damage tolerance. The pattern of damage development in these laminates, however, is not altered.

4. Interleafing significantly increases damage tolerance. It greatly reduces the size of internal delamination at all impact energy levels.

5. Interleafing alters the pattern of damage development. Although transverse cracks develop in the interlayered laminates at intermediate impact energy levels, they do not form delaminations at the ply interfaces.

Acknowledgments

This report represents a portion of the work accomplished under Air Force Contract F33615-84-C-5024. Appreciation is extended to H. S.

Schwartz of the Air Force Wright Aeronautical Labs/Materials Laboratory, the project monitor on that program. Special appreciation is also given to S. S. Kaminski and M. W. Warmkessel for all of their effort in developing these data.

References

[1] Greszczuk, L. G., "Response of Isotropic and Composite Materials to Particle Impact," *Foreign Object Impact Damage to Composites, ASTM STP 568*, American Society for Testing and Materials, Philadelphia, 1975, pp. 183–211.

[2] Starnes, J. H., Rhodes, M. D., and Williams, J. G., "Effects of Impact Damage and Holes on the Compression Strength of a Graphite/Epoxy Laminate," *Nondestructive Evaluation and Flaw Criticality for Composite Materials, ASTM STP 696*, American Society for Testing and Materials, Philadelphia, 1979, pp. 145–171.

[3] Labor, J. D., "Impact Damage Effects on the Strength of Advanced Composites," *Nondestructive Evaluation and Flaw Criticality for Composite Materials, ASTM STP 696*, American Society for Testing and Materials, Philadelphia, 1979, pp. 172–184.

[4] Starnes, J. H. and Williams, J. G., "Failure Characteristics of Graphite/Epoxy Structural Components Loaded in Compression," NASA Technical Memorandum 84552, National Aeronautics and Space Administration, Washington, DC, Sept. 1982.

[5] Palmer, R. J., "Investigation of the Effect of Resin Material on Impact Damage to Graphite/Epoxy Composites," NASA Contract Report 165677, National Aeronautics and Space Administration, Washington, DC, March 1981.

[6] Evans, R. E. and Masters, J. E., "A New Generation of Epoxy Composites for Primary Structural Applications: Materials and Mechanics," *Toughened Composites, ASTM STP 937*, American Society for Testing and Materials, Philadelphia, 1987, pp. 413–435.

[7] Krieger, R. B., "The Relation Between Graphite Composite Toughness and Matrix Shear Stress-Strain Properties," *Proceedings*, 29th National SAMPE Symposium, Technology Vectors, 3–5 Apr. 1984, Society for the Advancement of Material and Process Engineering, Covina, CA, pp. 1570–1584.

[8] Masters, J. E., "Development of Composites Having Improved Resistance to Delamination and Impact," Interim Report No. 3 (AFWAL Contract F33615-84-C-5024), American Cyanamid, Stamford, CT, July 1985.

[9] Byers, B. A., "Behavior of Damaged Graphite/Epoxy Laminates Under Compression Loading," NASA Contract Report 159293, National Aeronautics and Space Administration, Washington, DC, Aug. 1980.

[10] Williams, J. G. and Rhodes, M. D., "The Effect of Resin on the Impact Damage Tolerance of Graphite/Epoxy Laminates," NASA Technical Memorandum 83213, National Aeronautics and Space Administration, Washington, DC, Oct. 1981.

[11] Guynn, E. G. and O'Brien, T. K., "The Influence of Lay-Up and Thickness on Composite Impact Damage and Compression Strength," presented at AIAA/ASME/ASCE/AHS 26th Structures, Structural Dynamics and Materials Conference, Orlando, FL, April 1985.

[12] Donaldson, S. L., "Fractography of Mixed Mode I-II Fracture in Graphite/Epoxy and Graphite/Thermoplastic Unidirectional Composites," AFWAL-TR-84-4186, Air Force Wright Aeronautical Laboratories, Wright-Patterson Air Force Base, Ohio, June 1985.

[13] Russell, A. J. and Street, K. N., "Factors Affecting the Interlaminar Fracture Energy of Graphite/Epoxy Laminates," *Proceedings*, Fourth International Conference on Composite Materials, ICCM-IV, 25–27 Oct. 1982, Tokyo, Japan, The Metallurgical Society, Warrendale, PA.

[14] Bostaph, G. M. and Elber, W., "A Fracture Mechanics Analysis for Delamination Growth During Impact on Composite Plates," ASME Winter Conference, Feb. 1982, American Society of Mechanical Engineers, New York.

Metals

Feature Identification

Robert W. Gould[1] and Richard H. McSwain[2]

Fractographic Feature Identification and Characterization by Digital Imaging Analysis

REFERENCE: Gould, R. W. and McSwain, R. H., **"Fractographic Feature Identification and Characterization by Digital Imaging Analysis,"** *Fractography of Modern Engineering Materials: Composites and Metals, ASTM STP 948*, J. E. Masters and J. J. Au, Eds., American Society for Testing and Materials, Philadelphia, 1987, pp. 263–292.

ABSTRACT: Fracture surface analysis has been relegated to the fractographer, who has developed skills in visual interpretation of unique fracture features. This process is based on correlation of features obtained from laboratory-created fractures with known failure modes to specific fracture surface features. The result has been considerable subjectivity, with conflicting opinions sometimes being rendered by different fractographers. Recent approaches to expand objectivity in the process have been related to quantitative microscopy procedures and to digital imaging analysis, with both indicating that nonsubjective fractographic analysis can be extremely beneficial. This study has taken the approach of digital analysis of scanning electron fractographs through Fourier transform analysis of spatial frequency content. The most common fracture features—fatigue striations, dimples, intergranular facets, and cleavage facets—have been uniquely defined in terms of frequency components. The result has been a significant advance in objectivity for fracture feature identification and characterization.

KEYWORDS: fractography, fracture features, digital imaging, fatigue striations, dimple rupture, intergranular fracture, cleavage facets

Computer-Assisted Fractography

Fractography, as a critical part of failure analysis, relates fracture surface morphology to the mechanism of fracture. The fractographer uses a combination of experience and reference fractographs to identify specific fracture surface features. For fracture features that fit a recognized pattern, the process is straightforward. When fractures do not fit a recognized pattern, the

[1]Professor of Materials Science and Engineering, University of Florida, Gainesville, FL 32611.
[2]Senior Metallurgist, Naval Air Rework Facility, Materials Engineering Laboratory, Pensacola, FL 32508.

process may be extremely difficult and highly subjective. To reduce the subjectivity, an effort has been undertaken to develop a technique of computer-assisted fractographic analysis. Computerized imaging techniques have been applied in the past to numerous areas of technology, including electron microscopy for image enhancement. Based on the success of digital imaging analysis in these areas, an investigation was undertaken to examine the feasibility of applying digital Fourier transform techniques to fractographic imaging analysis.

Image Processing and Analysis

Image processing has as its goal the improvement of an image to aid in information extraction and interpretation. Image processing assumes that information of concern to the observer can be characterized in terms of the properties of perceived objects or patterns. The extraction of that information involves detection and recognition of image patterns, usually requiring considerable human interaction because of the complexity of the analysis and the lack of precise digital algorithms for automatic processing.

In contrast, imaging analysis is concerned with the description of images in terms of properties of objects or regions in the image and the relationship between them, resulting in a description of the input image. The description may relate to the properties of objects of interest, such as location, size, shape, relational structure, and directionality, usually segmented with parts determined by homogeneity with respect to a given gray level or to a geometric property. Regions of approximately constant gray level or approximately uniform texture may represent objects.

Image analysis in electron microscopy has been concerned primarily with improvement of image quality. Fourier transform imaging techniques have been applied in transmission electron microscopy not only for ultrahigh resolution imaging but also for analysis of periodic features. Most of the analysis has been performed optically, using the optical transform power spectrum version of the Fourier transform. The early basis for optical transform analysis of periodicities in electron micrographs was established by Taylor and Lipson [1,2]. Others have demonstrated that the optical transform is applicable for determination of spacings and angles in electron micrographs [3,4]. The optical transform has been used for removal of imaging defects such as defocusing and aberrations by optical reconstruction techniques [5]. Digital techniques of image reconstruction have also been developed for electron micrographs using the Fourier transform [6]. The digital Fourier transform is a proven technique for repeat feature analysis in two-dimensional images [7,8], and because fracture surfaces are typically a collection of repeat patterns, that is dimples, cleavage facets, fatigue striations, or intergranular facets, the Fourier transform is a logical selection for classification and characterization of fracture surface features.

Quantitative fractography has dealt with the translation of fracture features into parametric form. The first theoretical work of significance was performed by El-Soudani [9], and Chermant and Coster [10] published the first review of quantitative fractography. A second review by Coster and Chermant [11] provided further information concerning the areas of investigation in quantitative fractography. The simplest form of quantitative fractography has been striation counting in fatigue fractures to determine microscopic crack propagation rates [12–14]. Quantitative microscopy techniques have been used for the study of fatigue fracture using fundamental quantitative parameters [15]. Recent developments in quantitative fractography have involved stereological analysis of fracture surfaces using parallax measurements [16]. In general the quantitative parameters used to classify fracture surfaces have been the linear roughness index, the waviness index, the fractal dimension, the absolute mean curvature, the mean convex curvature, and the mean concave curvature [11,17–19]. The Fourier transform has been employed in a limited manner to characterize fracture surface sections taken perpendicular to a fracture plane [20]. This technique extracted frequency components corresponding to surface roughness and used those values to characterize the fracture. Fractal analysis techniques as well as the Fourier transform have been applied together for the characterization of fracture detail in a magnification range lying between the microscopic and macroscopic [21]. This approach has led to a correlation between the fractal dimension and fracture toughness.

Though fractographic techniques have become standardized and much effort has been applied to extracting quantitative information concerning a fracture surface, few practical means exist for fracture surface morphological pattern analysis. The goal of this work was to investigate the feasibility of the use of the two-dimensional Fourier transform as a means to characterize and classify fracture surface features from a perspective as typically seen by the fractographer.

Fractographic Imaging Analysis Technique

The approach of this work was to apply the two-dimensional Fourier transform to four primary classes of fracture features: dimples, cleavage facets, fatigue striations, and intergranular fracture, all from scanning electron fractographs. Instrument variables such as brightness contrast, tilt, working distance, and beam current were evaluated and carefully controlled to allow comparison of fractographs. Variations in these imaging conditions only affected the sensitivity of the technique. Because fracture features observed are obviously magnification-dependent, all fractures were examined at the same magnification, X250, to provide a basis for distinguishing the four classes of fracture from the Fourier transform spatial frequency power spectrum pattern. Other magnification selections would generate other characteristic feature patterns. The fractographs were produced from both laboratory and field-generated failures and were based on common structural alloys.

Theoretical Fractographic Models

Computer-generated fracture feature models were also prepared for digital imaging analysis to correlate simplified input features to the resulting Fourier transform power spectrum pattern. The four most pronounced features (dimples, cleavage facets, fatigue striations, and intergranular facets) were modeled as straight lines, curved lines, and multifaceted cellular structures. The fatigue model was composed of bands of parallel lines produced by a computerized plotter and made into a transparency for digital imaging analysis. Parallel lines with single curvature were also generated to simulate a bowed fatigue crack front. Parallel lines with double curvature and two spacings, 2.0 and 2.5 mm, were generated to simulate curved fatigue fronts. The multifaceted cellular models, which were ASTM grain-size charts, were used to simulate dimples, cleavage facets, and intergranular facets.

Image Processing Software

The conversion of a fractographic image into a Fourier transform power spectrum was performed by a sequence of operations. The image was digitized, an area was selected for analysis and windowed, the fast Fourier transform was calculated, and the input image and power spectrum were photographed. The data were further arranged into circular arrays to aid in computer display and sampling. The sampling of the Fourier transform power spectrum was performed to numerically characterize the transform pattern.

Digital Image Acquisition

The negative transparency of the scanning electron microscope (SEM) fractograph was digitized to a 256 by 256 pixel array for Fourier transform processing. The shape of the window used to limit a region of interest for Fourier transform imaging analysis was critical because the window is convolved with the Fourier transform of the input image data. A circular window produced the least amount of degrading effects in the final transform pattern. To eliminate the step created by the edge of the window, a softening function was applied to the data at the window interface. The window size was selected to contain typical regions of uniform fracture so that there were a sufficient number of repeat features in the input image to allow spatial frequency detection by the Fourier transform.

The fast Fourier transform program used to process the image data was the Cooley-Tukey algorithm software known as HARM. The processed image was stored as a 256 by 256 Fourier transform power spectrum, which was the sum of the squares of the real and imaginary parts of the Fourier transform, or the modulus squared. The power spectrum data were recorded on film, and the data were also converted into a ring format to aid in sampling.

A typical theoretical model and the resulting Fourier transform power spectrum are shown in Fig. 1. The theoretical input image contained parallel, equally spaced lines with single curvature. The resulting transform pattern contained the repeat frequency information from the input image. The centrally located white spot at the center of the transform pattern represents a background frequency of zero and is known as the d-c power spike. It is primarily due to the window function and the background gray level of the input image. The two outer large spots were due to the presence of the parallel lines in the input image. The distance of the lines from the d-c spike is directly correlated to the line spacing or repeat frequency in the input image. The two outer spots are nonsymmetrical and distorted horizontally, which indicated curvature of the parallel lines in the input image. The degree of curvature is reflected in the amount of distortion in the spots. The location of the two major spots along the vertical axis of the transform power spectrum indicates that the repeat pattern direction in the input image was also along the vertical axis. The two low-intensity spots on the vertical axis were due to the fact that some of the lines in the computer-generated input image had joined, resulting in halving of the repeat frequency and moving the Fourier transform power spectrum spots half the distance to the d-c spike. The high-intensity spots also have harmonics located twice their distance from the d-c spike, beyond the edge of the photograph. The harmonics have frequencies that are multiples of the fundamental frequency and thus have positions in the transform pattern that are multiples of the primary spot locations. The two spots located in the horizontal axis of the transform pattern are a result of perpendicular, vertical lines superimposed in the input image. The brightness of the spots in the transform pattern indicates the number of repeat features present in the input image. The greater the number of repeat features, the brighter the spot. Thus, visual analysis of the Fourier transform power spectrum is extremely valuable in extracting descriptive data concerning the input image features.

In order to view the transform power spectrum from an angular perspective, a rectangular three-dimensional (3D) surface plot was generated. The plot helps visualize the relative positions and heights of the spatial frequency data. Figure 2 shows the three steps in the production of the 3D surface plot. The top view is the transform power spectrum data in ring format, as viewed in the transform photograph. The central white spot represents the area removed, corresponding to the core of the d-c spike. The dark lobes and

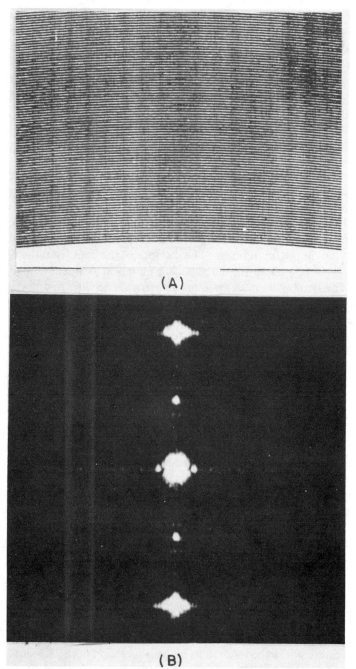

FIG. 1—*Theoretical fatigue model/Fourier transform pattern*: (A) *undigitized input image of theoretical fatigue as single curvature parallel lines, and* (B) *resulting Fourier transform power spectrum.*

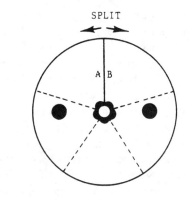

SPLIT

A B

EXPANDED INNER SURFACE

A

B

(RINGS)

COMPACTED OUTER SURFACE
(DISTORTED WEDGES)

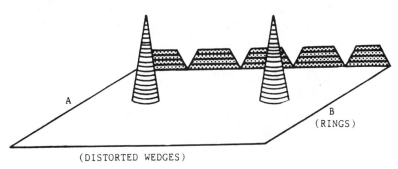

A

B
(RINGS)

(DISTORTED WEDGES)

FIG. 2—*Rectangular representation of three-dimensional surface plot of Fourier transform power spectrum data.*

spots represent data. The ring is divided into five sectors for simplicity. In order to produce the rectangular representation of the 3D surface plot, the ring data is split and opened, creating a rectangular surface as shown in the middle drawing of Fig. 2. This requires expansion of the inner sector area and compaction of the outer sector area, a process that is reasonable because of the positionally varying density of data within the sectors. Minor distortion results but does not interfere with visual interpretation of the data. If the middle drawing is then rotated back on a horizontal axis, it becomes a rectangular 3D surface plot of the Fourier transform power spectrum. The relative intensities of features in the spectrum now become height variations in the surface plot. The heights were not calibrated for specific numbers of repeat patterns, but serve as a means of comparing input images. The two outer rings of data were dropped from the surface plot because the ring-sampled data was taken from a square matrix, causing loss of data at the outer edges.

A radial plot, as shown in Fig. 3, was developed for sampling the Fourier transform power spectrum data to produce a directionally insensitive form of the data. Figure 3 shows a sample spectrum with two data spikes and a set

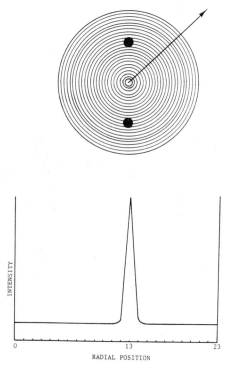

FIG. 3—*Radial plot sampling of Fourier transform power spectrum data.*

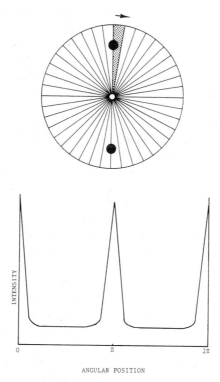

FIG. 4—*Wedge plot sampling of Fourier transform power spectrum data.*

of concentric rings. If the contents of each ring are summed and plotted with respect to radial position, a radial signature plot results. The radial plot, as shown at the bottom of Fig. 3, indicates the presence of data spikes at Ring 13. However, the plot provides no information about orientation of the data. The radial plot may also be displayed in the log-data format. The first difference of the log radial plot is presented in this paper and displays the slope of the curve, emphasizing inflections due to spots in the Fourier transform spectra.

The wedge plot was developed as a frequency insensitive technique for displaying the Fourier transform power spectrum data. The wedge plot, as shown in Fig. 4, provides a directional description of the input image that is independent of spatial frequency.

The combination of the Fourier transform power spectrum display and sampling patterns provides a means of classifying and characterizing the various types of input fractographic images.

Fractographic Image Feature Calibration

The two distinct classes of theoretical fracture feature models, parallel lines and multifaceted cellular patterns, were used to evaluate the ability of the digital Fourier transform to produce the expected transform power spectrum patterns. They were also used to calibrate the input image repeat spacing with the spot spacing for the transform pattern. Data from calibrated rectangular test grids at various magnifications as well as the theoretical fracture feature models at various magnifications were combined to establish a calibration curve for analysis of repeat features in the fractographic images.

A set of dual frequency straight parallel lines at a repeat spacing of 2 and 20 mm are shown in the input image of Fig. 5A. The resulting Fourier transform power spectrum is shown in Fig. 5B. For the two sets of lines in the input image, the repeat spacing along the horizontal axis resulted in a row of frequency spots in the transform pattern also along the horizontal axis. The finer lines in the input image resulted in the two highest intensity spots and their first harmonic, located twice the distance from the d-c spike. The coarse lines resulted in the row of multiple equally spaced spots, consisting of the principal spot adjacent to the d-c spike and multiple harmonics of the primary repeat frequency.

The parallel line input images were models of various types of fatigue fracture striations. The remaining theoretical models examined were composed of equiaxed cellular structures (ASTM grain-size charts) to simulate the equiaxed features that occur in dimple rupture, cleavage, and intergranular fractures.

Figure 6A shows the ASTM grain size 7 chart input image with an average repeat feature spacing of about 3.2 mm. The corresponding Fourier transform power spectrum, shown in Fig. 6B, indicated significant preferred orientation in the input image. The Fourier transform pattern indicated predominant parallel lines in the input image and had a repeat direction oriented approximately 2.6 rad counterclockwise from the horizontal axis in the input image. Visual examination of the input image at a low angle of incidence at the 2.6 -rad direction again revealed the predominant parallel lines. Thus, the Fourier transform power spectrum and associated sampling techniques formed a powerful tool for image analysis that revealed image features not readily apparent to the observer.

Calibration of the Fourier transform power spectrum was performed by analysis of the test grid patterns, the parallel lines, and the equiaxed cellular structures at a range of magnifications that extended over the limits of the transform power spectrum. The calibration was required to evaluate the feasibility of making quantitative measurements of fracture surface features. The photographs of the input image and transform patterns were all made to approximately the same magnification to allow measurements to be made directly from the photographs. The results from photographic measurements were compared to the transform sampling plots. Identical correlations resulted. Figure 7 shows the correlation of the input feature dimensions in millimetres with the transform spot pattern location expressed as ring

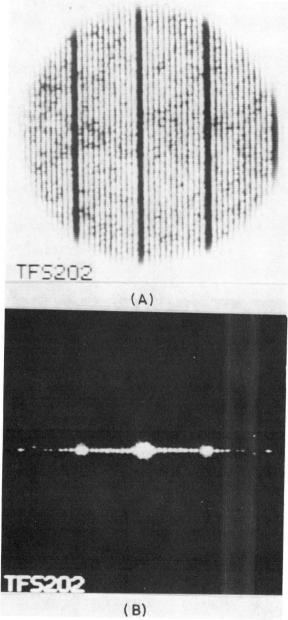

FIG. 5—*Parallel lines/Fourier transform pair:* (A) *input image of theoretical fatigue as two repeat frequency bands of straight parallel lines, and* (B) *resulting Fourier transform power spectrum.*

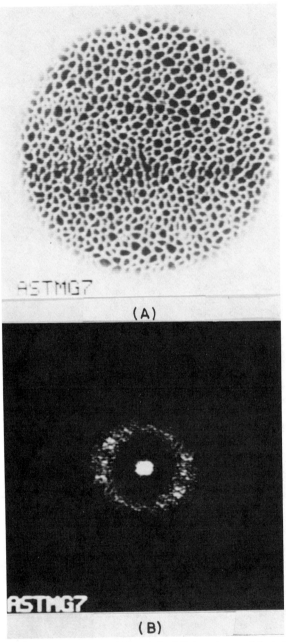

FIG. 6—*Cellular structure/Fourier transform pair*: (A) *input image of theoretical multifaceted cellular structure as ASTM grain size 7 chart, and* (B) *resulting Fourier transform power spectrum.*

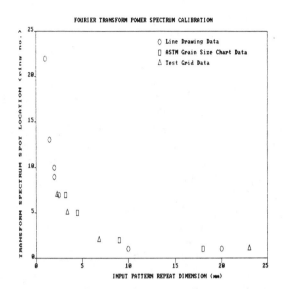

FIG. 7—*Correlation of input image repeat dimension to Fourier transform power spectrum spot separation.*

number for the transform pattern. The shape of the curve fits the expected hyperbolic shape as given by $S = K/X$, where X is the input spacing, K is a constant, and S is the output spacing. For very small input pattern dimensions the transform pattern has extremely high sensitivity; a change in input spacing from 1 to 2 mm causes the corresponding spot pattern to change from approximately 20 mm to approximately 10 mm. For very large input pattern dimensions, the transform pattern has extremely low sensitivity; a change in input spacing from 10 to 20 mm results in a spot location change from about 2 to 1 mm. The correlation curve in Fig. 7 showed the optimum range of input feature spacings to be from 1 to 10 mm.

Digital Fractographic Pattern Classification

The four primary fractographic features were evaluated with the Fourier transform to evaluate the potential of establishing spatial frequency signatures for each of the feature classes. For each fractograph, the input image, the resulting Fourier transform power spectrum, the surface plot, first difference of the radial plot, and wedge plot are presented. Each transform pattern and associated plots provide an objective description of the input image repeat feature pattern.

Microvoid Coalescence Dimples

The specimen that best typified tensile dimples resulting from microvoid coalescence was a 4340 quenched and tempered steel that failed in overload.

Figure 8A shows the input image composed of dimples of various sizes, and the corresponding Fourier transform power spectrum, shown in Fig. 8B, indicated a uniform structure with no strong indications of directionality or specific strong repeat frequency in the input image. The surface plot, Fig. 9, was also uniform. The radial difference plot, Fig. 10, was used to amplify the slope changes in the radial curve. The curve reached a constant negative slope at the second ring and remained constant to almost the last ring. The slope changes in the radial plot caused by the presence of repeat frequencies are shown as inflections in the first difference plot. The plot in Fig. 10 showed inflections at Rings 8, 10, and 20. Examination of the calibration curve with output pattern spacing expressed as ring number revealed an input image spacing of 2.4, 2.2, and 1.1 mm. The transform image also had low frequency spots near the d-c spike which corresponded to a spacing of 10 mm. The fine spacings (1 to 3 mm) were the approximate sizes of the dimples in the input image, and the coarse spacing (10 mm) was the approximate size of the pockets of dimples. The generally high slope shown by the radial difference plot reflected the presence of a full range of repeat sizes in the input image. The wedge plot, Fig. 11, also indicated the presence of orientation effects in the input image by the two frequency spikes with an initial angle of 1.04 rad. The spikes were located in the transform image, Fig. 8B, near the d-c spike. They were a result of the fracture ridges between pockets of dimples in the input image.

The features that appear to classify dimples are a uniform surface plot and a high slope in the radial plot, indicating the uniform presence of a range of frequency components. The ring plot and wedge plot showed no directional effects for equiaxed dimples.

Cleavage Planes

The specimen used to represent cleavage planes was a plain carbon steel specimen fractured in impact. The input image, shown in Fig. 12A, contained cleavage river patterns on flat cleavage facets. The resulting Fourier transform power spectrum, Fig. 12B, indicated the presence of coarse features by the presence of spots close to the central d-c power spike and also directionality in the input image as indicated by the grouping of spots in the transform pattern. The surface plot, Fig. 13, clearly demonstrated the directionality that was present. There were four major spikes, indicating two primary repeat directions. Figure 14, the first difference of the radial plot, showed specific inflections corresponding to an input image spacing ranging from 3.4 to 1.0 mm. Coarse features of approximately 10-mm spacing were also indicated by the spots in the transform power spectrum image close to the d-c spike. Each of the spacings corresponded to the spacings of the river patterns at certain positions. The radial difference plot showed a slowly rising slope. The wedge plot, Fig. 15, indicated the presence of multiple

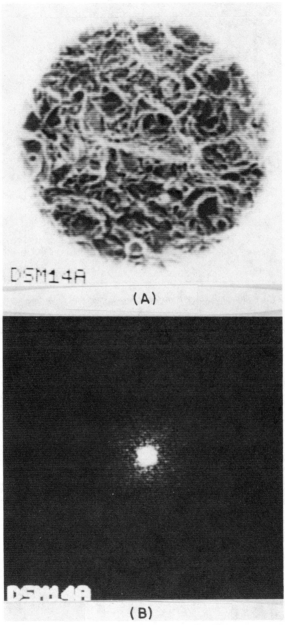

FIG. 8—*Tensile dimple image/Fourier transform pair*: (A) *input image of fine tensile dimples from a 4340 quenched and tempered steel, and* (B) *resulting Fourier transform power spectrum.*

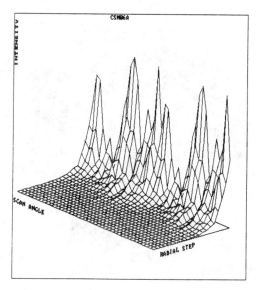

FIG. 9—*Surface plot of Fourier transform power spectrum from fine tensile dimples.*

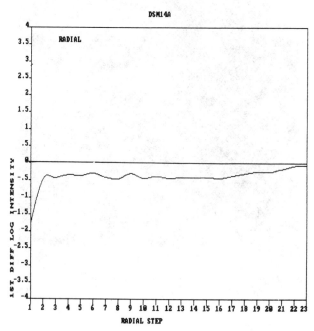

FIG. 10—*First difference of radial plot of Fourier transform power spectrum from fine tensile dimples.*

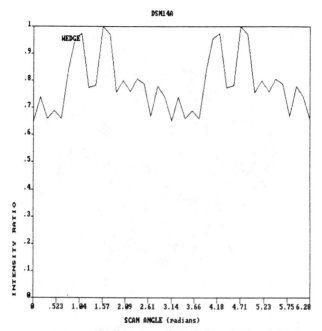

FIG. 11—*Wedge plot of Fourier transform power spectrum from fine tensile dimples.*

directional features. Spikes corresponded to the directions of the primary river pattern angles in the input image. Thus the Fourier transform analysis was able to detect the presence of river patterns and their angular orientation for a cleavage fracture.

The primary features for classifying cleavage fracture were large multiple spikes in the surface plot and wedge plot and a slowly rising negative slope in the radial plot.

Fatigue Striations

The specimen used to represent fatigue striations was a 2024 aluminum forging which failed in fatigue. The input image of the fatigue fracture is shown in Fig. 16A. The fatigue striations were equally spaced and had multiple curvature. The Fourier transform power spectrum, Fig 16B, showed the central d-c spike from the tapered window and two outer spikes that represented the fatigue striations. The two spikes were elongated horizontally, indicating the presence of curvature in the input image parallel lines. The surface plot, Fig. 17, showed two extremely distinct spikes. Measurement of the striation spacing in the input image revealed a repeat spacing of 4 mm. The radial difference plot, Fig. 18, showed multiple inflections with the fourth ring corresponded to a spacing of 4.0 mm, the striation spacing.

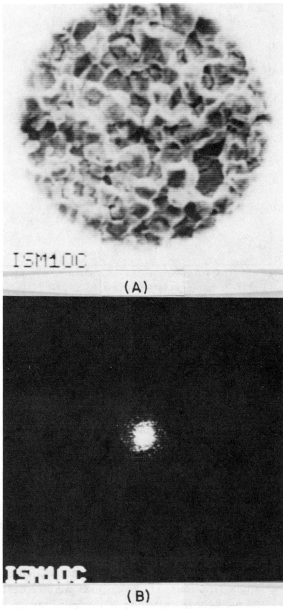

FIG. 12—*Cleavage facet image/Fourier transform pair: (A) input image of cleavage facets from a plain carbon steel, and (B) resulting Fourier transform power spectrum.*

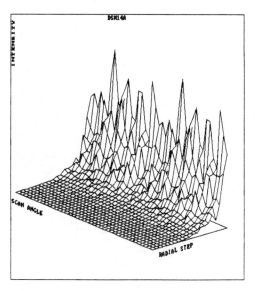

FIG. 13—*Surface plot of Fourier transform power spectrum from cleavage facets.*

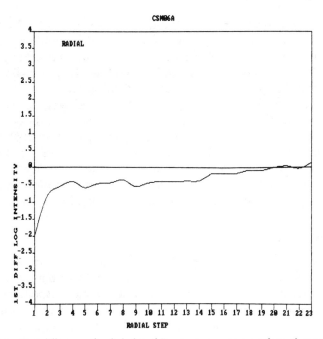

FIG. 14—*First difference of radial plot of Fourier power spectrum from cleavage facets.*

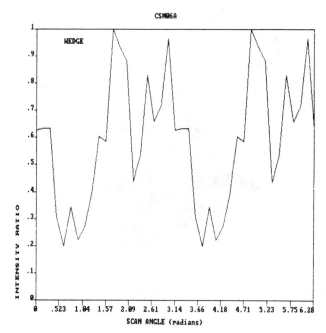

FIG. 15—*Wedge plot of Fourier transform power spectrum from cleavage facets.*

The other rings corresponded to features with spacings of 2.5 to 1.4 mm. These fine spacings were not readily apparent in the input image and were probably due to harmonic spots in the transform pattern caused by the large primary spikes. The wedge plot, Fig. 19, showed the two principal spikes and the spread in the spikes due to curvature of the striations.

The Fourier transform power spectrum for fatigue was the most unique for all the fracture modes and was the easiest to classify by the dual spots in the transform pattern or dual spikes in the surface plot. Fatigue also showed strong curvature in the radial plot due to the absence of a full range of frequency components. The Fourier transform had the ability to detect the presence of striations, to automatically measure the striation spacing and shape, and to isolate frequency components produced by random loading.

Intergranular Facets

The specimen used to represent intergranular fracture was a 4340 quenched and tempered steel forging fractured by hydrogen-assisted cracking. The input image is shown in Fig. 20*A*, and the corresponding Fourier transform power spectrum is shown in Fig. 20*B*. The transform pattern indicated the presence of coarse features in the input image as evidenced by the collection of spots close to the d-c power spike. The transform pattern

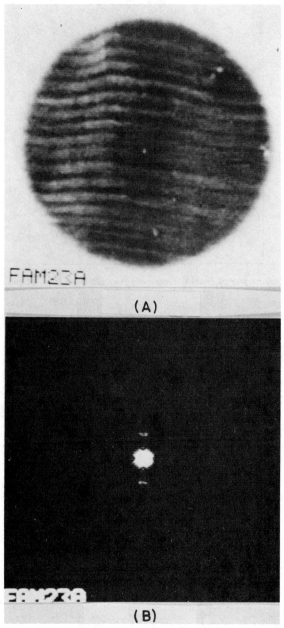

FIG. 16—*Fatigue striation image/Fourier transform pair:* (A) *input image of fatigue striations from 2024 aluminum forging, and* (B) *resulting Fourier transform power spectrum.*

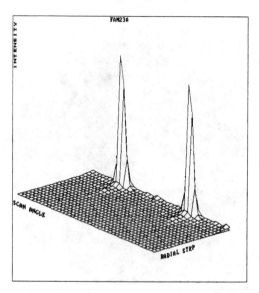

FIG. 17—*Surface plot of Fourier transform power spectrum from fatigue striations.*

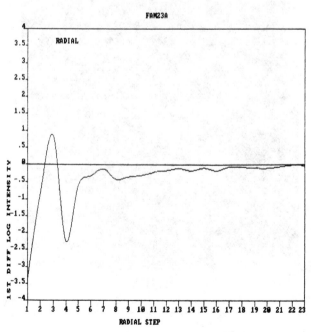

FIG. 18—*First difference of radial plot of Fourier transform power spectrum from fatigue striations.*

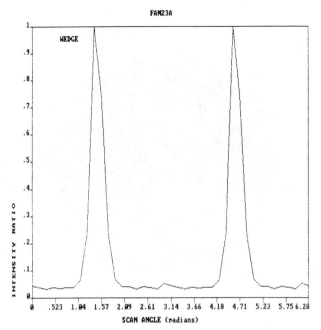

FIG. 19—*Wedge plot of Fourier transform power spectrum from fatigue striations.*

also indicated directionality by the collection of spots at the top and bottom of the d-c power spike. The directionality was probably due to grain elongation in the forging. The surface plot, Fig. 21, indicated no strong orientation effects or repeat features. The radial difference plot, Fig. 22, showed a slowly changing negative slope with minor inflections. The size of the minor inflections indicated a limited number of repeat units for the size classes. A small inflection was present which corresponded to an input repeat pattern spacing of about 5 mm. Low-frequency components were contained in the transform pattern near the d-c spike corresponding to an input image spacing of 10.0 mm. The intergranular facets in the input image had a projected diameter of approximately 10.0 mm. Other facets in the input image had projected diameters of approximately 5 mm. Smaller facets were also present as indicated by the outlying inflections in the radial difference curve. The wedge plot, Fig. 23, showed little indication of directionality.

The features used to classify intergranular fracture were a uniform pattern in the surface plot, moderate curvature of the radial plot, and a few outlying inflections in the radial difference plot. Intergranular fracture transform features were similar to dimple rupture fracture features. A primary difference was the lower slope of the radial plot due to fewer high frequency components present in the intergranular fracture images. The

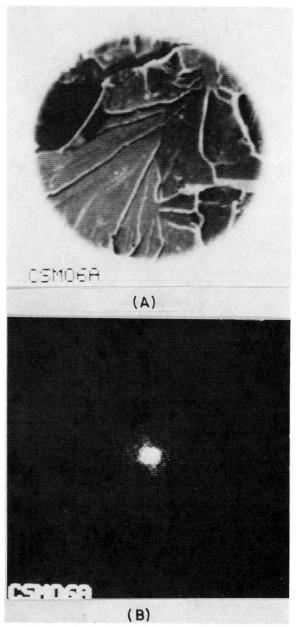

FIG. 20—*Intergranular facet image/Fourier transform pair:* (A) *input image of intergranular fracture from hydrogen-assisted cracking in anxle, and* (B) *resulting Fourier transform power spectrum.*

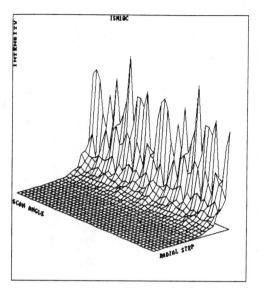

FIG. 21—*Surface plot of Fourier transform power spectrum from intergranular hydrogen-assisted cracking facets.*

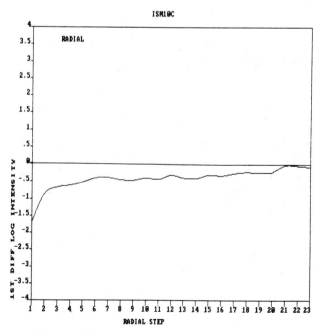

FIG. 22—*First difference of radial plot of Fourier transform power spectrum from intergranular hydrogen-assisted cracking facets.*

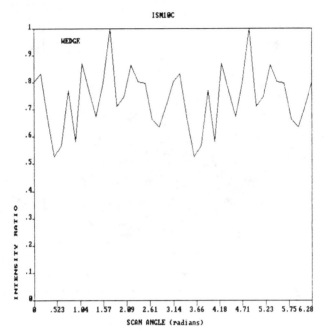

FIG. 23—*Wedge plot of Fourier transform power spectrum from intergranular hydrogen-assisted cracking facets.*

Fourier transform analysis was able to determine the size of facets, the proportion of facets within a given size class, and the presence of grain elongation.

Digital Fractographic Pattern Characterization

The two-dimensional Fourier transform has the potential for fractographic feature classification of specific patterns and for characterization of descriptive information concerning a certain pattern. The characterization can produce information on feature orientation, feature size, and quantity of features of a given size, class, and spacing.

A shear dimple pattern, shown in Fig. 24*A*, from a plain carbon steel that failed in torsional overload was also analyzed. The resulting Fourier transform power spectrum, Fig. 24*B*, showed a spread of spots oriented in a direction that corresponded to the edges of the shear dimples. The surface plot, Fig. 25, showed two primary broad spikes that corresponded to the shear dimple edges. The spikes in the surface plot indicated the directionality in the input image due to the shear dimples. The location of the spikes corresponded to the orientation of the repeat patterns formed by the edges of the shear dimples. The heights of the peaks indicated strong orientation

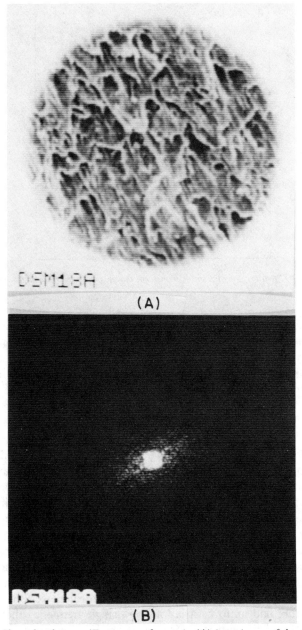

FIG. 24—*Shear dimple image/Fourier transform pair: (A) input image of shear dimples from a plain carbon steel, and (B) resulting Fourier transform power spectrum.*

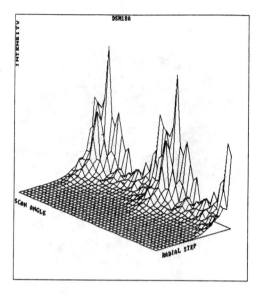

FIG. 25—*Surface plot of Fourier transform power spectrum from shear dimples.*

effects in the input image. The Fourier transform analysis of the shear dimples illustrated the power of the technique to characterize fracture features within a given class. The sampled Fourier transform power spectrum was capable of revealing the degree of orientation and orientation direction of the shear dimples.

Summary

Digital fractographic imaging analysis using the Fourier transform has been demonstrated to be a useful technique for classifying and characterizing fractographic images. For four primary fracture modes, microvoid coalescence, cleavage, fatigue, and decohesive rupture, the digital imaging analysis technique was able to distinguish and classify the spatial frequency character of each type of fracture surface for a given specified magnification. The microvoid coalescence fracture had a wide range of repeat frequency components related to dimple size and a uniform distribution of frequency content. The cleavage fracture had river patterns that created a narrow range of directional repeat frequency effects in the transform power spectrum. The fatigue fracture had uniform striations of a single frequency that produced distinct spots in the Fourier transform pattern. The decohesive rupture fractures produced intergranular facets with a range of facet sizes that typically lacked higher frequency component content. It also showed minor directional effects due to elongated grain shape.

The Fourier transform was able to extract useful information from subclasses of fractographs, such as shear dimples. Size classes of repeat features were determined and the extensiveness of the size class was evaluated. Orientation effects were easily determined using the wedge sampled Fourier transform pattern, and the direction of orientation was evaluated for many of the fractures. The technique was even able to detect orientation effects in ASTM grain size charts, which was completely unexpected. The digital fractographic imaging analysis technique is not expected to replace the fractographer, but to assist the fractographer in making qualitative and quantitative assessments of fracture surface features.

Digital fractographic imaging analysis using the two-dimensional Fourier transform offers a new and creative approach to fracture surface feature analysis that is worthy of further investigation. It has the ability to draw the fractographer's attention to subtle characteristics in a fracture surface pattern, to classify the pattern, and to extract quantitative information concerning the nature of the pattern. With the further development of digital imaging analysis applications for SEM, digital fractographic imaging analysis should become an instrumentation option readily available to the fractographer.

Acknowledgments

This work was supported by Contract N000-14-83-K-2035 from the Naval Air Systems Command, Washington, D.C. Assistance from Alan Gregalot of the University of Florida and equipment support from the Department of Engineering Sciences at the University of Florida and Naval Air Rework Facility, Pensacola, Florida, is appreciated.

References

[1] Taylor, C. A. and Lipson, H., "Optical Methods in Crystal Structure Determination," *Nature*, Vol. 167, 1951, p. 809.
[2] Taylor, C. A. and Lipson, H., *Optical Transforms*, G. Bell and Sons, Ltd., London, 1964.
[3] Markham, R., Frey, S., and Hills, G. J., "Methods for the Enhancement of Image Detail and Accentuation of Structure in Electron Micrographs," *Virology*, Vol. 20, 1963, p. 88.
[4] Klug, A. and Berger, J. E., "An Optical Method for the Analysis of Periodicities in Electron Micrographs and Some Observations on the Mechanism of Negative Staining," *Journal of Molecular Biology*, Vol. 10, 1964, p. 564.
[5] Erickson, H. P. and Klug, A., "Measurements and Composition of Defocusing and Aberrations by Fourier Processing of Electron Micrographs," *Philosophical Transactions of the Royal Society of London*, Vol. 261, Series B, 1971, p. 105.
[6] Crowther, R. A. and Klug, A., "Structural Analysis of Macromolecular Assemblies by Image Reconstruction from Electron Micrographs," *Annual Review of Biochemistry*, Vol. 44, 1975, p. 161.
[7] Lendaris, G. G. and Stanley, G. L., "Diffraction-Pattern Sampling for Automatic Pattern Recognition," *Proceedings of the IEEE*, Vol. 58, No. 2, 1970, Institute of Electrical and Electronics Engineers, New York, p. 198.

[8] Pincus, H. J., "Optical Diffraction Analysis in Microscopy," in *Advances in Optical and Electron Microscopy*, V. E. Cosslett and R. Barer, Eds., Academic Press, New York, NY, 1978.

[9] El-Soudani, S. M., "Theoretical Basis for Quantitative Analysis of Fracture Surfaces," *Metallography*, Vol. 7, 1974, p. 271.

[10] Chermant, J. L. and Coster, M., "Review Quantitative Fractography," *Journal of Materials Science*, Vol. 14, 1979, p. 509.

[11] Coster, M. and Chermant, J. L., "Recent Developments in Quantitative Fractography," *International Metals Reviews*, Vol. 28, 1983, p. 228.

[12] Abelkis, P. R., "Use of Microfractography in the Study of Fatigue Crack Propagation Under Spectrum Loading," in *Fractography in Failure Analysis, ASTM STP 645*, B. M. Strauss and W. H. Cullen, Jr., Eds., American Society for Testing and Materials, Philadelphia, PA, 1978, p. 213.

[13] Madeyski, A. and Albertin, L., "Fractographic Method of Evaluation of the Cyclic Stress Amplitude in Fatigue Failure Analysis," in *Fractography in Failure Analysis, ASTM STP 645*, B. M. Strauss and W. H. Cullen, Jr., Eds., American Society for Testing and Materials, Philadelphia, PA, 1979, p. 73.

[14] Au, J. J. and Ke, J. S., "Correlation Between Fatigue Crack Growth Rate and Fatigue Striation Spacing," in *Fractography and Materials Science, ASTM STP 733*, L. N. Gilbertson and R. D. Zipp, Eds., American Society for Testing and Materials, Philadelphia, PA, 1981, p. 202.

[15] Rhines, F. N., "Quantitative Microscopy and Fatigue Mechanisms," in *Fatigue Mechanisms, ASTM STP 675*, J. T. Fong, Ed., American Society for Testing and Materials, Philadelphia, PA, 1979, p. 22.

[16] Kobayashi, T., Irwin, G. R., and Zhang, X. J., "Topographic Examination of Fracture Surfaces in Fibrous Cleavage Transition Behavior," in *Fractography of Ceramic and Metal Failures, ASTM STP 827*, J. J. Mecholsky, Jr. and S. R. Powell, Jr., Eds., American Society for Testing and Materials, Philadelphia, PA, 1984, p. 234.

[17] Underwood, E. E. and Chakrabortty, S. B., "Quantitative Fractography of a Fatigued Ti-28V Alloy," in *Fractography and Materials Science, ASTM STP 733*, L. N. Gilbertson and R. D. Zipp, Eds., American Society for Testing and Materials, Philadelphia, PA, 1981, p. 337.

[18] Mandelbrot, B., *Fractals, Form, Chance, and Dimension*, W. H. Freeman and Co., San Francisco, CA, 1977.

[19] Banerji, K. and Underwood, E. E., "Quantitative Analysis of Fractographic Features in a 4340 Steel," *Acta Stereologica*, Vol. 2, No. 1, 1983, p. 65.

[20] Passoja, D. E. and Psioda, J. A., "Fourier Transform Techniques—Fracture and Fatigue," *Fractography and Materials Science, ASTM STP 733*, L. N. Gilbertson and R. D. Zipp, Eds., American Society for Testing and Materials, Philadelphia, PA, 1981, p. 355.

[21] Mandelbrot, B. B., Passoja, D. E., and Paullay, A. J., "Fractal Character of Fracture Surfaces of Metals," *Nature*, Vol. 308, 1984, p. 721.

Metals I: Ferrous Alloys

Jagannathan Sankar,[1] *David B. Williams,*[2] *and Alan W. Pense*[2]

Fractography of Pressure Vessel Steel Weldments

REFERENCE: Sankar, J., Williams, D. B., and Pense, A. W., **"Fractography of Pressure Vessel Steel Weldments,"** *Fractography of Modern Engineering Materials: Composites and Metals, ASTM STP 948,* J. E. Masters and J. J. Au, Eds., American Society for Testing and Materials, Philadelphia, 1987, pp. 295–316.

ABSTRACT: An experimental program was carried out to interpret the effect of submerged arc flux compositions on the Charpy toughness and fracture behavior of typical pressure vessel steel weldments. Three different fluxes with varying basicity—namely, Linde Grade 80, Linde Grade 0091, and Oerlikon OP76—were selected for the program. The steel weldments used were C-Mn-Si, $2\frac{1}{4}$Cr-1Mo, and A543 Class 1. It has been found that silicon-rich inclusions are harmful to toughness and that Grade 80 welds with the presence of high amounts of these inclusions result in lower toughness compared to Grade 0091 or OP76. Of the three fluxes, OP76 develops a much cleaner weld with very few silicon-rich inclusions, resulting in extremely tough weld. Characterization of toughness at different positions of the weld shows sensitivity of the toughness tests on the weld metal to impact test position.

KEYWORDS: scanning electron microscopy/energy dispersive spectrometry, weld metal, fractography, mechanical properties, structure/property relationships

In recent years, for pressure vessel service, steels of reasonable strength, 415 MPa (60 ksi), with relatively high toughness are used due to the design requirements. These steels, often of heavy section, are welded by the submerged arc (SA) process since it is technically and economically advantageous for welding heavy sections. In SA welding a granulated flux is used that shields the molten weld metal from the atmosphere and further acts as a slag, permitting desired (or undesired) slag/metal reactions to take place during the welding process. The SA, in common with other welding procedures, subjects both the base metal that is being joined and the weld metal region to severe thermal fluctuations, particularly in multipass welds, which are quite common in pressure vessel steels. These factors, in combination

[1]Assistant professor, Mechanical Engineering, North Carolina Agricultural and Technical State University, Greensboro, NC 27411.
[2]Professor, Metallurgy and Materials Engineering, Lehigh University, Bethlehem, PA 18015.

with the normal solidification process, result in a weldment that often contains severe residual stresses and a microstructure that is a result of many complex interacting variables. Furthermore, it is a common practice to subject such welds to long postweld heat treatments or stress relief annealing (SRA) processes to avoid any fabrication cracking and subsequent distortion or fracture. However, this again affects the microstructure and mechanical properties of the weld and that of the base plate [1,2]. In general, SRA causes the weld metal to lose strength and gain toughness, but in some instances decreases both strength and toughness [3,4].

Another important variable that affects the weld properties, especially toughness, is the composition of welding flux. The final chemistry of the weld metal is partially determined by the slag-metal reactions that are in turn influenced by the flux composition and the wire-plate compositions [5]. A useful concept when dealing with the effect of flux on weld metal toughness is the basicity index (B). The basicity index formula can be written after Tuliani et al. [6] as

$$B = \frac{CaO + MgO + BaO + SnO + K_2O + Li_2O + CaF_2 + \frac{1}{2}MnO + FeO}{SiO_2 + \frac{1}{2}(Al_2O_3 + TiO_2 + ZrO_2)}$$

(1)

When B is less than 1, the flux is classified as acid; 1 to 1.5 is neutral, 1.5 to 2.5 is semibasic, and greater than 2.5 is basic.

The effect of slag basicity on the partitioning of manganese (Mn) and silicon (Si) between metal and slag is shown in Fig. 1. Previous works [2,7] have shown that a high Si concentration at low B-values can lead to a lowering of the weld metal toughness. Dissolved oxygen also seems to have a disastrous influence on impact strength [8]. The general trend of the weld metal oxygen content in terms of the basicity index of the flux is shown in Fig. 2 after Eagar [9]. It is now a well-accepted fact that high oxygen content results in high amounts of inclusions in the weld metal [8–11], thus decreasing the notch toughness of the weld [12–15]. The failure mechanism is one of void formation by the decohesion of the inclusion particle/matrix interface followed by the failure of the interparticle region. Chin [14] reported a model for the toughness of welds. He suggested that the low Charpy V-notch toughness is associated with a discontinous mode of crack propagation where inclusions act as preformed cracks defining a path of least resistance. Figure 3 shows the concept of the discontinuous mode of crack propagation. The word "discontinuous" is used, since the main crack propagates by coalescence with the small crack immediately ahead of its root; the whole process is repeated many times as the fracture spreads across the whole specimen.

The other metallurgical factors which may play a role in determining weld

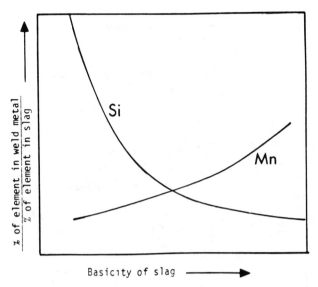

FIG. 1—*Effect of slag basicity on the partition of Mn and Si between metal and slag.*

metal toughness are [7,16,17]: (1) the microstructure of the columnar grains; (2) the microstructure of the recrystallized zone developed in the previous pass as a result of heating during the laying of a subsequent bead in a multipass weld; (3) an alloying or impurity element in the solid solution; (4) the chemical nature and distribution of precipitates; (5) the chemistry of the inclusions that may be trapped in the weld metal during solidification, particularly in multipass weldments. It is appropriate to mention here that

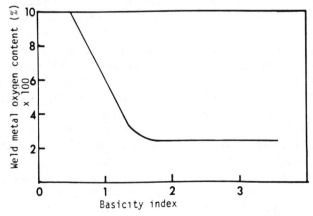

FIG. 2—*General trend of the weld metal oxygen content as a function of flux basicity.*

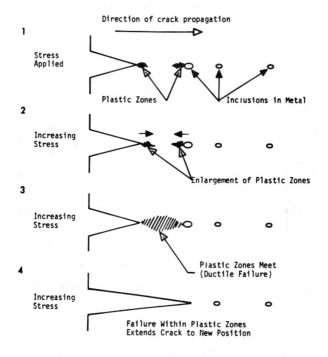

FIG. 3—*Model of discontinuous crack propagation in steel welds.*

these variables are not independent and may be influenced by flux itself for a particular heat input.

In recent years electron microscopy has begun to play a major role in the investigation of the microstructure and fracture surfaces of welds. In particular, the use of X-ray microanalysis coupled with fracture surface investigation using scanning electron microscopy (SEM) has facilitated in scientifically correlating the mechanical behavior of weldments to the presence of various microstructures, submicron precipitates, and inclusions. In the present investigation an attempt was made to interpret the effect of submerged arc flux compositions on the Charpy toughness and fracture behaviors of typical pressure vessel steel SA weld metals. This type of research is needed since higher toughness steels fabricated using the SA process are in increasing demand in the pressure vessel industry. The paper describes the fracture surface/microanalysis of pressure vessel steel weld metals and shows how X-ray energy dispersive microanalysis in combination with SEM fractography can provide insight into hitherto unexplained toughness behavior of these weld metals.

Definition of the Materials

Three different fluxes with varying basicity index were selected for the program. The SA fluxes were Linde Grade 80, Linde Grade 0091, and Oerlikon OP76, Grade 80 being the least basic of the three and OP76 the most basic. The compositions of the fluxes along with their basicity index are given in Table 1. The steel weldments used were mild steel (0.09C-1.2Mn-0.3Si), low-alloy steel ($2\frac{1}{4}$Cr-1Mo), and A543 Class 1 (3.27Ni-1.73Cr-0.47Mo). Table 2 gives the welding details of the work. All the weldments were SRA treated except A543 before Charpy V-notch toughness characterization. Table 3 gives the SRA treatment details and the thickness of the plates used. In Table 3 all the 25 mm (1-in.) plates (C-Mn-Si and $2\frac{1}{4}$Cr-1Mo, both G80 and G0091 fluxes) were fabricated at the Franklin Institute Research Laboratories while the other two weldments ($2\frac{1}{4}$Cr-1Mo/OP76 and A543) were made at the Chicago Bridge and Iron Co.

Test Program and Results

Toughness Characterization

The Charpy V-notch toughness characterization of all the 25 mm (1-in.) plates (C-Mn-Si and $2\frac{1}{4}$Cr-1Mo, both G80 and G0091 fluxes) were done at the Franklin Institute Research Laboratories [18]. The toughness characterization of $2\frac{1}{4}$Cr-1Mo/OP76 and the A543 Class 1/G0091 flux were carried out at the Lehigh University.

In all the 25-mm (1-in.) plates (C-Mn-Si and $2\frac{1}{4}$Cr-1Mo both G80 and G0091 fluxes), the notch of each specimen was perpendicular to the plate surface and centered in the weld metal. The top of each specimen was located 6.4 mm ($\frac{1}{4}$ in.) below the plate surface. Figure 4 shows the Charpy data obtained. Examination of these results indicates that the weld metal's toughness can vary significantly depending on the type of flux used in the fabrication. The G80 flux invariably gives a weld of inferior toughness compared to G0091, irrespective of whether the weld plate is C-Mn-Si or $2\frac{1}{4}$Cr-1Mo. For the C-Mn-Si weld metal, the G80 flux shows an average ambient temperature toughness of $\sim$90 J ($\sim$70 ft · lb), while the G0091 flux has $\sim$260 J ($\sim$193 ft · lb). As for the transition temperature concerned, the G80 (C-Mn-Si) has a higher one ($-18°C$) compared to the G0091 ($\sim -29°C$). The transition temperature here is the temperature at which the material changes from ductile to brittle failure. Similarly, for the $2\frac{1}{4}$Cr-1Mo weld metal, the G80 shows an average upper shelf toughness of $\sim$90 J ($\sim$70 ft · lb) compared to 170 J ($\sim$130 ft · lb) of the G0091 flux. Here again in the $2\frac{1}{4}$Cr-1Mo weld the G80 indicates a higher transition temperature of $\sim$35°C compared to G0091 (18°C).

TABLE 1—*Chemical analysis of the fluxes (wt. %).*

Flux	SiO_2	Al_2O_3	TiO_2	Fe_2O_3	MnO	CaO	CaF_2	MgO	Na_2O	K_2O	Basicity Index, B
G80	38.0	15.0	...	1.5	7.0	22.0	6.0	10.0	...	...	0.93
G0091	32.57	3.99	5.02	0.99	0.48	41.92	10.02	0.29	4.45	0.15	1.55
OP76	Unavailable due to proprietary										Much higher than G0091

TABLE 2—*Welding details of the work.*

Base Plate	Weldment	SA Flux	Wire	Wire dia, mm (in.)	Heat Input, KJ/mm (KJ/in.)	Preheat and Interpass Temperature, °C (°F)	No. of Passes
A516-70	C-Mn-Si	G80 G0091	Linde 36	3.18 (1/8)	3 (75)	149 (300)	14 18
A387-22	$2\frac{1}{4}$Cr-1Mo	G80 G0091 OP76	Linde U571	3.18 (1/8)	3 (75)	149 (300)	15 19 59
A543 Class 1	Ni-Cr-Mo	G0091	Armco XW-28	3.18 (1/8)	2 (50)	93 to 149 (200 to 300)	146

TABLE 3—*Various weldments used in the present study.*

Weldment	SA Flux	SRA Condition	Thickness of Plate, mm (in.)
C-Mn-Si	G80	100 h at 605°C and furnace cool	25 (1)
C-Mn-Si	G0091	100 h at 605°C and furnace cool	25 (1)
$2\frac{1}{4}$Cr-1Mo	G80	100 h at 690°C and furnace cool	25 (1)
$2\frac{1}{4}$Cr-1Mo	G0091	100 h at 690°C and furnace cool	25 (1)
$2\frac{1}{4}$Cr-1Mo	OP76	100 h at 690°C and furnace cool	63 ($2\frac{1}{2}$)
A543 Class 1	G0091	As welded (0 SRA treatment)	150 (6)

In order to understand more clearly the effect of basicity on the toughness characteristics, $2\frac{1}{4}$Cr-1Mo SRA weld metal fabricated using Oerlikon OP76 flux, a flux which is more basic than Linde G0091, was studied. Toughness was characterized at two different positions of the weld, one near the top passes and the other near the bottom thickness passes of the weld. This was done to find variations, if any, in the toughness of the weld in terms of the weld position of the specimens. The obtained test results are presented in Fig. 4. It can be seen that the OP76 produces the toughest $2\frac{1}{4}$Cr-1Mo weld metal, followed by G0091 and G80. The $2\frac{1}{4}$Cr-1Mo/OP76 flux, when characterized at the bottom thickness weld passes, shows an average ambient

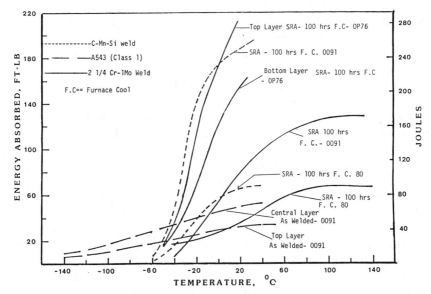

FIG. 4—*Charpy V-notch data showing the variation in energy absorbed to fracture as a function of SA flux used, for C-Mn-Si (mild steel) weld metal, $2\frac{1}{4}$Cr-1Mo (low alloy) steel weld metal, and A543 Class 1.*

temperature toughness of ~ 219 J (~ 162 ft · lb) compared to ~ 284 J (~ 210 ft · lb) of the top thickness level. The weld also shifts to a lower transition temperature when characterized near the top thickness passes compared to bottom level passes. A shift of $\sim 13°$C is observed here.

Finally, A543 Class 1 SA weldment fabricated using Linde G0091 flux was characterized for its Charpy toughness. Previous investigation [19] showed that these newer generation plates when welded using G0091 developed stronger weld metal. The idea here is to understand the toughness characteristics of this newer weld metal. The Charpy experimental data are presented in Fig. 4. Here again for G0091, just like OP76 flux, the toughness was characterized at two different positions of the weld, one near the top passes and the other near the center thickness layers of the weld. It is thought that this type of study will reveal the effect of flux, if any, on the toughness values as the number of beads increases in a multipass SA weldment. It can be seen from Fig. 4 that when the specimens are taken near the top weld passes the values are somewhat low, the ambient temperature toughness being in the order of 46 J (~ 34 ft · lb). When specimens are obtained near the center thickness passes of the weld, the toughness of these specimens is 50% higher, averaging about 65 J (49 ft · lb) in the ambient temperature range.

Fractography of Charpy Fracture Surfaces

Since the variation in toughness between the G80 and G0091 is maximum near the upper shelf (Fig. 4) for both mild and low-alloy steels, Charpy specimens tested under this condition were investigated to determine their mode of fracture. Initially the Charpy specimens tested near the upper shelf transition were selected. In the case of C-Mn-Si weld metal this lies near 5°C and for $2\frac{1}{4}$Cr-1Mo near 71°C. Here the G80 samples (for both weldments) showed 50% shear fracture, while the G0091 Charpy specimens failed in a completely ductile fashion. Consequently, another investigation was carried out where the G0091 Charpy specimens (for both welds) which failed in 50% shear mode ($-29°$C for C-Mn-Si and 21°C for $2\frac{1}{4}$Cr-1Mo) were characterized so that they could be compared with that of the G80. All the fracture surfaces were ultrasonically cleaned with acetone and were observed immediately under SEM to avoid contamination. In-situ X-ray energy dispersive spectrometry (EDS) was used in all SEM studies to observe the microchemistry.

C-Mn-Si Weld Metals (G80 and G0091)—Figure 5a-d shows representative areas of the fracture surface from the notched end to the other end of the Charpy of C-Mn-Si weld metal–G80 flux tested at 5°C. The fracture starts with microvoid coalescence near the notched end of the Charpy due to inclusion/matrix interface failure. From Figs. 5a-b and 6 it can be seen that the fracture surface contains inclusions of high Si and Mn content [along with traces of aluminum (Al) and sulfur (S) (Fig. 7)]. Observations indicate

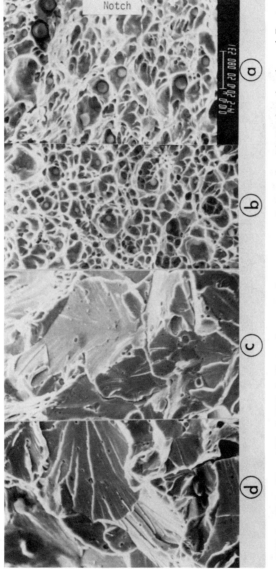

FIG. 5—*SRA mild steel (Grade 80 flux): Charpy fracture surfaces representative areas from the notch to the other end. Fracture starts with low energy "equiaxed dimple" morphology followed by cleavage fracture. Test temperature: 5°C.*

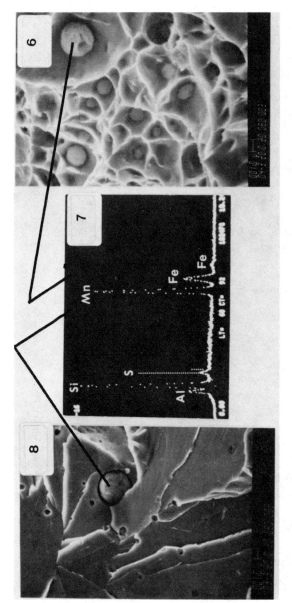

FIG. 6-8—(6) Higher magnification image near the notched end of the fracture (SRA mild steel—G80). (7) EDS spectrum typical of inclusions observed here in SRA mild steel—G80 flux. (8) High magnification image from the cleavage area of the fracture (SRA mild steel—G80 flux).

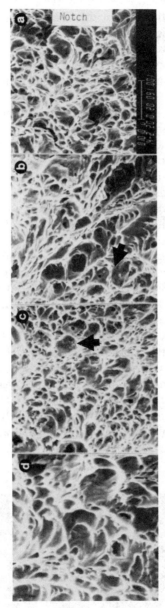

FIG. 9—SRA mild steel (Grade 0091 flux): Charpy fracture surface showing representative areas from the notch to the other end. Clearly the fracture process is more ductile than in Fig. 5. Test temperature: 5°C.

that the final fracture (about half of the total fracture surface) occurred by cleavage mechanisms (Figs. 5c-d and 8). On the other hand, mild steel weld metal of the G0091 flux tested at 5°C produced a more ductile failure from the notched end to the other end of the Charpy (Fig. 9a-d). The fracture surface contains fine particles at the bottom of some voids, although occasionally some inclusions are observed. The EDS analysis carried out on these particles indicated that they contain essentially iron (Fe) and Mn (Fig. 10). Occasional coarse inclusions showed mainly Si and Mn with traces of Al and titanium (Ti).

Figure 11a-d represents the SEM images obtained for the G0091 flux weld metal Charpy with 50% shear fracture. Here, also, the fracture starts with microvoid coalescence near the notched end followed by cleavage failure similar to Fig. 5. The only difference in the fracture topography is the amount and size of the Si-rich inclusions observed, the G80 showing higher and larger sized inclusions compared to the G0091. Further, the G0091 sample contains fine particles at the bottom of some voids of composition Fe and Mn. An EDS spectrum typical of the inclusions observed in mild steel–G0091 flux is given in Fig. 12.

$2\frac{1}{4}Cr$-$1Mo$ Weld Metals (G80 and G0091)—Studies conducted on the $2\frac{1}{4}$Cr-1Mo Charpy fracture surfaces show a similar trend to the C-Mn-Si specimens (Figs. 13–16). Specifically, G80 fracture started with a microvoid coalescence followed by cleavage, and the fracture surface contains inclusions of high Si and Mn content (Fig. 13).

In contrast, G0091 flux weld Charpy tested at 71°C failed in a ductile fashion (Fig. 14a-d) with the occasional presence of inclusions. Only precipitates of the composition shown in Fig. 15 are observed in some voids.

Comparative studies conducted on the G0091 flux weld Charpy with 50% shear fracture with those from G80 showed that the G80 fracture contains a higher amount of coarse inclusions compared to G0091. An EDS spectrum typical of the inclusions observed in the fracture surface of the $2\frac{1}{4}$Cr-1Mo G0091 flux is given in Fig. 16.

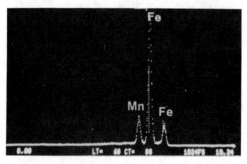

FIG. 10—*EDS spectrum typical of precipitates observed in SRA mild steel–G0091 flux (Fig. 9).*

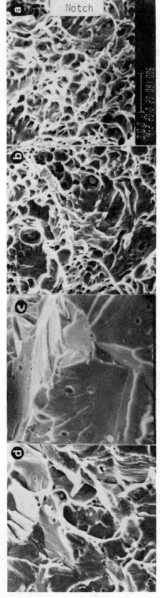

FIG. 11—*SRA mild steel (Grade 0091 flux): Charpy fracture surface showing representative areas from the notch to the other end. Fracture surface contains fewer and smaller inclusions compared to Fig. 5. Test temperature: −29°C.*

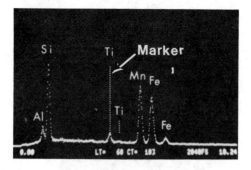

FIG. 12—*EDS spectrum typical of inclusions observed in SRA mild steel–G0091 flux (Fig. 11).*

$2\frac{1}{4}Cr$-*1Mo Weld Metal (OP76 Flux)*—For the fracture surface investigation here, Charpy specimens tested at 5°C were selected for both top weld thickness and bottom thickness layers. The bottom layer Charpy specimens showed the fracture to start with microvoid coalescence near the notched end followed with cleavage fracture (slightly more than half of the total fracture surface). On the other hand, the top layer Charpy specimens produced a more ductile fracture with little cleavage area. The EDS analysis of the particles observed in these fracture surfaces shows that they contain molybdenum (Mo), some chromium (Cr), and Fe. Between the two levels, the bottom level Charpy specimens showed slightly more inclusions content compared to the top layer Charpy specimens. In general, irrespective of the weld position of the Charpy specimens, the OP76 flux resulted in a weld of few Si-rich inclusions compared to either G80 or G0091 ($2\frac{1}{4}$Cr-1Mo).

A543 Class 1 Weld Metal (G0091 Flux)—The SEM images of the A543 Charpy fracture surfaces are shown in Figs. 17 and 18. Figure 17 represents areas of the fracture surface observed for the ambient temperature Charpy specimens studied near the top passes of the A543 (G0091), as welded material. High inclusion content along with large microvoids can be observed.

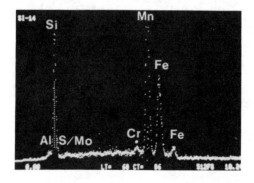

FIG. 13—*EDS spectrum typical of inclusions observed in SRA $2\frac{1}{4}$Cr-1Mo, G80 flux.*

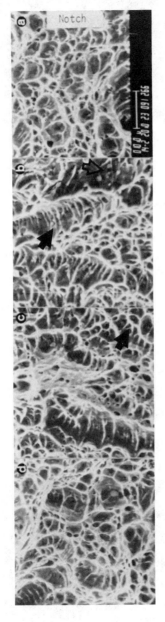

FIG. 14—SRA $2\frac{1}{4}$ Cr-1Mo (Grade 0091 flux): Charpy fracture surface showing representative areas from the notch to the other end. Clearly the fracture process is ductile. Test temperature: 71°C.

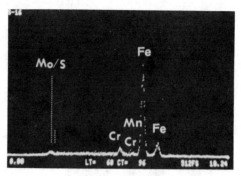

FIG. 15—*EDS spectrum typical of the precipitates observed in SRA $2\frac{1}{4}Cr$-$1Mo$, G0091 flux* (*Fig. 14*).

The EDS work on the inclusions (arrowed) indicates that they are complex silicates (Fig. 19) similar to the ones observed in C-Mn-Si weld metal (Fig. 12). On the other hand, Fig. 18 shows very fine particles and fine microvoids for the Charpy specimens taken near the center thickness layers. The precipitates show composition of Al and Fe and some complex precipitates containing small amounts of Al, Si, Cr, Mn, nickel (Ni), and Fe.

Discussion

From the various results it can be summarized that some of the critical factors responsible for the observed notch toughness characteristics (in all different weld metals) are:

1. Inclusions—their distribution, size, and chemistry.
2. Precipitates—their distribution, size, and chemistry.

In all specimens the size of both slag inclusions and precipitates observed

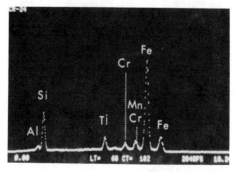

FIG. 16—*EDS spectrum typical of inclusions observed in SRA $2\frac{1}{4}Cr$-$1Mo$, G0091 flux.*

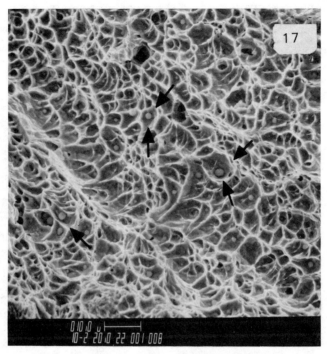

FIG. 17—*SEM image of the Charpy fracture surface (A543/G0091) observed for the specimen taken near the top layers. Inclusions are arrowed (2 KJ/mm, as-welded state). Test temperature: 26°C. Energy absorbed: 43 J (32 ft · lb).*

is such that bulk SEM/EDS data invariably come from both the particle and the surrounding matrix. This was a major problem with the small micro-precipitates and inclusions especially when they were located well inside the microvoids.

C-Mn-Si Weld Metal (G80 and G0091)

Figures 5*a-b*, 6, and 7 show that G80 weld contains a high density of large-sized (1–3 μm) inclusions with large amounts of Si and Mn as well as Al and S, and may possibly be complex silicates. Similar results were also found by the authors in another parallel program [*4*] when observed directly on the polished and etched surface of these specimens. A typical etched surface image of this specimen is given in Fig. 20*a,b* for discussion purpose. The electron probe microanalysis (EPMA) conducted on this other program showed that the inclusions have oxygen with no oxygen in the matrix, indicating that the Si-rich inclusions are probably silicates. Previous works [*7,12,14*] have shown that inclusions, especially in the form of silicates,

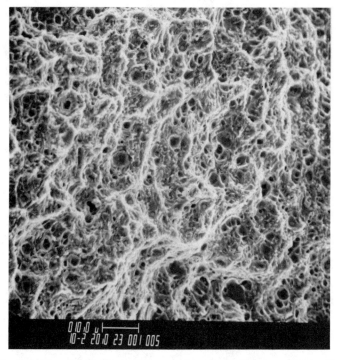

FIG. 18—*SEM image of the Charpy fracture surface observed for the specimen taken near the center thickness layers (2 KJ/mm, as-welded condition). Test temperature: 26°C. Energy absorbed: 70 J (52 ft · lb) A543/G0091.*

represent a path of least resistance for crack propagation and that the resistance further decreases as the size and number of inclusions increase.

It is generally accepted that ductile fracture occurs by nucleation of voids due to decohesion of second phase particles and the matrix followed by void coalescence [20]. Fissures along the inclusion/matrix boundary in G80 weld in Figs. 8 and 20 indicate a weak interface, probably resulting in easy void formation such as seen in Figs. 5a-b and 6. In fact, from Fig. 20 there is evidence for the existence of cavities formed by the shearing off of inclusions, confirming that the inclusion/matrix interface is weak. The presence of Si-rich inclusions in large amount is considered to embrittle the G80 weld. Smith [21] showed that inclusion cracking or decohesion can initiate a cleavage-type crack. These processes are considered to be the reason for the observed low energy to fracture (Figs. 4 and 5–8) of the G80 weld.

On the other hand, the G0091 weld metal has mainly Fe-Mn precipitates with few, small Si-rich inclusions (Figs. 9–12). Taylor and Farrar [7] showed that small particles will deform plastically rather than crack and that the particle/matrix decohesion is difficult. This is considered to be the reason for the better toughness characteristics of the G0091 SRA weld metal (Fig. 4).

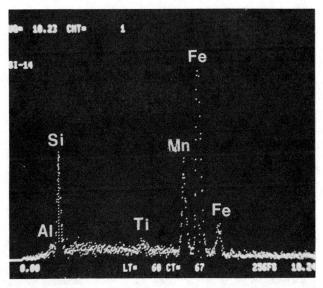

FIG. 19—*EDS spectrum typical of the inclusions observed in Fig. 17. This is similar to Fig. 12, the composition typical of the inclusions observed on the fracture surface of the C-Mn-Si/G0091 SRA weld.*

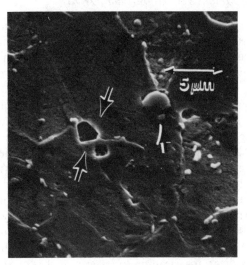

FIG. 20—*SEM image of SRA mild steel weld metal–G80 flux. Polished and etched surface. Precipitates along the grain boundaries are visible but several coarse slag inclusions or voids where inclusions have pulled away from the surrounding matrix are clearly visible (arrowed).*

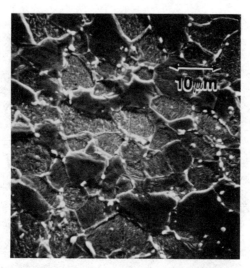

FIG. 21—*SEM image of SRA mild steel weld metal, G0091 flux. Polished and etched surface. Many small grain boundary precipitates are visible but no coarse inclusions are observed.*

A photomicrograph of the polished and etched surface of this specimen is shown in Fig. 21. Small grain boundary precipitates are visible but no coarse inclusions are observed.

$2\frac{1}{4}Cr$-$1Mo$ Weld Metal (G80, G0091, and OP76)

In a similar manner to C-Mn-Si weld metal, G80 flux Charpies compared here to G0091 show high amounts of large-sized Si-rich inclusions (Figs. 13–16). As discussed in the previous section, these Si-rich inclusions contribute to the lower toughness characteristics of the G80 weld compared to G0091 (Fig. 4).

It has been demonstrated [6,8–11] that an increase in the basicity index of the flux generally produces a tougher material due to low nonmetallic inclusions in weld metal. From Table 1 and as well as from previous work [22], it can be seen that G80 flux contains larger amounts of oxides, for example silica (SiO_2), manganese oxide (MnO), and alumina (Al_2O_3) compared with G0091, leading to a lower basicity index: 0.93 for G80 and 1.55 for G0091 as calculated using the standard basicity formula (Eq 1). Consequently, G80 weld metals carry a large amount of inclusions, especially in the form of complex silicates, thus adversely affecting the notch toughness. With lower amounts of MnO and Al_2O_3 in the G0091 flux, it is thought that Si acts as an effective deoxidizer, thus leading to lesser amounts of Si-containing precipitates and inclusions, resulting in a tougher weldment.

The work carried out on the OP76 flux confirms the previously stated reasonings. Oerlikon OP76 flux with higher basicity than G0091 flux results in a weld of very few Si-rich inclusions, thus achieving high toughness values (Fig. 4). It is considered that the variation in the amounts of inclusions between the top and bottom thickness passes of the OP76 weld contributes to their observed toughness variations between the two levels. The bottom thickness weld passes with higher amounts of Si-rich inclusions (probably due to the impurities on the base plate and back-up plate surfaces before welding), resulting in a lower toughness weld compared to the top passes.

A543 Class 1 (G0091)

Similar to OP76, the work carried out on the A543 Class 1 (G0091) weld at two different weld positions suggests that the weld microcleanliness is dependent on position in the weld joint, and this in turn influences weld metal transition temperature and upper shelf energy. Higher cleanliness welds tend to have lower oxygen contents as well as fewer inclusions and thus have lower transition temperatures and higher upper shelf energies. This is consistent with the results seen in Figs. 4 and 17–19.

It can be summarized from the OP76 and A543 work that inclusion content is very high and toughness low at the initial weld passes due to impurities on the base plate and back-up plate surfaces before welding. It seems that as the number of passes increases the inclusion content of the weld also improves as the inclusions keep moving to the top after each pass. It is thought that this same process finally results in high inclusion content in the final passes of the weld due to the buildup of inclusions. The low toughness values observed in A543/G0091 when characterized in top thickness passes are probably due to this phenomenon.

Conclusions

From the investigation of the toughness and fracture characteristics of C-Mn-Si, $2\frac{1}{4}$Cr-1Mo, and A543 Class 1 submerged arc weld metals fabricated using Linde Grade 80, Linde Grade 0091, and Oerlikon OP76 fluxes, the following has been concluded:

1. Both C-Mn-Si and $2\frac{1}{4}$Cr-1Mo SRA weld metals show poorer toughness and higher transition temperature when welded with Linde G80 flux compared to G0091 or Oerlikon OP76. Of the three fluxes studied, the OP76, due to its higher basicity, produces the toughest weld metal, followed by G0091 and G80.

2. Linde Grade 80 flux welds contain a large number of silicate inclusions compared to Linde G0091 or OP76 welds. These silicate inclusions seem to have a deleterious effect on Charpy toughness. The presence of Si-rich inclusions in large amounts is considered to embrittle the SRA–Grade 80 flux

weld, thus resulting in lower toughness and higher transition temperature compared to G0091 or Oerlikon OP76.

3. The toughness of the weld metal depends on the impact test position. The weld metal when characterized near the initial (bottom thickness) or final (top thickness) weld passes will tend to have more submicron silicate inclusions, resulting in a higher transition temperature and lower upper shelf energy compared to the center thickness layers.

Acknowledgments

The funding for this program was provided through the Fabrication Division of the Pressure Vessel Research Committee.

References

[1] Billy, J., Johansson, T., Loberg, B., and Easterling, K. E., *Metals Technology*, Vol. 7, 1980, p. 67.

[2] Dorshu, K. E. and Stout, R. D., *Welding Journal*, Vol. 40, 1961, p. 97s.

[3] Bosansky, J., Porter, D. A., Astrom, H., and Easterling, K. E., *Scandinavian Journal of Metals*, Vol. 6, 1977, p. 125.

[4] Sankar, J., "The Effect of Submerged Arc Welding Variables on the Structure and Mechanical Behavior of Pressure Vessel Steel Weldments," Ph.D thesis, Lehigh University, Bethlehem, PA, 1982.

[5] Wittstock, G. G., *Welding Journal*, Vol. 55, 1976, p. 733.

[6] Tuliani, S. S., Boniszewski, T., and Eaton, N. F., *Welding and Metal Fabrication*, Vol. 37, 1969, p. 327.

[7] Taylor, L. G. and Farrar, R. A., *Welding and Metal Fabrication*, Vol. 43, 1975, pp. 305–310.

[8] Palm, J. H., *Welding Journal*, Vol. 51, 1972, p. 358s.

[9] Eagar, T. W., *Welding Journal*, Vol. 57, 1978, p. 76s.

[10] North, T. H., Bell, H. B., Nowicki, A., and Craig, I., *Welding Journal*, Vol. 57, 1978, p. 63s.

[11] Lewis, W. J., Faulkner, G. E. and Rieppel, P. J., *Welding Journal*, Vol. 40, 1961, p. 337s.

[12] Eaton, N. F., Glover, A. G., and McGrath, J. T., *Proceedings of Fracture 1977*, Waterloo, Canada, Vol. 1, 1977, p. 751.

[13] Palm, J. H., *Welding and Metal Fabrication*, Vol. 44, 1976, p. 135.

[14] Chin, L. L. J., *Welding Journal*, Vol. 48, 1969, p. 290s.

[15] Widgery, D. J., *Welding Journal*, Vol. 55, 1976, p. 57s.

[16] Bennett, A. P. and Stanley, P. J., *British Welding Journal*, Vol. 13, 1966, pp. 59–74.

[17] Peck, J. V. and Witherell, C. E., *Welding Journal*, Vol. 42, 1963, pp. 193s–201s.

[18] Dorshu, K., "The Effects of Long Time Stress Relieving on the Structures and Properties of Pressure Vessel Steel Weld Metals," Technical Report P-C4649-3, Franklin Institute, Philadelphia, PA, 1978.

[19] Pense, A. W., "The Fracture Behavior of A737 Microalloyed Steel and A543 Class 1 Steel Weldments," Quarterly Report No. 2, Pressure Vessel Research Committee, Fabrication Division, New York, NY, 1980.

[20] Dieter, G. E., *Mechanical Metallurgy*, Chapter 7, McGraw-Hill, New York, 1976.

[21] Smith, E., *Conference Proceedings*, Scarborough, England, Iron and Steel Institute, London, England, 1971, p. 37.

[22] Sankar, J. and Williams, D. B., "Microstructural Characterization of Pressure Vessel Steel Weld Metals Using Electron Optical Techniques," Status Report No. 3, Pressure Vessel Research Committee, Fabrication Division, New York, 1979.

Golam M. Newaz[1]

A Fractographic Investigation of Fatigue Damage in Carburized Steel

REFERENCE: Newaz, G. M., "A Fractographic Investigation of Fatigue Damage in Carburized Steel," *Fractography of Modern Engineering Materials: Composites and Metals, ASTM STP 948*, J. E. Masters and J. J. Au, Eds., American Society for Testing and Materials, Philadelphia, 1987, pp. 317–333.

ABSTRACT: A fatigue study was conducted to determine the performance of unnotched and notched carburized American Iron and Steel Institute (AISI) 8620H steel specimens and to evaluate the corresponding fatigue damage. Completely reversed, constant amplitude, load-controlled tests were conducted using cylindrical carburized steel specimens with stress concentrations of 1.00, 1.46, and 2.22. All tests were conducted at a frequency of 2 Hz under an ambient laboratory environment. Fatigue data plotted as stress amplitude versus mean fatigue life show that, at long life, both sets of notched specimens have similar performance. At shorter lives, the specimens with stress concentration of 1.46 exhibit superior performance compared with the specimens having a stress concentration of 2.22, as expected. The performance of smooth specimens is significantly better compared with the notched specimens at all lives.

Fractographic investigation conducted using scanning electron microscopy revealed the nature of initiation and the subsequent propagation of fatigue cracks in these specimens. At shorter lives, the crack initiation in smooth specimens was found to be surface-initiated at one or more sites on the surface, which then circumferentially converged and propagated through the carburized case into the core. The fracture surface of the case was smooth and flat, while in the core it was nonuniform and coarse. At higher lives, crack initiation in the smooth specimen originated at inclusions within the core. The subsurface crack initiation site showed convergence of radial lines toward a focal point, the site of a nonmetallic inclusion. Fractographs of notched specimens which failed at shorter lives show damage similar to that in smooth specimens. At longer lives, clear evidence of separation between the case and the core was found, with some evidence of subsurface initiation from nonmetallic silicon oxide inclusions having dimensions of 20 to 50 μm.

KEYWORDS: fatigue life, carburized steel, fatigue damage, fractography, nonmetallic inclusion, subsurface crack initiation

[1]Principal research scientist, Mechanics Section, Battelle Memorial Institute, Columbus, OH 43201.

Carburizing is a selective surface-hardening process utilized for enhancing fatigue performance of certain engineering components made of steel. There are two major applications, namely, (1) members with a stress gradient at the surface and (2) members experiencing surface contact loading that are subject to wear. A common example of such a member is a gear. Improvements obtained from carburizing are due to the increased strength and hardness of the case accompanied by development of compressive residual stresses in the surface layers.

In analyzing the monotonic deformation and fracture behavior of carburized bend specimens, Krotine et al. [1] and McGuire et al. [2], in two closely related studies, adopted the approach that a surface-hardened member is a composite consisting of a high-strength, low-ductility case and a lower strength, higher ductility core. Landgraf and Richman [3] utilized this concept in evaluating the cyclic deformation and fatigue resistance of carburized members. Their work centered upon the evaluation of total component behavior as related to the individual behavior of the case and the core. From their axial fatigue results, they found that in low-cycle lives the fatigue cracks were surface initiated and that in high-cycle lives they were subsurface initiated. Landgraf and Richman [3,4] also studied the bending fatigue behavior of carburized steel.

Magnusson and coworkers [5–7] investigated a wide range of problems related to fatigue crack propagation. Also, a considerable amount of effort by these researchers rested in the area of the metallurgical aspect of case carburizing. It may be stated quite justifiably that the metallurgical aspects of the problem are quite complex. An excellent account is given by Parrish [8].

Case hardening gives rise to large residual stresses due to the volume expansion of martensite. In a complex manner, these depend primarily on the cooling rate, carburizing parameters, transformation characteristics, and elevated temperature properties of the steel. The large residual stresses can either be detrimental and cause quench cracking or they can be advantageous and improve the properties of the hardened steel. Ebert [9] studied the role of residual stresses in the mechanical performance of case-carburized steels. His work assesses the difficulties in defining the quantitative influence of residual stresses. Residual stresses may relax due to plastic strain generated by fatigue loading; Morrow and coworkers [10] investigated the relaxation of residual stresses in both soft and hard steels. They proposed a design guideline to account for the change in residual stresses and adopted the approach that residual stress could be treated as mean stress. Burke and Lawrence [11] have shown that mean stress relaxation may be described by a power function. According to their relation, the mean stress does not change if the member experiences elastic strain only.

Although various studies have been conducted to understand the fatigue behavior of carburized members as discussed earlier, reports of coherent studies on the evolution of fatigue damage in carburized members have not been found. Reported here are results from efforts undertaken, in the context of fatigue performance of smooth and notched carburized specimens, to determine the location of damage initiation as a function of fatigue life. Scanning electron photomicrography was extensively utilized to provide valuable information in this respect.

Material and Specimens

Circumferentially notched axial specimens were prepared from radial sections of forged, normalized gear blanks (30.48 cm diameter, 5.08 cm thick) of AISI 8620H steel with a grain size of ASTM No. 9. Material composition and mechanical and fatigue properties are given in Tables 1, 2, and 3. Three different sets of specimens with stress concentrations ($K_t = 1.00$, $K_t = 1.46$, and $K_t = 2.22$) were prepared. The monotonic and tension cyclic stress-strain response of the core material is shown in Fig. 1. All specimen dimensions and related features are presented in Fig. 2.

TABLE 1—*Material compositions and processing.*

Material	C	S	P	Mn	Si	Cr	Ni	Mo
Core	0.22	0.027	0.024	0.81	0.15	0.56	0.43	0.15
Case	0.72	0.029	0.016	0.78	0.17	0.64	0.49	0.17

PROCESSING

Core: Pseudocarburized at 927°C, reheated to 843°C, quenched in oil, and tempered at 149°C for 1 h.

Case: Carburized at 927°C in an endothermic atmosphere for 10 h, carbon potential of 0.90 to 1.00%, process repeated six times. Hardened in a neutral salt pot at 843°C for 15 min, quenched in oil, refrigerated for 4 h at −84°C, air cooled, and tempered at 163°C for 2 h.

Carburized: Carburized at 899°C in an endothermic atmosphere with a dew point of −3.9°C for 0.5 h, cooled to 843°C, quenched in oil, and tempered at 177°C for 1 h.

TABLE 2—*Mechanical properties of simulated case, simulated core, and carburized materials.*

Material	Case	Core	Carburized
Modulus of elasticity, E, $\times 10^5$ MPa ($\times 10^3$ ksi)	1.93 (28.0)	1.98 (28.8)	1.95 (28.3)
0.2% offset yield strength, MPa (ksi)	. . .	1200 (174)	1200 (174)
Ultimate tensile strength, S, MPa (ksi)	1600 (232)	1510 (219)	1586 (230)
Percent reduction in area, % RA	. . .	42.2	2.6
True fracture strength, σ_f, MPa (ksi)	1600 (232)	2034 (295)	1669 (242)
True fracture ductility, ε_f	0.0085	0.549	0.026
Strain hardening exponent, n	. . .	0.094	0.129
Strength coefficient, K, MPa (ksi)	. . .	2151 (312)	2668 (387)

NOTE: RA = reduction in area.

TABLE 3—*Fatigue properties of simulated case, simulated core, and carburized materials.*

Material	Case	Core	Carburized
Cyclic yield strength, 0.2% offset, MPa (ksi)	. . .	1082 (157)	1213 (176)
Cyclic strain hardening exponent, n'	. . .	0.261	0.244
Cyclic strength coefficient, K', MPa (ksi)	. . .	2220 (322)	2875 (417)
Fatigure strength coefficient, σ'_f, MPa (ksi)	1931 (280)	2510 (364)	2344 (340)
Fatigue ductility coefficient, ε'_f	. . .	0.053	0.027
Fatigue strength exponent, b	-0.109	-0.116	-0.097
Fatigue ductility exponent, c	. . .	-0.444	-0.398
Transition fatigue life, $2N_t$, reversals	. . .	79	15

Carburization and Related Processes

Both sets of notched and unnotched specimens were gas carburized at 899°C in an endothermic atmosphere with a dew point of 39°C for 0.5 h, cooled to 843°C, quenched in oil, and tempered 177°C for 1 h. Threaded sections of all the specimens were copper-plated to an approximately 0.025-mm-thick layer prior to carburizing in order to suppress carburization of these sections and prevent embrittlement at the root of the threads due to the carburizing process. The copper layers were stripped off electrochemically prior to testing.

A layer of approximately 0.04 mm was polished from the surface using γ-alumina(Al_2O_3) polishing cloth of 0.5-μ grit for all specimens. No evidence of microcracking was observed as determined using an optical microscope at X200 magnification.

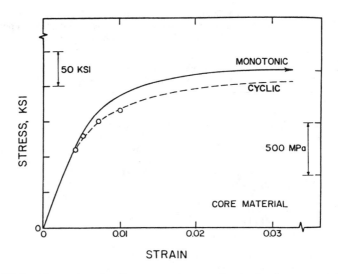

FIG. 1—*Monotonic and cyclic stress-strain curves for simulated core material.*

Experimental Program

Microhardness Measurement

Several radial hardness traverses were obtained at the notch root of notched carburized specimens using diamond pyramid indentations at 100-g load level. The hardness profiles for the two different notched members ($K_t = 1.46$ and $K_t = 2.22$) were nearly identical. A hardness profile for the $K_t = 2.22$ member is shown in Fig. 3. From the profiles, the case thickness may be estimated to be approximately 0.4 mm. This was intended to eliminate the effects of surface decarburization and internal oxidation. Hardness values typically vary from 46 to 51 Rockwell C in the core and from 58 to 64 Rockwell C in the case. At the case-core interface region there is a sharp drop in hardness values as the core is approached, and this feature is well depicted in the measured hardness profiles. These hardness values compare well with the case and core hardness values obtained by Waraniak [12] for smooth carburized cylindrical specimens with identical carburizing and heat treatments.

Residual Stress Measurements

The measurement of residual stress in a small notched specimen poses considerable problems. The small notch is the primary constraint. Residual stress was measured through the case and part of the core using the X-ray diffraction technique on the shallower notched specimen ($K_t = 1.46$). Only the longitudinal stress profile was obtained. The residual stress profile is shown in Fig. 4. The curve through the measured points is a second order parabolic fit. The obtained profile compares very well with the smooth specimen profile reported by Waraniak [12] in terms of surface and core peak values as well as the general nature of the stress distribution. Based on this corroboration, it will be assumed that the residual stress profile in the sharper notched specimen ($K_t = 2.22$) will be similar for the same case depth.

Fatigue Testing Method

Constant amplitude, completely reversed, load-controlled tests were conducted on a Materials Test Systems (MTS) closed-loop servocontrolled hydraulic test system. The test frequency was 2 Hz. Tests were conducted at a room temperature of 22.8°C (73°F) and 50% humidity to 2×10^7 reversals if failure did not occur; such a specimen was considered a runout. A minimum of four specimens were tested at each load level.

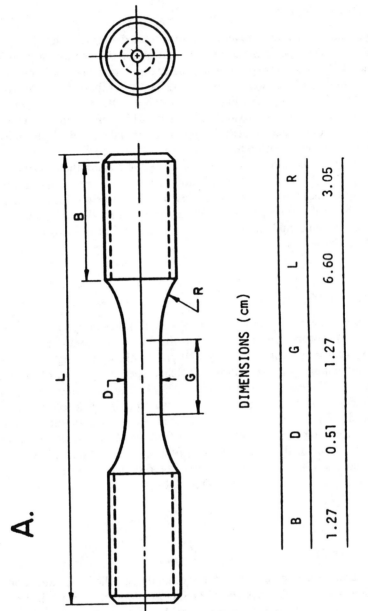

A.

DIMENSIONS (cm)

B	D	G	L	R
1.27	0.51	1.27	6.60	3.05

FIG. 2(A)—*Smooth specimen* ($K_t = 1.00$) *dimensions.*

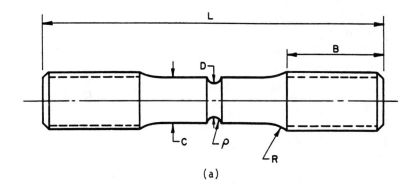

(a)

B.

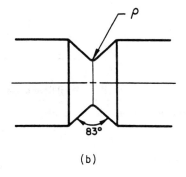

Detailed notch
root dimensions.
All other
dimensions as
in (a).

(b)

DIMENSIONS (cm)

Specimen	B	D	ρ	L	R	C
a	1.27	0.51	0.20	6.60	2.54	0.91
b	1.27	0.51	0.05	6.60	2.54	0.91

FIG. 2(B)—*Dimensions of cylindrical notched specimens:* (a) $K_t = 1.46$ *and* (b) $K_t = 2.22$.

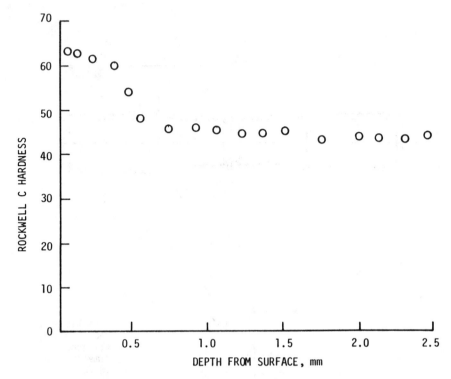

FIG. 3—*Hardness profile of carburized* $K_t = 2.22$ *specimen.*

Scanning Electron Microscopy

Failed specimens from each set were examined fractographically to determine the nature of fatigue damage. Special efforts were made to determine the fatigue damage initiation sites as a function of fatigue life.

Results and Discussion

Fatigue behavior of the unnotched and notched samples are shown in Fig. 5. The performance of unnotched specimens is substantially better than the notched specimens at all lives. The notched specimens ($K_t = 1.46$ and $K_t = 2.22$) show some difference in performance at low cycles. However, the fatigue performance at higher cycles for the two different notched specimens are similar. This result is quite interesting in that the specimens with higher stress concentration perform as well as the ones with lower stress concentration, indicating that the failure processes are similar.

In studying the fractographs, the smooth specimen ($K_t = 1.00$) shows relatively planar surface characteristics at low cycles ($2N_f \sim 1000$) (Fig. 6). Dis-

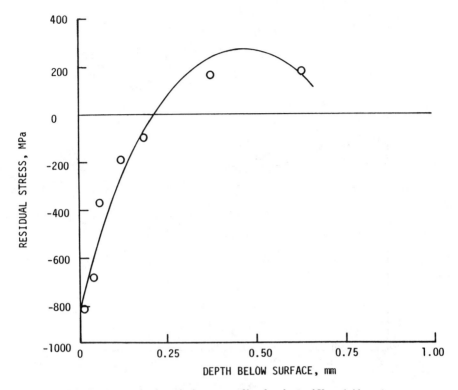

FIG. 4—*Longitudinal residual stress profile of carburized* $K_t = 1.46$ *specimen.*

tinction between the case and the core can be easily established. Final fracture can be attributed to surface-initiated cracking; the smooth case is an indication of fatigue crack growth through this layer. At intermediate life ($2N_f \sim 1.0 \times 10^5$), fatigue damage appears evident in the core. However, in this life regime, inclusion-originated fatigue cannot be established conclusively; both surface cracking and internal inclusion-originated damage compete for dominant failure process and may occur simultaneously. At high cycles, the fatigue damage is clearly due to inclusion-originated fracture (Fig. 7). Similar observations had been made by Landgraf and Richman [3] for long-life fatigue in smooth carburized members.

The fracture process in $K_t = 1.46$ specimens is very similar to that observed in $K_t = 2.22$ specimens, so that the $K_t = 2.22$ specimen failure process will be used to illustrate fatigue damage in notched carburized specimens as undertaken in this study.

For notched specimens, fracture at low cycle-life exhibits a smooth appearance (Fig. 8). This feature seems to be similar to that observed for unnotched specimens. At long life, the fatigue failure mode is changed to one where case-core, interface-initiated damage is predominant (Figs. 9 and

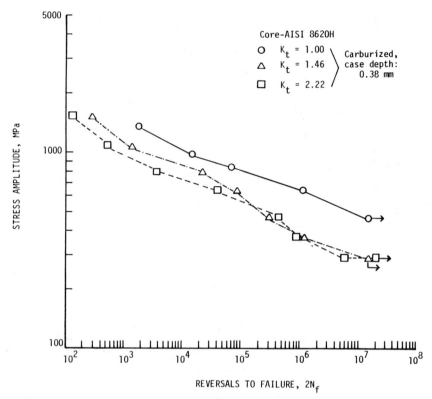

FIG. 5—*Combined fatigue data (points are mean of log* $(2N_f)$ *of specimens tested at the same load level).*

10A). The case-core separation is circumferential in nature and is essentially a step fracture; the damage initiation site at the interface appears to be a shear plane. Fatigue damage in the form of striations can be observed in such planes (Fig. 10B). At long life, in some instances, an inclusion-originated fatigue crack was observed near the case-core interface as shown in Fig. 11.

One of the most interesting failure processes observed was case-core separation in notched specimens under long-term fatigue loading. This can be attributed to a triaxial state of stress at the case-core interface, worsened by the tensile residual stress just inside the core. In elastic loading conditions as expected at lower stress levels, the core residual tension stress acts as the mean stress, while compressive residual stresses in the case prove beneficial by shifting the overall stress severity to the case-core interface or inside the core. (Further discussion of stress severity at the case-core interface in relation to fatigue damage initiation is provided by Newaz [*13*].) At high imposed stress levels the residual stress relaxes due to plastic deformation in the specimen, resulting in less shifting of the stress severity from the case. In this

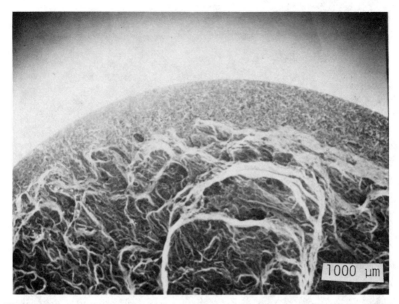

FIG. 6—*Scanning electron photomicrograph showing relatively planar fracture surface of case and core* ($K_t = 1.00$, $2N_f \sim 1000$).

situation, imposed stresses would be expected to govern the failure by surface-initiated cracking, which has been clearly observed.

A major driving force for carburizing structural parts is to enhance fatigue life by shifting fatigue damage from surface to subsurface locations. Subsurface fatigue damage is normally encountered at nonmetallic inclusions below the surface in high-cycle, axially loaded, carburized specimens [4,5]; this is also found to be a primary mode of failure in contact fatigue [14]. Initiation and propagation of fatigue cracks are the result of cyclic stress, which are locally intensified by the shape, size, and distribution of nonmetallic inclusions.

The mechanism of crack initiation at subsurface nonmetallic inclusions is less studied than surface-initiated fatigue cracks. It can be suggested that the nonmetallic inclusion creates a significant stress concentration such that (1) localized plastic flow takes place and (2) a fatigue crack starts to propagate from this localized plastic deformation with no detectable plastic flow on a macroscopic scale.

Kunio et al. [15] made some interesting observations of early fatigue crack growth in martensitic steel, especially with respect to inclusion-originated fatigue. Poor adhesion between matrix and inclusion lead to the formation of an inclusion pit, which served as a simple stress raiser. The fatigue crack originated at the periphery of this inclusion pit at an angle of 45° to the

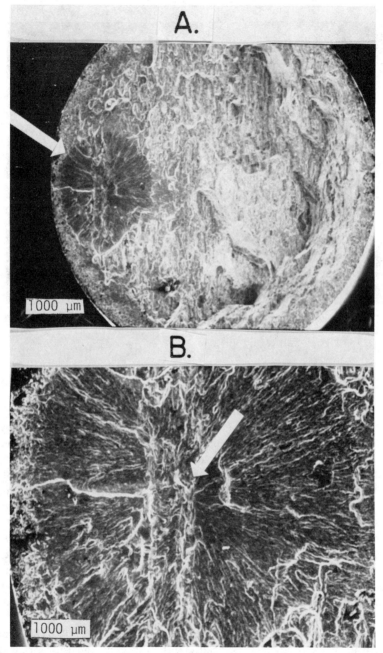

FIG. 7—(A) *Scanning electron photomicrograph showing inclusion-controlled fatigue crack initiation as pointed by the arrow* ($K_t = 1.00$, $2N_f \sim 0.8 \times 10_6$). (B) *Enlarged view of the crack initiation site.*

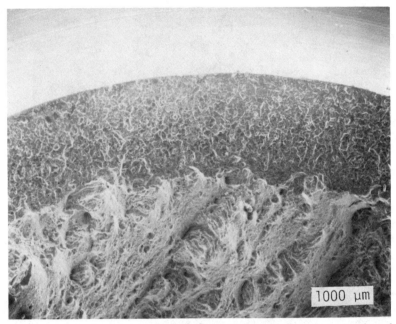

FIG. 8—*Scanning electron photomicrograph showing relatively planar fracture surface of case and core* ($K_t = 2.22$, $2N_f \sim 340$).

FIG. 9—*Scanning electron photomicrograph showing case-core separation* ($K_t = 2.22$, $2N_f \sim 0.7 \times 10^6$).

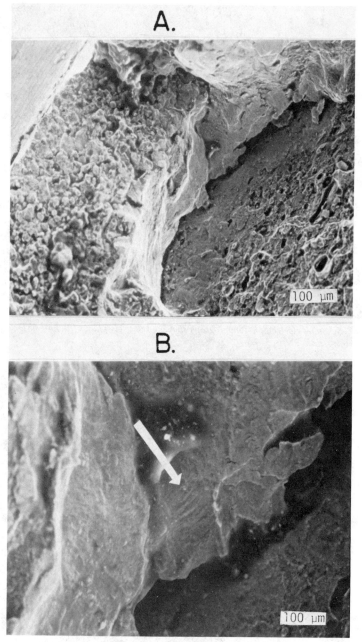

FIG. 10—(A) *Scanning electron photomicrograph showing case-core interfacial separation* ($K_t = 2.22$, $2N_f \sim 0.9 \times 10^6$). (B) *Enlarged view of interfacial separation showing fatigue striation marks.*

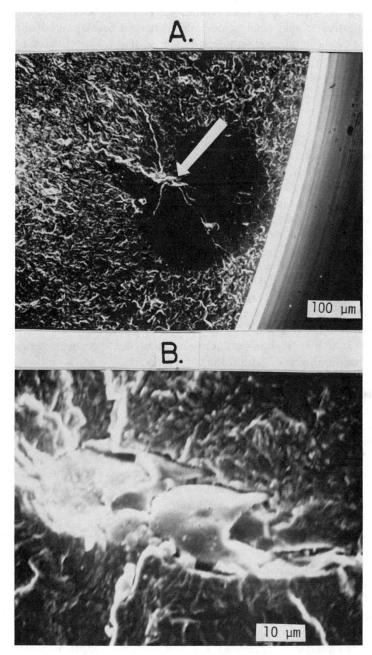

FIG. 11—(A) *Scanning electron photomicrograph showing inclusion-controlled fatigue crack initiation* ($K_t = 2.22$, $2N_f \sim 1.2 \times 10^6$). (B) *Nonmetallic inclusion at the crack initiation site.*

principal stress direction. Their findings suggest that the early growth (stage one) of fatigue cracks from nonmetallic inclusions develop out of slip bands and are of shear rather than tensile mode, so that the mechanism of initiation and early growth are identical to what is observed at the surface in ductile materials.

Summary

Based on the fatigue behavior and fractographic investigation undertaken in this study, the following remarks can be made.

1. At shorter lives, crack initiation in unnotched specimens ($K_t = 1.00$) was found to be surface initiated. Initiation occurred at one or more sites on the surface, and subsequent crack growth occurred radially through the carburized case. This is consistent with the fact that loading at high stresses may induce plastic strain within the specimen and will account for relaxation of residual stress. Hence the magnitude of compressive residual stress in the case may be reduced and beneficial effects may be lost. The fracture surface of the case was smooth and planar and the core surface was nonuniform and coarse.

2. At long life, crack initiation in the unnotched specimens was found to be inclusion-originated within the core. Subsurface crack initiation showed convergence of radial lines toward a focal point, the site of a nonmetallic inclusion.

3. Fractographs of notched specimens ($K_t = 1.46$ and 2.22) which failed at shorter lives showed a similar type of damage as in the unnotched specimens.

4. At long life in notched specimens, clear evidence of separation at the case-core interface was found. These separations were circumferentially oriented around the case-core interface. In some cases, subsurface initiation at nonmetallic inclusion was also observed.

5. Subsurface fatigue crack initiation was typically found to occur at nonmetallic inclusions with dimensions of 20 to 50 μm for the specimens used in this study.

6. Fatigue data, plotted in the form of stress amplitude versus mean fatigue life, showed similar performance of notched specimens ($K_t = 1.46$ and 2.22) at long life. This is attributed to case-core interface initiation and the similar overall multiaxial state of stress present at the interface in both notched conditions. At shorter lives, where surface initiation controls fatigue response, specimens with $K_t = 1.46$ exhibited superior performance compared with specimens with $K_t = 2.22$. The performance of smooth specimens was significantly better compared with the notched specimens at all lives.

Acknowledgment

The author acknowledges the encouragement of Professor Herbert T. Corten, Department of Theoretical and Applied Mechanics, University of Illinois at Urbana-Champaign, during the course of this study.

References

[1] Krotine, F. T., McGuire, M. F., Ebert, L. J., and Troiano, A. R., "The Influence of Case Properties and Retained Austenite on the Behavior of Carburized Components," *Transactions*, American Society for Metals, Vol. 621, 1969, pp. 829–838.

[2] McGuire, M. F., Troiano, A. R., and Ebert, L. J., "Phase Transformation Effects on the Bending Stress Distributions in Carburized Steel Components," *Transactions*, Series D, American Society of Mechanical Engineers, Vol. 93, 1971, pp. 699–707.

[3] Landgraf, R. W. and Richman, R. H., "Fatigue Behavior of Carburized Steel," in *Fatigue of Composite Materials, ASTM STP 569*, American Society for Testing and Materials, 1975, pp. 130–144.

[4] Landgraf, R. W. and Richman, R. H., "Some Effects of Retained Austenite on the Fatigue Resistance of Carburized Steel," *Metallurgical Transactions A*, Vol. 6A, May 1975, pp. 955–964.

[5] Magnusson, L., "Low Cycle Behavior of Case Hardened Steel," Mechanisms of Deformation and Fracture, Interdisciplinary Conference, Lulea, Sweden, 1978, K. E. Easterling, Ed., Pergamon Press, Oxford, England, 1979, pp. 105–110.

[6] Magnusson, L. and Ericsson, T., "Initiation and Propagation of Fatigue Cracks in Carburized Steel," Heat Treatment '79, Internationl Conference, Birmingham, England, The Metals Society, London, England, 1980, pp. 202–206.

[7] Magnusson, L., Ericsson, T. and Hildenwall, B., "Cyclic Behavior of Carburized Steel," Scandinavian Symposium in Material Science, Lulea, Sweden, 1980.

[8] Parrish, B., "The Influence of Microstructure on the Properties of Case-Carburized Components," American Society for Metals, Metals Park, OH, 1980.

[9] Ebert, L. J., "The Role of Residual Stresses in the Mechanical Performance of Case Carburized Steels," *Metallurgical Transactions*, Vol. 9A, Nov. 1978, pp. 1537–1551.

[10] Morrow, J., Ross, A. S., and Sinclair, G. M., "Relaxation of Residual Stresses Due to Fatigue Loading," *SAE Transactions*, Society of Automotive Engineers, Warrendale, PA, Vol. 68, 1960, pp. 40–48.

[11] Burke, J. D. and Lawrence, F. V., Jr., "The Effect of Residual Stresses on Weld Fatigue Life," Fracture Control Report No. 29, College of Engineering, University of Illinois, Urbana, IL, 1977.

[12] Waraniak, J. M., "Cyclic Deformation and Fatigue Behavior of Carburized Steel," M.S. thesis, Department of Mechanical and Industrial Engineering, University of Illinois, Urbana, IL, May 1980.

[13] Newaz, G. M., "Axial Cyclic Response of Unnotched and Notched Carburized Cylindrical Member under Constant Amplitude Completely Reversed Loading," Fracture Control Program Report No. 41, University of Illinois, Urbana, IL, Nov. 1981.

[14] Littman, W. E. and Widner, R. L., "Propagation of Contact Fatigue from Surface and Subsurface Origins," *Transactions*, Series D, American Society of Mechanical Engineers, Vol. 88, No. 3, Sept. 1966, pp. 624–636.

[15] Kunio, T., Shimizu, M., Yamada, K., Sakura, K., and Yamamoto, T., "The Early Stage of Fatigue Crack Growth in Martensitic Steel," *International Journal of Fracture*, Vol. 17, No. 2, April 1981, pp. 111–119.

Sushil K. Bhambri[1]

Fracture Morphology of 13% Chromium Steam Turbine Blading Steel

REFERENCE: Bhambri, S. K., "**Fracture Morphology of 13% Chromium Steam Turbine Blading Steel,**" *Fractography of Modern Engineering Materials: Composites and Metals, ASTM STP 948*, J. E. Masters and J. J. Au, Eds., American Society for Testing and Materials, Philadelphia, 1987, pp. 334–349.

ABSTRACT: To understand micromechanisms of steam turbine blade failures, a variety of fracture morphology was generated in 13Cr-Mo-V steel by controlled laboratory tests. The primary variables were microstructural condition and environment. Micromechanisms operative during fracture under impact and fatigue loading have been presented and discussed.

KEYWORDS: 13% chromium steels, impact, fatigue fracture, corrosion fatigue, microstructure, micromechanisms of fracture

Steam turbine blade failures form a significant proportion of the power generation plant outages. The 13% chromium class of martensitic stainless steels are commonly employed in steam turbine blading applications. Modified 13% chromium steel with molybdenum and vanadium additions is used for high-temperature stages at steam inlet in high-pressure and intermediate-pressure turbines. These alloying elements impart heat-resistant properties to the 13% chromium steels. In steam turbine blading applications, these steels are employed in hardened and tempered microstructural condition. However, microstructural deviations may result due to inadequate control during manufacturing and processing.

Investigations related to fracture morphology of failed components play an important role in identifying the cause of failure. A variety of micromechanisms of fracture have been observed in turbine blade failures, such as fatigue striations, beach marks, and intergranular fracture. While striations and beach marks may be due to fatigue, the causes leading to fatigue failure

[1]Laboratory manager—Metallurgy, Corporate R & D, Bharat Heavy Electricals, Ltd., Hyderabad, India.

may be different. Likewise, intergranular fracture may result due to various causes such as temper embrittlement of the material, presence of corrosive environment, hydrogen embrittlement, and caustic embrittlement. Of these, temper embrittlement and hydrogen embrittlement relate to material defect and are accompanied by low-impact energy values. The other causes are related to the operating environments. The influence of sodium chloride (NaCl)-saturated solutions on the fatigue behavior of 13% chromium steel has been extensively reported. However, the effect of small concentrations of NaCl has not received much attention. The present study, therefore, explores the influence of smaller concentrations of NaCl on the fracture morphology of fatigue failures. Recent studies of Doig et al. [1] show that stress corrosion cracking of 13% chromium steels in an alkaline solution of sodium hydroxide (NaOH) results in intergranular fracture when the material is tempered to a hardness level above 280 Brinnell Hardness Number (BHN). Blading materials, on the other hand, are tempered to hardness levels below 270 BHN. Therefore, stress corrosion cracking investigations have not been included here.

The nature and magnitude of stress have a significant influence on the fracture morphology. In a fatigue fracture, for example, the stress amplitude or the stress intensity factor range would determine the fracture mode. Striations are not formed during fatigue fractures as a rule, and other fracture modes such as ductile tearing, cyclic cleavage, and intergranular separation may operate [2,3]. The knowledge of micromechanisms of fracture is, therefore, assimilated through laboratory tests simulating operational conditions and possible deviations thereto. The influence of various combinations of microstructural condition, stress, and environment on fracture morphology has been investigated in this study.

Experimental Procedure

The 13Cr-Mo-V steel has the chemical composition given in Table 1. The material was received in the form of bars of 75 by 46-mm section which were hardened and tempered (solutionized at 1000°C and tempered at 700°C). The mechanical properties of the as-received material (Heat Treatment A) are given in Table 2. A part of the material was subjected to other heat treatment cycles shown in Table 2. The resultant microstructures and the mechanical properties obtained for the different microstructural conditions are given in Table 2.

TABLE 1—*Chemical composition of 13Cr-Mo-V steel investigated.*

C	Mn	Si	S	P	Cr	Mo	V	Ni
0.22	0.42	0.35	0.01	0.015	13.00	0.42	0.22	0.42

TABLE 2—*Heat treatment, resultant microstructure, and mechanical properties.*

Designation	Heat Treatment and Microstructure	0.2% Proof Stress, (MPa)	Tensile Strength, (MPa)	Elongation, %	Charpy V-notch Impact, Energy, (J)
(A)	1000°C—1h, AC 700°C—5h, AC Tempered martensite with fine coalescence of carbides in ferrite	686	823	18	60
(B)	1000°C—1h, AC 800°C—5h, AC Overtempered structure of carbide precipitates in ferrite	451	706	29	116
(C)	1000°C—1h, AC 535°C—5h, AC Tempered martensite with grain boundary impurity segregation and carbide precipitation	764	872	7	15
(D)	1000°C—1h, AC Autotempered martensite	784	990	14	7
(E)	1000°C—1h, FC to 600°C AC to 25°C Alloy pearlitic network structure in ferrite	431	700	9	9
(F)	1300°C—1h, AC 700°C—5h, AC Prior austenite grain boundary network of delta ferrite in tempered martensite	674	820	15	45

NOTE: AC = air cooled; FC = furnace cooled.

Impact tests were carried out on Charpy V-notch specimens in accordance with the ASTM Methods for Notched Bar Impact Testing of Metallic Materials (E 23-82). An instrumented Tinius-Olsen impact testing machine was used for these tests. Fatigue crack growth tests were conducted on a 250 KN-MTS servohydraulic testing machine employing compact-tension specimens according to ASTM Test Method for Constant-Load-Amplitude Fatigue Crack Growth Rates Above 10^{-8} m/Cycle (E 647-83). A constant stress ratio of 0.1 was maintained during fatigue crack growth tests. The conventional fatigue tests were carried out on rotary bending fatigue testing machines of Carl-Schenck make. The stress ratio for these rotary-bending fatigue tests was -1.0. Corrosion fatigue tests were also conducted on rotary-bending fatigue testing machines.

The fracture surfaces of all specimens were examined in a Cambridge Stereoscan, S-150, scanning electron microscope.

Results

Microstructures

The microstructures obtained in each heat-treated condition are shown in Fig. 1. Various phases and their morphology observed in the microstructures are described in following paragraphs.

The microstructure (Fig. 1a) resulting from 700°C tempering (Heat Treatment A) of martensite formed during air cooling from austenitizing temperature (this is an air-hardening type of steel) comprises fine coalescence of carbides precipitated out in a matrix of ferrite. Gooch [4] has identified these precipitates to be $M_{23}C_6$ carbides in the form of irregular particles or as needles in the martensite lath boundaries.

Tempering at 800°C (Heat Treatment B) resulted in an overtempered structure (Fig. 1b), and the coarse carbides are clearly resolved by metallurgical microscope. The carbides are $M_{23}C_6$ with a near spheroidal shape. The martensitic lath structure has apparently disappeared.

The resultant microstructure for 535°C tempered material (Heat Treatment C) is shown in Fig. 1c. In this microstructural condition, impurity elements are segregated on prior austenite grain boundaries. Stable carbides are M_7C_3 and $M_{23}C_6$ precipitated along grain boundaries and grain interiors [5].

Air cooling of 13% chromium steel from the austenitizing temperature (Heat Treatment D) results in martensite formation. The structure obtained (Fig. 1d) is an autotempered martensite with a small amount of dispersed carbides. These carbides have been identified as M_3C and form from austenite during cooling in the range of 600 to 800°C [4].

Furnace cooling from austenitizing temperature (Heat Treatment E) resulted in a network type of structure with alloy pearlite along prior austenite grain boundaries in a ferritic matrix (Fig. 1e). Austenitizing at 1300°C (Heat Treatment F) resulted in formation of delta ferrite at prior austenite grain boundaries (Fig. 1f).

Schematics of the various microstructures are presented in Fig. 2. The

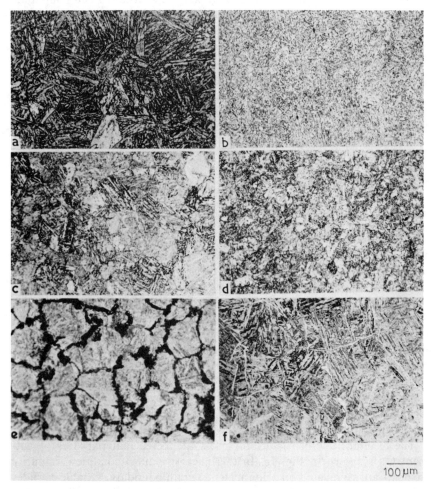

100 μm

FIG. 1—*Microstructures introduced in 13Cr-Mo-V steel by heat treatment*: (a) *tempered at 700°C*, (b) *tempered at 800°C*, (c) *tempered at 535°C*, (d) *air cooled from 1000°C*, (e) *furnace cooled from 1000°C*, (f) *air cooled from 1300°C*.

various microstructural features are: dispersion of carbides in the grain interiors as well as along grain boundaries (Fig. 2a—fine carbides, and 2b—coarse carbides), a structure wth embrittled grain boundaries (Fig. 2c), an autotempered martensite structure (Fig. 2d), a network structure of alloy pearlite along prior austenite boundaries with small incursions of carbide into grain interiors of ferrite (Fig. 2e), and a grain boundary delta ferrite (Fig. 2f).

FIG. 2—*Schematic representation of microstructures introduced in 13Cr-Mo-V steel:* (a) *dispersion of fine carbides,* (b) *dispersion of coarse carbides,* (c) *grain boundary embrittlement,* (d) *autotempered martensite,* (e) *alloy pearlite network,* (f) *grain boundary delta ferrite.*

Fracture Morphology

Impact Fractures

Impact fracture mechanisms observed in different microstructural conditions are shown in Fig. 3. Materials tempered at 700 and 800°C (Heat Treatments A and B) showed a macroscopic ductile impact fracture with appreciable shear lip. The micromechanism of fracture was, however, transgranular mixed mode with dimples at the initiating notch (Fig. 3a) and quasicleavage at the midsection. Dimples dominated gradually as the triaxial stress relaxed, and, simultaneously, crack tip plasticity increased with crack extension.

Impact fracture of the 535°C tempered specimen (Heat Treatment C) was completely intergranular (Fig. 3b). The air-cooled specimen (Heat Treatment D) fractured in a brittle manner and depicted quasicleavage mode (Fig. 3c).

The furnace-cooled structure (Heat Treatment E) fractured in cleavage mode under impact (Fig. 3d). The material austenitized at 1300°C (Heat Treatment F) with delta ferrite at prior austenite grain boundaries showed an intergranular mode of fracture (Fig. 3e). However, the grain faces revealed cleavage and dimple mechanisms of fracture at higher magnifications (Fig. 3f).

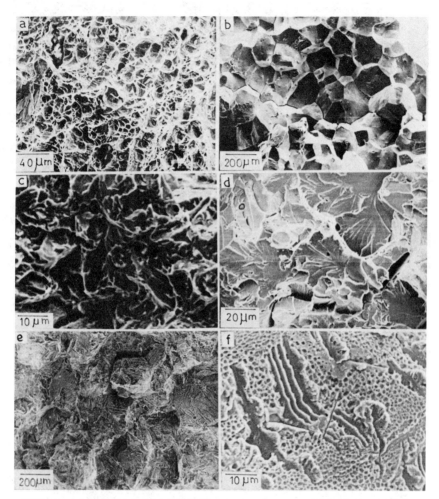

FIG. 3—*Scanning electron fractographs showing fracture mechanisms during impact failures:*
(a) *dimple mode in 700°C/800°C tempered structures,* (b) *intergranular mode in 535°C tempered structure,* (c) *quasicleavage mode in martensite,* (d) *cleavage mode in pearlite network/ferrite structure,* (e) *intergranular fracture in material with grain boundary delta ferrite,* (f) *same as in* (e) *at higher magnification showing transcrystalline fracture features on a grain face.*

Fatigue Fractures

The representative fatigue fracture surfaces for the lower, middle, and upper parts of the intermediate regime of the fatigue crack growth rate curve are shown in Fig. 4. A typical example of fatigue crack growth rate (da/dN) versus stress intensity factor range (ΔK) curve is shown in Fig. 5. The micromechanisms operative in materials tempered at 700 and 800°C (Heat

Treatments A and B) are transgranular, primarily by the striation formation and ductile tearing. The fractographs presented in Fig. 4a show ductile tearing. Striations are found to be poorly defined as earlier presented by this author [6]. In specimens air cooled from austenitizing temperature (Heat Treatment D), intergranular facets appeared along with striations. The proportion of the intergranular part of the total fracture varied with the stress intensity factor range (Fig. 4b). The variation of intergranular fracture in this specimen with ΔK was bell shaped as shown in Fig. 6. Intergranular facets also appeared during fatigue crack growth in material tempered at 535°C (Heat Treatment C) (Fig. 4c). These specimens showed a larger percentage of intergranular fracture in comparison with air-cooled specimens (Fig. 4c). The variation in proportion of intergranular fracture as a function of stress intensity factor range was, however, not bell shaped, and intergranular facets appeared all along the upper protion of the intermediate regime (Fig. 6). In the furnace-cooled material (Heat Treatment E), the fatigue crack growth process involved formation of striations with the occurrence of microcleavage (Fig. 4d). The material containing delta ferrite at prior austenite grain boundaries (Heat Treatment F) also showed intergranular facets during fatigue crack propagation (Fig. 4e).

The specimens tempered at 700°C (Heat Treatment A), fatigue tested in air under a completely reversed cycle, have revealed well-defined striations (Fig. 7a). Introduction of a mild corrosive environment, 0.01% NaCl aqueous solution, resulted in severe pitting at successive crack front positions near the crack initiation location, but the crack growth mechanism has remained transgranular (Fig. 7b). Increasing the concentration of NaCl in an aqueous solution to 3.0% through 0.1 and 1.0% resulted in the appearance of intergranular facets (Figs. 7c and 7d). Up to this level (3.0%) of chloride-induced aggressiveness in environment, fracture was in a mixed trans- and intercrystalline mode. Debris of corrosion products on fracture surfaces were occasionally observed in these fractures.

Discussion

The influence of variable parameters, such as microstructure and environment, on the fracture morphology under impact and high-cycle fatigue has been examined.

Impact Fractures

The observed transgranular ductile-brittle fracture in material tempered at 700 and 800°C is in agreement with the fracture mode reported for 13% chromium steels tempered above 650°C [5]. The dominance of dimples in the fracture mode of a 800°C tempered specimen has resulted due to ease in microvoid formation with enhanced ductility. The triaxial stresses at the

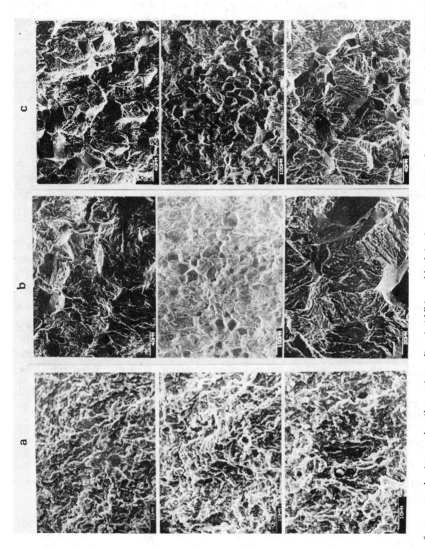

FIG. 4—*Fatigue fracture mechanisms at low (bottom), medium (middle), and high (top) stress intensity factor range values for:* (a) *700°C/800°C tempered structure,* (b) *air cooled from 1000°C,* (c) *535°C tempered structure,* (d) *furnace cooled from 1000°C,* (e) *air cooled from 1300°C.*

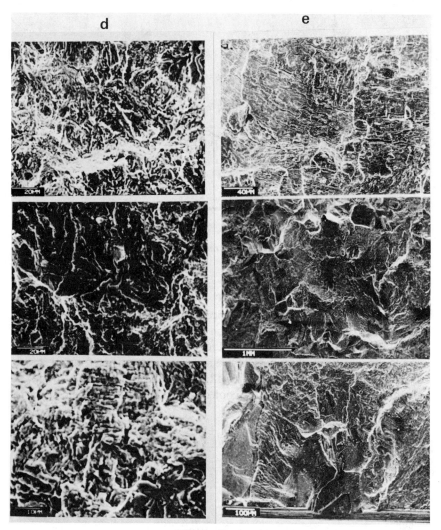

FIG. 4—*Continued.*

notch tip set in large plastic strains in the ductile material, resulting in decohesion of the particle-matrix interface. A mixed size of these dimples suggests that both second-phase precipitate particles and inclusions have been the sites for microvoid initiation.

Intergranular fracture observed in 535°C tempered material also conforms to the results reported by other authors on 13% chromium steel tempered within the temper embrittlement range. Thirteen percent chromium steels are susceptible to temper embrittlement within 475 to 550°C [5]. Mechanisms of

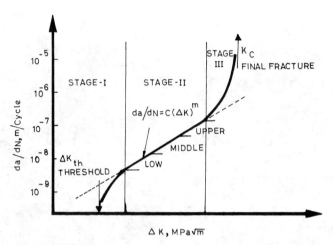

FIG. 5—*Schematic representation of fatigue crack growth rates plotted as a function of stress intensity factor range.*

grain boundary embrittlement have been studied extensively [7], and it is generally agreed that both segregation of impurity elements and carbide precipitates at grain boundaries reduce their cohesion such that grain boundaries are the weakest part of the microstructure. Therefore, the crack advances along grain boundaries, resulting in intergranular separation.

Autotempered martensite in air-cooled material has a high yield strength

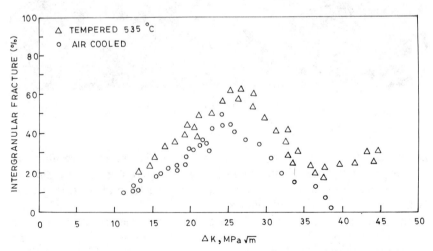

FIG. 6—*Variation of intergranular fracture percent as a function of stress intensity factor range in air-cooled (Heat Treatment D) and 535°C tempered structure (Heat Treatment C).*

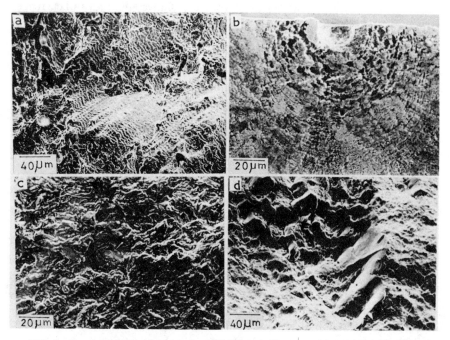

FIG. 7—*Corrosion fatigue fractures for 700°C tempered microstructure under completely re-versed cycles*: (a) *air*, (b) *0.01% NaCl aqueous solution*, (c) *0.1% NaCl aqueous solution*, (d) *1.0% NaCl aqueous solution*.

level which facilitates a high triaxial stress state at the notch front. Strain concentrations are also developed at the boundaries between martensitic laths, resulting in reduction of plane-strain ductility. A brittle fracture, there-fore, triggers in quasicleavage mode since tear ridges are formed at the carbides present in the lath boundaries. The appearance of the dimple mode of fracture as the crack front approached edges is related to the loss in severity of the triaxial stress state.

The impact fracture in furnace-cooled material with a network structure of alloy carbides along prior austenite boundaries revealed river patterns, a characteristic feature of cleavage fracture. The applied stress in this case exceeded the cleavage fracture stress in the ferritic phase, leading to a frac-ture by cleavage. A small proportion of dimples resulted from microvoids formed at alloy carbides ahead of the crack front as it approached the alloy pearlitic region.

The impact fracture morphology in the material containing delta-ferrite phase depended on the distribution and configuration of the ferrite phase. Presence of delta ferrite at the prior austenite grain boundaries rendered the latter weaker, and hence fracture occurred by intergranular separation. At higher magnification, however, the grain facets showed dimples and cleavage fracture modes.

Fatigue Fracture

The fatigue fractures in materials tempered at 700°C (Heat Treatment A) following hardening have revealed a transgranular mode of crack propagation with striation formation and ductile tearing. The crack growth has occurred by the plastic blunting process proposed by Laird and Smith [8]. However, the presence of carbides ahead of the crack tip resulted in ductile tearing during this process. The voids nucleated at these carbide particles prior to the crack front intersecting them. Thus, the interference of carbide particles has resulted in poorly defined fatigue striations at the stress ratio of 0.10 considered in this study. An increase in ductile tearing is observed with increased stress intensity factor range.

In material tempered at 800°C (Heat Treatment B), ductile tearing at carbide particles is more pronounced, while the crack growth mechanism continues to be the striation formation mechanism. These observations are, however, converse to the fracture mode reported by Ishii et al. [9] in a 13% chromium steel. Ishii et al. observed intergranular facets in materials tempered at 600 and 750°C.

The appearance of intergranular facets with the striations on the fracture surface of specimens air cooled from austenitizing temperature conforms to the fracture morphology reported by Kobayashi et al. [10], Lui and Le May [11], and Crooker et al. [3] for fatigue crack growth in martensitic structure. The proportion of intergranular facets in fatigue fracture is known to decrease with an increasing degree of tempering of martensite [10], as a result of increasing tempering temperature, till they completely disappear. The absence of intergranular facets in materials tempered at and above 700°C in the present study is in agreement with this.

The occurrence of intergranular fracture with striations during fatigue crack propagation in a material tempered at 535°C is analogous to that observed in a martensitic structure. The intergranular separation in this material condition, however, increased continuously with the increasing stress intensity factor range since the static mode of fracture is also intergranular.

The appearance of intergranular facets in fatigue fractures in martensitic structure (Heat Treatment D) and in 535°C tempered structure (Heat Treatment C) is attributed to the ratio of plastic zone size to prior austenite grain diameter. For plastic zone size less than grain diameter, the crack growth occurs by slip plane decohesion (Fig. 8a). When plastic zone size equals the grain diameter, the intergranular cracking sets in along favorable oriented grain boundaries (Fig. 8b). The underlying basic mechanism is one of dislocation pileups at grain boundaries. Grain boundary embrittlement in material tempered at 535°C (Heat Treatment C) further facilitates inter-

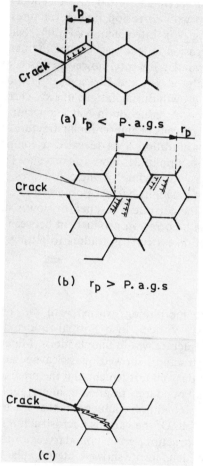

FIG. 8—*Schematic representation of micromechanisms of fracture:* (a) *transgranular fracture by striation formation,* (b) *intergranular fracture when crack tip plastic zone equals prior austenite grain diameter,* (c) *cyclic cleavage.*

granular separation. Grain boundary strengthening and gradual dispersion of carbides into grain interiors in an alloy pearlitic network structure resulted in transcrystalline mode of fatigue crack growth. The incursions of carbides prevented plastic strain concentrations from occurring at the pearlite/ferrite interface. Secondary fissures during fatigue crack growth have formed due to low toughness of material in this microstructural condition. The observed cyclic cleavage in ferritic structure has also been reported extensively [2]. Microcleavage occurred when, within the crack tip plastic zone, the fracture stress exceeded the cleavage stress locally on favorably oriented habit planes (Fig. 8c).

Fatigue fracture of specimens tested in air (under completely reversed cycle) for comparison with corrosion fatigue fractures has reavealed well-defined striations (Fig. 7a). Introduction of 1.0% NaCl aqueous solution environment in a corrosion fatigue test resulted in intergranular facets. The proportion of intergranular fracture, however, was a small fraction of the fracture surface. Ishii et al. [9] and Ebara et al. [12] have also reported a mixed mode of crack growth during fatigue of 13% chromium steel in a 3% NaCl aqueous solution. A reduction in NaCl concentration in the aqueous solution reduced the percent of intergranular fracture. Specimens tested at the 0.01% NaCl concentration level revealed a completely transgranular fracture mode at stress amplitudes close to corrosion fatigue strength level. In the slow crack growth regime of these fractures, extensive pitting at successive crack front positions has been observed (Fig. 7b). The reduction in aggressiveness of the corrodent resulted in slower near threshold crack growth rates, and, with corrosion products in between the crack faces, the crack tip region acted like a crevice, leading to pitting.

Conclusions

1. Impact fracture morphology varied with the initial microstructural condition in 13Cr-Mo-V steel. In a normal-tempered microstructure the micromechanism of fracture was in dimple mode. Materials with martensite and pearlite/ferrite structures showed quasicleavage and cleavage mode of fracture. Grain boundary embrittlement and the presence of delta ferrite on prior austenite grain boundaries resulted in intergranular fracture.

2. Microstructural condition influenced the fracture morphology of fatigue fractures as well. Of the various microstructures introduced, materials with martensitic structure, grain boundary embrittlement, and a grain boundary network of delta ferrite showed intergranular facets. Material with alloy pearlitic net structure in ferrite showed the appearance of microcleavage in addition to striations.

3. Corrosion fatigue fracture morphology was influenced by the concentration of NaCl in aqueous solution. At a 0.01% NaCl concentration, fracture occurred in transcrystalline mode with severe pitting at the slow crack growth regime. Intergranular facets appeared when the NaCl concentration in aqueous solution exceeded 0.01%.

Acknowledgement

The author wishes to thank the general manager, Corporate Research and Development Division of BHEL for permission to present this paper. Acknowledgment is also given to K. Rajanna, Y. Sitaramaiah, and S. V. Reddy for their assistance in experimental work.

References

[1] Doig, P., Chastell, D. J., and Flewitt, P. E. J., *Metallurgical Transactions A*, Vol. 13A, 1982, p. 913.

[2] Benson, J. P. and Edmonds, D. V., *Metal Science*, Vol. 12, 1978, p. 223.

[3] Crooker, T. W., Hasson, D. F., and Yoder, G. R., in *Fractography-Microscopic Cracking Process*, *ASTM STP 600*, American Society for Testing Materials, Philadelphia, 1976, p. 205.

[4] Gooch, D. J., *Metal Science*, Vol. 16, 1982, p. 79.

[5] Prabhu Gaunker, G. V., Huntz, A. M., and Lacomb, P., *Metal Science*, Vol. 14, 1980, p. 241.

[6] Bhambri, S. K. and Rama Rao, P., *Advances in Fracture Research*, Vol. 3, Valluri et al., Eds., Pergamon Press, Oxford, U.K., 1984, p. 1751.

[7] McMohan, C. J., Jr., Briant, C. L., and Banerji, S. K., *Metallurgical Transactions A*, Vol. 10A, 1979, p. 1729.

[8] Laird, C. and Smith, G. C., *Philosophical Magazine*, Vol. 7, 1962, p. 847.

[9] Ishii, H., Sakakibara, Y., and Ebara, R., *Metallurgical Transactions A*, Vol. 134, 1982, p. 1521.

[10] Kobayashi, H., Marakami, R. and Nakazawa, H., *Fracture Mechanics and Technology*, Vol. 1, Noordhoff, Amsterdam, 1979, p. 205.

[11] Lui, M. W. and Le May, I., *Microstructural Science*, Vol. 8, Elsevier, Amsterdam, 1980, p. 341.

[12] Ebara, R., Kai, T., and Inoue, K., *Corrosion Fatigue Technology*, *ASTM STP 642*, American Society for Testing and Materials, Philadelphia, 1978, p. 155.

T. Alan Place,[1] *T. Srinivas Sudarshan,*[2] *Cindy K. Waters,*[2] *and M. R. Louthan, Jr.*[3]

Fractographic Studies of the Ductile-to-Brittle Transition in Austenitic Stainless Steels

REFERENCE: Place, T. A., Sudarshan, T. S., Waters, C. K., and Louthan, M. R., Jr., **"Fractographic Studies of the Ductile-to-Brittle Transition in Austenitic Stainless Steels,"** *Fractography of Modern Engineering Materials: Composites and Metals, ASTM STP 948*, J. E. Masters and J. J. Au, Eds., American Society for Testing and Materials, Philadelphia, 1987, pp. 350–365.

ABSTRACT: Austenitic stainless steels are not generally thought to undergo ductile-to-brittle transitions as the temperature is lowered to 77 K or to be susceptible to hydrogen embrittlement. However, some austenitic steels do undergo a classic ductile-to-brittle transition, and are susceptible to hydrogen embrittlement even at the high strain rates used in an impact test. In this study the fracture behavior of Tenelon, 21-6-9, and 304L was investigated by testing hydrogen charged and uncharged Charpy specimens at temperatures ranging from 77 to 298 K. The fracture mode was determined with the aid of optical and scanning electron microscopy and X-ray analysis. Fracture initiation in the 21-6-9 and 304L was generally by ordinary dimple zone formation, though the Tenelon developed a brittle initiation mode at low temperatures. All of the steels showed a change from single to multiple crack initiation as the temperature was reduced. Propagation was by microvoid coalescence at higher temperatures, with a strong orientation effect in evidence where inclusions provided weak interfaces. A ductile-to-brittle transition was observed in both the Tenelon and the 21-6-9 steels. In the case of Tenelon, this transition resulted from {111} plane separation, and the reduction in toughness in 304L and 21-6-9 was associated with strain localization and formation of large microvoids. The presence of hydrogen did not introduce any new fracture modes but did enhance the embrittlement mechanisms which were promoted by low temperatures.

KEYWORDS: stainless steel, mechanical properties, fracture (metals), impact testing, fractography

[1]Professor, Department of Mechanical Engineering, University of Idaho, Moscow, Idaho 83843.

[2]Former graduate student, Department of Materials Engineering, Virginia Polytechnic Institute, Blacksburg, VA 24061.

[3]Professor, Department of Materials Engineering, Virginia Polytechnic Institute, Blacksburg, VA 24061.

Previous work [1–7] on the deleterious effects of hydrogen on the low temperature mechanical properties of austenitic stainless steels has been principally concerned with tensile properties and crack growth behavior. These studies have shown that the tendency for hydrogen embrittlement increases as strain rate and temperature are decreased. Despite this recognition of the importance of strain rate and low temperatures, little work has been done on the effects of hydrogen on impact properties of austenitic steels at low temperatures.

It is often assumed by designers that austenitic stainless steels do not exhibit a ductile-to-brittle transition. However, studies have clearly demonstrated that some austenitic steels are susceptible to low-temperature embrittlement. It is essential, therefore, if stainless steels are to be considered for low-temperature structural applications, that their tendency for brittle fracture be investigated. Hydrogen is also known to have a deleterious effect on austenitic stainless steels, although there is disagreement about the mechanism of hydrogen embrittlement. Previous studies [8–14] have attributed the embrittlement to the interaction of hydrogen with grain boundary impurities, to hydrogen effects on plastic deformation processes, to dislocation transport of hydrogen, to hydrogen interactions on martensite formation, to hydrogen diffusion in stress gradients, or to hydrogen interactions at interfaces and voids. Many of the suggested models require the redistribution of hydrogen during the test. Such redistribution either results in localized high hydrogen concentrations or the interaction of hydrogen with some microstructural feature produced during the test. During slow-strain-rate tests at ambient temperatures, dislocation transport may contribute to this redistribution. However, at the high strain rates produced in impact tests, especially when they are conducted at very low temperatures, it is unlikely that any hydrogen could be swept in by the dislocations. The purpose of the present study was to investigate the effects of hydrogen on the impact behavior of three austenitic stainless steels tested over a wide range of temperatures, in both the hydrogen charged and uncharged conditions.

Procedure

The investigation was made using samples of Tenelon, 21-6-9, and 304L stainless steels. Some specimens were hydrogen charged to saturation before the impact tests were conducted.

The specimens were machined from as-received hot cross-rolled plates. Tension specimens were prepared with a gauge length of 5 cm and a gauge diameter of 0.6 cm. The crosshead speed during the tension test of 0.13 cm/min produced a strain rate of 423 μs^{-1}. Standard Charpy V-notch specimens were cut from the Tenelon plate in the T-S and T-L orientations and from the 304L and 21-6-9 plates in the L-T orientations. Testing was conducted in accordance with ASTM Methods for Notch Bar Impact Testing of

TABLE 1—*Gaseous hydrogen charging conditions.*

	Tenelon	21-6-9	304L
Pressure, MPa	0	0	0
	69	13.8	13.8
	. . .	138	138
Temperature, K	645	573	573
Time, days	28	45	45

Metallic Materials (E 23-82). The impact machine was instrumented so that load-time and energy-time traces were recorded. Some specimens were tested in the as-received condition and others after charging to saturation with gaseous hydrogen under the various conditions listed in Table 1. The charged specimens were stored at 273 K until they were tested. Tenelon Charpy specimens were broken at temperatures ranging from 77 to 298 K, and the 21-6-9 and 304L specimens were tested at 77 and 298 K.

Optical microscopy was performed on the as-received materials and on some of the broken Charpy specimens. Scanning electron microscopy was carried out on the fracture surfaces of broken Charpy specimens. X-ray analysis and single-surface stereographic analysis were also performed on the fractured specimens.

Results

The chemical composition of the steels used in this study are shown in Table 2. The tensile properties are shown in Table 3. The 304L and 21-6-9 had equiaxed microstructures with few inclusions or orientation effects. The Tenelon, on the other hand, contained a number of inclusions strung out in the predominant direction of rolling.

TABLE 2—*Chemical compositions of Tenelon, 21-6-9, and 304L steels in weight percent.*

Element	Tenelon	21-6-9	304L
Cr	17.44	19.34	19.21
Ni	0.22	5.89	11.35
Mn	15.31	8.80	1.73
C	0.1 max	0.04	0.024
N	0.5	0.29	0.053
Si	0.53	0.32	0.46
P	< 0.02	0.017	0.023
S	0.02	0.013	0.02
Mo	. . .	. . .	0.12
Cu	. . .	. . .	0.06
Co	. . .	. . .	0.09
O	. . .	0.007	. . .
Fe	Balance	Balance	Balance

TABLE 3—*Tensile properties of Tenelon, 21–6–9, and 304L steels.*

Material	Test Temperature, K	0.2% Proof Stress, MPa	UTS, MPa	Elongation		Reduction in Area, %
				Uniform, %	Total, %	
Tenelon	350	675	1265	48	59	76
	273	830	1480	50	58	68
	198	1050	1960	59	66	50
	78	1735	1780	19	19	8
21–6–9	298	568	775		34	75
304L	298	254	595		39	80

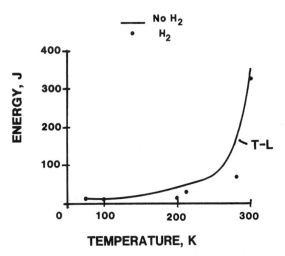

FIG. 1—*Charpy energy versus temperature for T-L orientation Tenelon specimens. The line is for uncharged specimens. Points are for charged specimens. T-S orientation specimens showed similar behavior.*

A plot of Charpy energy versus temperature for the T-L orientation Tenelon specimens is shown in Fig. 1. The T-S plot was similar. The solid line in Fig. 1 was drawn through six points obtained from uncharged specimens. The solid circles on the graph show the points for specimens charged with hydrogen for 28 days at 645 K and 69 MPa. These points were slightly below the solid line as was the case for the T-S orientation specimens. Ease of crack propagation was estimated from the tearing slope, which was calculated from the inverse of the slope of the load-time trace in the propagation region. Larger values of the tearing slope therefore imply more energy absorption in propagation. The tearing slope for the T-L orientation specimens increased sharply in the ductile-to-brittle transition range (Fig. 2), and the T-S orientation specimens behaved in much the same manner.

Analysis of the load-time traces for the Tenelon specimens yielded the load-temperature plots shown in Figs. 3, 4, and 5. Despite the presence of elongated inclusions in the Tenelon, orientation seemed to make little difference, so these plots show the combined data for both the T-L and T-S orientations. Figure 3 shows the variation in yield and maximum loads with change in temperature for the uncharged specimens. The yield and maximum loads were the same at all temperatures below about 214 K. The lines drawn through the data for the uncharged specimens were transposed onto Figs. 4 and 5, which show the data points for the hydrogen-charged specimens. Figure 4 shows the yield load data and Fig. 5 shows the maximum load data. Almost all the data for the hydrogen-charged specimens fell below the lines.

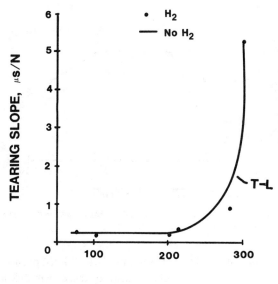

FIG. 2—*Tearing slope versus temperature for T-L orientation Tenelon specimens. Larger slope values represent more energy absorption during propagation. The line is for uncharged specimens. Points are for charged specimens. T-S orientation specimens showed similar behavior.*

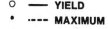

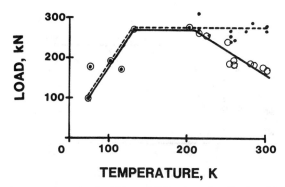

FIG. 3—*Load-temperature plot for T-L and T-S orientation Tenelon Charpy specimens in uncharged condition. Yield and maximum load data were derived from load-time traces.*

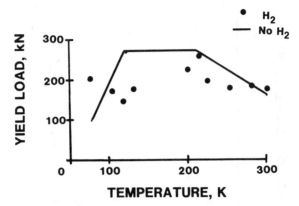

FIG. 4—*Yield load versus temperature for Tenelon Charpy specimens. The solid line is from Fig. 3; points are from hydrogen-charged specimens.*

The limited data obtained for 21-6-9 and 304L did not permit similar plots to be made for these materials. The load-time traces for 298 and 77 K are shown on the same scale in Fig. 6 for 21-6-9 and Fig. 7 for 304L. It can be seen that in both steels, lowering the temperature elevates the yield load and maximum load and reduces the tearing slope (makes propagation easier). Hydrogen was observed to have little effect on the yield and maximum loads but has an appreciable effect on the propagation energy. The effect of hydrogen charging pressure (and therefore hydrogen content) and temperature on total Charpy energy for both 21-6-9 and 304L is shown in Fig. 8.

The fracture processes underlying the changes just described were thoroughly characterized by scanning electron microscopy. In all three steels, fracture initiation at room temperature was associated with formation of a dimple rupture zone. Below about 214 K, however, this single initiation process gave way to multiple initiation in which a number of cracks could be

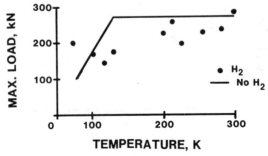

FIG. 5—*Maximum load versus temperature for Tenelon Charpy specimens. The solid line is from Fig. 3; points are for hydrogen-charged specimens.*

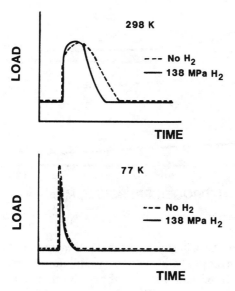

FIG. 6—*Load-time traces from instrumented impact tests on 21-6-9.*

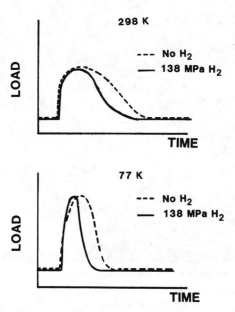

FIG. 7—*Load-time traces from instrumented impact tests on Type 304L.*

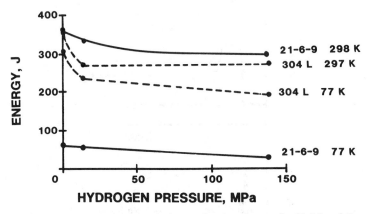

FIG. 8—*Charpy energy versus gaseous hydrogen-charging pressure for 21-6-9 and Type 304L.*

seen at the tip of the Charpy V-notch. A few measurements of the total dimple zone width showed wide variations and no apparent decline in width with decreasing temperature. Below about 130 K, a third fracture mechanism was observed in Tenelon. Multiple initiation still occurred, but fracture commenced by a brittle mechanism (dark areas at crack tip in Figs. 9 and

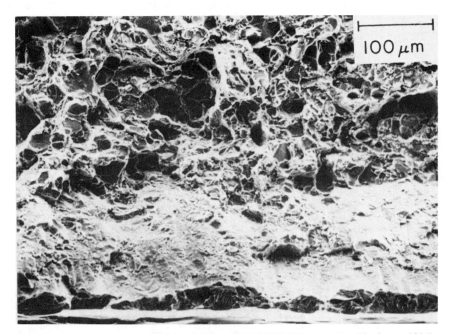

FIG. 9—*Crack initiation in Tenelon Charpy specimen, hydrogen charged and broken at 130 K. Notch at bottom/dark area is brittle crack initiation and then relatively smooth dimple rupture crack initiation zone. Rougher crack propagation zone is at top.*

10). This brittle mechanism gave way to typical dimple zone formation (top of Fig. 10), which in turn gave way to a different mechanism for propagation (top of Fig. 9). All three zones are visible in Fig. 9. The brittle crack initiation mechanism was not observed in 21-6-9 or 304L.

The mode of the crack propagation depended on the temperature. On the upper shelf the predominant mode was dimple rupture, although pronounced splitting occurred ahead of the main crack front in the Tenelon. The splits can be seen on the side of the Charpy specimen shown in Fig. 11. The splitting phenomenon diminished as the temperature was reduced, and there was little evidence of splitting below about 214 K. The transition region in Tenelon was characterized by the onset and increase of fracture along what appeared to be specific crystallographic planes. As the test temperature was reduced from 255 to 214 K, the area fraction of the flat facets seen in Fig. 12 increased from less than 10% at 255 K to about 70% at 214 K. X-ray studies of these fracture surfaces showed that the relative {111} peak intensity was far greater than that found in polished specimens of the same orientation. Single surface stereographic analysis confirmed that the facets were of the {111} orientation. The area fraction of facets (brittle fracture) increased along the lower shelf from about 70% at 214 K to about 90% at 131 K. Below 131 K the area fraction remained at about 90%.

Reducing the test temperature from 298 to 77 K reduced the toughness of 304L, but not drastically. Despite multiple crack initiation at 77 K, the yield

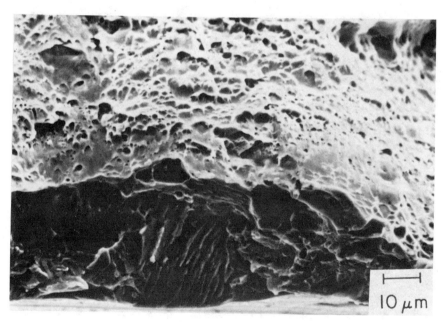

FIG. 10—*Same as Fig. 9, but showing brittle crack initiation zone at higher magnification.*

FIG. 11—*Side of a Tenelon Charpy specimen, showing crack arrester orientation splitting ahead of the main crack front. Hydrogen charged and tested at 255 K.*

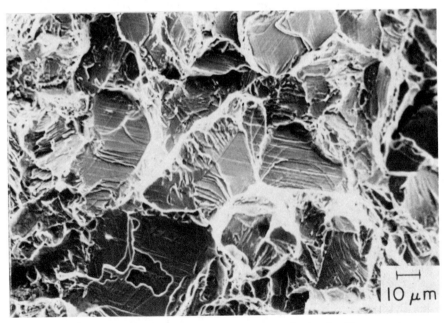

FIG. 12—*Flat {111} plane facets on the fracture surface of a Tenelon Charpy specimen.*

and maximum loads were much higher than at 298 K. The propagation energy was less and was associated with large shallow dimples (Fig. 13). The same trends were observed in 21-6-9, only in this case the propagation energy was drastically reduced. The fracture propagation mode at 298 K was by coalescence of fine microvoids. At 77 K the fracture mode was still by microvoid coalescence, although at this lower temperature two sizes of microvoids were apparent (Fig. 13). There were microvoids of more than 10 μm diameter surrounded by a network of microvoids of about 1 μm diameter. The large microvoids were sometimes separated by a chisel point fracture. This effect was much more pronounced than in 304L. Neither the 304L nor the 21-6-9 exhibited fracture along the {111} planes, which was observed in Tenelon.

Hydrogen charging did not cause any new fractographic features to form but had the same effect as a reduction in temperature. At a given temperature, a charged specimen of any of the three steels would exhibit more evidence of embrittlement than an uncharged specimen. Charged specimens showed multiple initiation, brittle initiation, {111} plane parting, or lower tearing slopes at a slightly higher temperature than uncharged specimens.

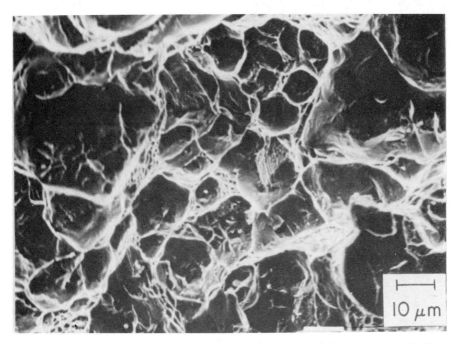

FIG. 13—*Large shallow dimples on the fracture surface of a 21-6-9 Charpy specimen. Smaller microvoids are around large ones. Uncharged. Broken at 77 K.*

Discussion

The toughness of all three steels at 298 K was very high. The inclusions in Tenelon gave rise to pronounced splitting at the inclusion-matrix interfaces, although there was not much difference in toughness between the crack divider (T-L) and crack arrester (T-S) orientations. The elevation of yield and maximum loads with decrease in temperature in these steels has been observed in other studies [2,17,18]. The effect is less pronounced in 304L than in 21-6-9 or Tenelon. The ductile-to-brittle transition behavior in Tenelon can be described with the aid of the load-temperature schematic shown in Fig. 14. The decline in toughness in the transition region is associated with a sharp drop in propagation energy, which was related to a dramatic increase in {111} plane parting. No {111} plane parting was observed at 298 K. About 70% of the fracture surface area was formed by parting at 214 K. The separation on {111} planes in all three austenitic steels has also been observed by other workers [2,6,17]. West and Louthan [2] observed that such interfacial separation was not observed in 21-6-9 at strength levels below about 1100 MPa. This supports the observation in the present study that no interfacial separation occurs in 21-6-9 with a tensile strength of 775 MPa. Even at test temperatures on the lower shelf, the Charpy energy of Tenelon continues to decline, and this was associated with the onset of multiple crack initiation at the notch tip and a further increase in {111} plane

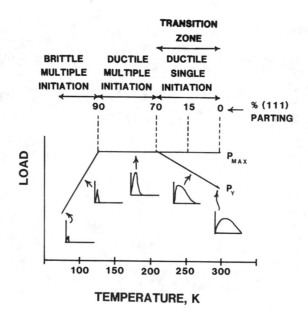

FIG. 14—*Change in fracture modes with temperature in Tenelon.*

parting to such an extent that about 90% of the fracture at 130 K is along the {111} plane. There was no further increase in {111} plane separation as the temperature was reduced from 130 to 77 K, but there was further decline in the Charpy energy. The final decline was due to a sharp drop in fracture load, which was caused by the onset of a brittle crack initiation mechanism. These brittle fracture initiation areas may be the result of strain-induced martensite produced at temperatures below 130 K. Many other workers have observed martensite formation in all three of these steels [17,19,21,22]. Hanninen et al. [19,20] observed surface cracking related to martensite formation in charged 304. Surface cracking in hydrogen-charged specimens of these steels tested in tension was previously shown to be intergranular, although no such intergranular cracking was observed in the present high strain rate study.

The presence of a notch has been shown to increase the tendency for surface cracking in 21-6-9 [17] tension specimens, which may partly explain the onset of multiple crack intiation at low temperatures. The formation of martensite, however, requires shear, and this would tend to be restricted in the triaxial stress region just remote from the notch tip. The outer surface of the notch, however, would be in a state of plane stress, and so it should be possible to form martensite in the near surface region, whereas martensite formation may be suppressed or severely limited in the triaxial stress region a little further below the surface. This would explain martensite-assited crack initiation in Tenelon at low temperatures but no apparent effects of martensite on crack propagation. Apparently conditions were not suitable for the same mechanism to operate in 304L or 21-6-9.

The lower toughness in 304L and 21-6-9 at 77 K was related to the formation of large microvoids, which were most pronounced in 21-6-9. Reducing the temperature or introducing hydrogen appears to lead to strain localization which favors the growth of a few microvoids rather than numerous small ones. Such strain localization has been observed previously [1,7,15,16]. The only fractographic change coinciding with the loss in toughness in 21-6-9 at the low temperature was formation of the large voids. West and Holbrook [17] also observed a substantial increase in microvoid size associated with charged 304L specimens as compared to uncharged. Thompson [23], however, has shown reductions in toughness in 309S stainless steel to be due to enhanced microvoid formation at smaller inclusions than those showing interfacial separation in uncharged specimens. Hydrogen promoted void formation over a wider range of inclusion sizes, thus causing a reduction in average void size. In the present case, it appears that hydrogen promotes strain localization at certain interfaces, so that some of the voids become very large in comparison to the others. The influence of hydrogen in promoting growth of voids but not their initiation has been observed in spheroidized eutectoid steels [24].

Summary

The fracture behavior of three austenitic stainless steels was investigated over the temperature range of 77 to 298 K in the as-received and hydrogen-charged conditions. Reductions in toughness were observed when the temperature was reduced or if hydrogen was present in the material. The reductions in Tenelon and 21-6-9 were appreciable. The effect of hydrogen was similar to that resulting from a reduction in temperature. The embrittlement mechanisms included multiple crack initiation, brittle crack initiation and $\{111\}$ plane parting in the case of Tenelon, and multiple crack initiation and strain localization in the case of 304L and 21-6-9. The strain localization resulted in the production of large microvoids.

References

[1] Louthan, M. R., Jr., Caskey, G. R., Jr., Donovan, J. A., and Rawl, D. E., Jr., *Materials Science and Engineering*, Vol. 10, 1972, p. 357.

[2] West, A. J. and Louthan, M. R., Jr., *Metallurgical Transactions A*, Vol. 13A, 1982, p. 2049.

[3] Donovan, J. A., *Metallurgical Transactions A*, Vol. 7A, 1976, p. 145.

[4] Odegard, B. C., Brooks, J. A., and West, A. J., *Effect of Hydrogen on Behavior of Materials*, A. W. Thompson and I. M. Bernstein, Eds., American Institute of Mining, Metallurgical, and Petroleum Engineers, New York, 1976, p. 116.

[5] Fidelle, J. P., Bernardi, R., Broudeur, R., and Rapin, M., in *Hydrogen Embrittlement Testing, ASTM STP 543*, American Society for Testing and Materials, Philadelphia, 1974, p. 221.

[6] Caskey, G. R., Jr., "Hydrogen Induced Brittle Fracture of Type 304L Stainless Steel," in *Fractography and Materials Science, ASTM STP 733*, American Society for Testing and Materials, Philadelphia, 1981, p. 86.

[7] West, A. J. and Louthan, M. R., Jr., *Metallurgical Transactions A*, Vol. 10A, 1979, p. 1675.

[8] Kane, R. D. and Berkowitz, B. J., *Corrosion*, Vol. 36, 1980, p. 29.

[9] Hanninen, H. and Hakkarainen, T., *Metallurgical Transactions A*, Vol. 10A, 1979, p. 1196.

[10] Sridhar, N., Kargol, J. A., and Fiore, N. F., *Scripta Metallurgica*, Vol. A, 1980, p. 225.

[11] Kock, G. H., *Corrosion*, Vol. 35, 1979, p. 73.

[12] Briant, C. L., *Metallurgical Transactions A*, Vol. 10A, 1979, p. 181.

[13] Kennedy, J. R., Adler, P. N., and Schulte, R. L., *Scripta Metallurgica*, Vol. 14, 1980, p. 299.

[14] Cialone, H. and Asaro, R. J., *Metallurgical Transactions A*, Vol. 10A, 1979, p. 367.

[15] Thompson, A. W., in *Effect of Hydrogen on Behavior of Materials*, The Metallurgical Society of American Institute of Mining, Metallurgical, and Petroleum Engineers, New York, 1976, p. 467.

[16] Hirth, J. P. and Onyewuenyi, O. A., *Environmental Degradation of Engineering Materials in Hydrogen*, M. R. Louthan, Jr., R. P. McNitt, and R. D. Sisson, Eds., Virginia Polytechnic Institute, Blacksburg, VA, 1981, p. 133.

[17] West, A. J. and Holbrook, J. H., in *Hydrogen Effects in Metals*, I. M. Bernstein and A. W. Thompson, Eds., American Institute of Mining, Metallurgical, and Petroleum Engineers, New York 1981, p. 607.

[18] Holbrook, J. H. and West, A. J., in *Hydrogen Effects in Metals*, I. M. Bernstein and A. W. Thompson, Eds., American Institute of Mining, Metallurgical, and Petroleum Engineers, 1981, p. 655.

[19] Hannula, S.-P., Hanninen, H., and Tahtinen, S., *Metallurgical Transactions A*, Vol. 15A, 1984, p. 2205.

[20] Hanninen, H., Hannula, S.-P, and Tahtinen, S., in *Environmental Degradation of Engineering Materials in Hydrogen*, M. R. Louthan, R. P. McNitt, and R. D. Sisson, Eds., Virginia Polytechnic Institute, Blacksburg, VA, 1981, p. 347.

[21] Hyzak, J. M., Rawl, D. E., and Louthan, M. R., Jr., *Scripta Metallurgica*, Vol. 15, 1981, p. 937.

[22] Caskey, G. R. in *Environmental Degradation of Engineering Materials in Hydrogen*, M. R. Louthan, R. P. McNitt, and R. D. Sisson, Eds., Virginia Polytechnic Institute, Blacksburg, VA, 1981, p. 283.

[23] Thompson, A. W. in *Effect of Hydrogen on Behavior of Materials*, A. W. Thompson and I. M. Bernstein, Eds., American Institute of Mining, Metallugical, and Petroleum Engineers, New York, 1976, p. 467.

[24] Carber, R. and Bernstein, I. M. in *Environmental Degradation of Engineering Materials*, M. R. Louthan, Jr. and R. P. McNitt, Eds., Virginia Polytechnic Institute, Blacksburg, VA, 1977, p. 463.

A. O. Ibidunni[1]

Fractography in the Failure Analysis of Corroded Fracture Surfaces

REFERENCE: Ibidunni, A. O., **"Fractography in the Failure Analysis of Corroded Fracture Surfaces,"** *Fractography of Modern Engineering Materials: Composites and Metals, ASTM STP 948,* J. E. Masters and J. J. Au, Eds., American Society for Testing and Materials, Philadelphia, 1987, pp. 366–379.

ABSTRACT: Fractographic information retrieval from oxidized or corroded fracture surfaces is very important in failure analysis. In order to determine the mode of failure, the surface oxides must be removed without destroying the original surface morphology of the fracture. Over the years, several methods have been utilized in the removal of surface oxides. However, chemical dissolution of the oxides coupled with ultrasonic cleaning has been the most effective. The effectiveness of two commercially available proprietary products and of orthophosphoric acid in the removal of surface oxides without destroying the fracture surfaces is discussed. Three case histories, in which each of the chemicals has been utilized, are presented. The fractographic information obtained in each case was supported by metallography. Orthophosphoric acid, when coupled with ultrasonic cleaning, is very effective in the removal of oxides from fracture surfaces.

KEYWORDS: corrosion, oxidation, failure analysis, fatigue, weld defect, oxide remover, metallography, fractography

In most material failure analyses, the identification of the mode of failure is important in the determination of the cause of failure. Therefore, the mode of failure or modes of failure must be accurately identified. Fractography, the study of fracture surfaces, has been widely applied by failure analysts because it provides the necessary information needed in the identification of failure modes. This fractographic information is generally obtained by using the scanning electron microscope (SEM).

[1]Formerly, materials engineer, Systems Engineering Associates, 7349 Worthington-Galena Rd., Columbus, OH 43085. Presently, member of technical staff, AT&T Bell Laboratories, 1600 Osgood Street, North Andover, MA 01845.

Fractographic information can be easily obtained from clean fresh fractures. However, since most failures occur in the field during operations, the fracture surfaces of these failures are not always fresh and clean. Material failures in severe operating conditions, for example, at high temperatures or in corrosive environments, may provide little or no fractographic information because of corrosion. On the other hand, failures in mild operating conditions, such as at room temperature and with lubricants, can present the same problem if adequate precautions are not taken to prevent long-term exposure.

The formation of oxides on the fracture surface has been one of the most difficult problems the failure analyst has had to confront during the analyses of failures, because these surface oxides hinder fractographic information retrieval. Another problem is the removal of these oxides from the fracture surface. When these oxides are properly removed, the fractographic information on the original fracture can be obtained. However, severe surface oxidation can destroy the original fracture surface and make fractographic information retrieval impossible.

The ideal oxide removal process must assure complete removal or dissolution of the surface oxide without dissolving or attacking the underlying alloy. Over the years, several oxide removal methods have been developed. Madeyski's [1] hydrogen reduction method is probably the most ideal. However, this method is very involved and requires special apparatus and high temperatures.

Another method which has been proven to be close to ideal is known as "Endoxing" [2,3]. This method works extremely well on ferrous materials when ultrasonic cleaning and electrochemical reduction of the surface oxides are combined in a solution of Endox. However, it has been demonstrated to produce effective results only on ferrous materials. Endox 214 is a commercially available alkaline proprietary product containing sodium cyanide and manufactured by Enthone, Inc. The use of this oxide removal method is not recommended by this author because of the high toxicity of the alkaline deoxidizing solution.

A variety of other oxide removal methods utilizing chemical attack and ultrasonic cleaning have also been developed. Some of these methods can produce excellent results if adequate precautionary measures are taken. Concentrated and diluted orthophosphoric acid is a chemical which has been used successfully; it is recommended in a technical report prepared by McDonnell Douglas for the Air Force Materials Laboratory [4]. Two other commercial proprietary products which have also been used successfully are "Bransonic Ultrasonic Oxide Remover" and "Oakite 33"; they are manufactured by Bransonic Cleaning Equipment Co. of Connecticut and Oakite Products, Inc. of New Jersey. The chemical composition of the Bransonic solution is unknown, while Oakite 33 contains phosphoric acid, solvents, and biodegradable products.

Procedures

Sample preparation should be in accordance with standard metallurgical procedures. The oxidized fracture should be degreased using suitable degreasers such as acetone or ether [5]. Should identification of the corrosion product be necessary, the specimen should be examined with energy dispersive X-ray spectroscopy (EDS) prior to oxide removal. SEM micrographs documenting the condition of the fracture surface should also be taken at this time.

The degreased specimen should be inserted in the selected cleaning solution after the ultrasonic cleaner has been activated. Proprietary products should be mixed with the correct amount of water recommended by the manufacturer. For best results, the sample should be frequently withdrawn from solution and rinsed in distilled water. The wet sample should then be rinsed in acetone and then dried. After drying, the fracture surface should be observed with a low-power optical microscope to determine the extent of rust removal.

Once most of the loose oxides have been removed and the underlying alloy can be observed on most of the fracture surface, cleaning should be stopped. Further attempts to remove tight adherent oxides could result in alloy dissolution and pitting of the fracture surface.

The rate of oxide removal increases with the concentration of the cleaning solution. The probability of pitting any exposed fracture surface also increases with the concentration of cleaning solutions. Orthophosphoric acid has very low pitting tendencies when diluted down to 50% concentration. This acid also removes surface oxides as effectively as the proprietary products which contain surfactants.

Case Histories

Three case histories are presented in this section to show the effects of each of the chemicals utilized in removing surface oxides.

Stainless Steel Aircraft Manifold

A stainless steel manifold which had a dent on one of the pipes (Fig. 1) cracked in service. The crack propagated along the welded joint and radially across the dent. The manifold was fabricated from a titanium stabilized stainless steel type 321 (Table 1). The tight adherent oxide covering the fracture surface was ultrasonically removed in Bransonic solution. Fractographic studies using the SEM revealed intergranular fracture, as shown in Fig. 2a and 2b. Some of the exposed grains were covered with layered precipitates or films. Metallography revealed the presence of grain boundary precipitates and internally distributed precipitates. The effectiveness of this

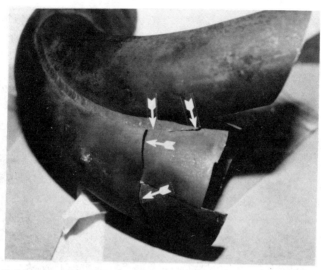

FIG. 1—*Stainless aircraft manifold which cracked along a welded seam and across the dent in one of the pipes (arrows).*

cleaning solution can be inferred from the micrographs in Fig. 2c through 2e showing a minimum oxide thickness of 35 μm and the absence of the morphological characteristics of oxidized stainless steel.

Figure 2c shows intergranular precipitates throughout the 0.89-mm (0.35-in.)-thick stainless steel tube, which was covered with a tight adherent oxide scale. This micrograph was observed in the heat-affected zone of the

TABLE 1—*Chemical composition of aircraft manifold.*

Element	Amount, wt. %
Fe	69.
Cr	17.9000
Ni	9.5000
Mn	1.5800
C	0.0800
Ti	0.9000
W	0.0200
V	0.0700
Nb	0.0310
Mo	0.2200
Al	0.1400
P	0.0270
S	0.0170
Si	0.3800
Cu	0.1500
Sn	0.0110
Co	0.1400
B	0.0007

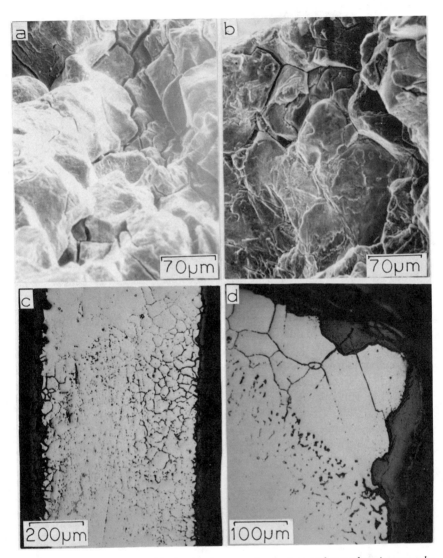

FIG. 2—(a) *SEM micrograph showing intergranular fracture and secondary intergranular cracking.* (b) *SEM micrograph of intergranular fracture. Note carbide film or flames on alloy grains.* (c) *Grain boundary carbides in stabilized stainless steel.* (d) *Grain boundary carbide along fracture surface. Note carbide in oxide layer.*

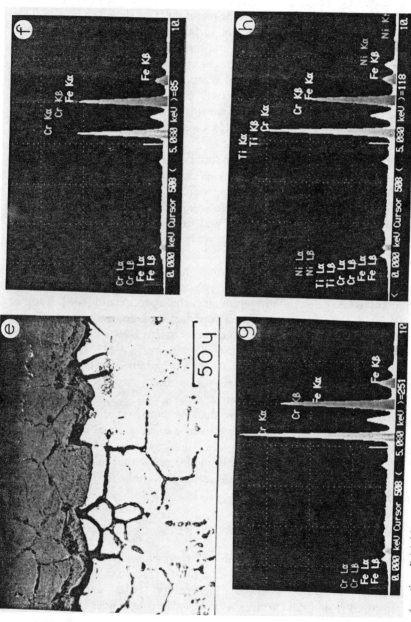

FIG. 2—(continued) (e) Microstructure of stainless steel manifold showing adherent chromia scale with precipitates along prior alloy grain boundaries. Note metallic phase within scale and grain boundary precipitates in alloy. (f) EDS spectrum of scale close to the scale gas interface. (g) EDS spectrum of scale between the scale interfaces. (h) EDS spectrum of scale close to the scale/metal interface.

welded tubes. Grain boundary and internal precipitates can also be observed in Fig. 2d at the fracture surface. In this micrograph, at the upper right corner, a clearly defined grain boundary precipitate can be observed within the surface oxide. A micrograph of the manifold ID is shown in Fig. 2e. This micrograph revealed a tight adherent surface oxide scale which contained darker grain boundary precipitates and metallic stringers or islands along prior alloy grain boundaries. Grain boundary precipitates were also present, ahead of the scale, in the unoxidized alloy.

Identification of these grain boundary precipitates were not conducted because of the limited scope of this investigation. However, thermodynamic and high-temperature oxidation analyses were performed in order to evaluate stable phases.

An efficient aircraft engine will produce combustion products rich in carbon dioxide (CO_2), and other gaseous species will be present in small quantities. Depending on the fuel, sulfur-, vanadium-, and lead-containing species can be present in addition to nitrogen compounds. This exhaust gas containing multioxidants, especially oxygen and carbon, will contact the stainless steel manifold, which attains temperatures between 873 and 1073 K (600 and 800°C) during operation.

In this temperature range, iron chromium alloys operating in a dual oxidizing environment (carbon and oxygen) have been known to form stable chromium oxide scales and intergranular carbide precipitates [6–8]. Another known fact is the formation of grain boundary carbides in the heat-affected zones of welded stainless steels with carbon content higher than 0.02 weight %, even in the presence of strong carbide stabilizers [9–11]. The formation of these carbides is known to reduce the chromium content at the grain boundaries, resulting in excessive corrosion along these boundaries. This process is generally known as sensitization.

In most environments containing dual oxidants, especially oxygen and carbon, the formation of a complete surface oxide scale of chromium has been known to stifle the growth of grain boundary carbides by hindering the diffusion of carbon. However, any carbide formed at the grain boundary will be displaced by the advancing oxide scale. The Gibbs free energy for this reaction

$$4M_{23}C_6 + 69O_2 \rightarrow 46M_2O_3 + 24\underline{C} \qquad (1)$$

is negative when M is chromium and $\underline{C}$ represents dissolved carbon [12]. This reaction will therefore generate carbon to react with chromium ahead of the scale. The kinetics of this favorable reaction is fast, and ultimately, in the sensitized region, all carbides will be replaced by oxides [8].

The oxygen activity at the chromia scale/gas interphase was probably lower than that necessary for the formation of the spinel ($FeCr_2O_4$) because the brownish or reddish color of this phase was not observed during metallography [13]. Furthermore, the presence of a metallic phase (Fe) within the chromia scale (Fig. 2e) is also indicative of oxygen activity lower than that necessary for the oxidation of iron or the formation of a spinel ($P_{O_2} \ll 10^{-3}$ atm). EDS was performed on the scale. The spectrum shown in Fig. 2f reveals the presence of iron and chromium in this scale; however, the amount of chromium increased upon approaching the scale/metal interface (Fig. 2g). Seybolt [13] has confirmed the existence of a rhombohedral solid solution of ferric oxide (Fe_2O_3) and chromium sequioxide (Cr_2O_3) and showed that this oxide phase can exist in equilibrium with iron.

In a complex low-oxygen-activity environment, such as the one inside an aircraft manifold, the displacement of carbides formed by sensitization or inward diffusing carbon is expected as just indicated [1]. However, the presence of a darker phase within the chromia scale (see Fig. 2d and 2e) can be interpreted as either the inability of this displacement reaction to take place for kinetic reasons, which is in direct opposition to the observation of Page et al. [8], or the formation of a new phase (probably an oxycarbide of chromium), or from Fig. 2h, one of the 18 phases of chromium titanate (Cr_2TiO_5) or chromium nitrides. The scale was also identified as a chromia phase during metallography by its characteristic greenish color under polarized light. The grain boundary precipitates were definitely darker than carbides and clearly identifiable within the chromia scale. Although there are no known thermodynamic data on oxycarbides of chromium, its existence in a low-oxygen-activity environment has been suspected.[2]

The existence of an oxycarbide phase with the formula $CrO_{1.13}O_{0.12}$ has been reported [14]. The reported oxygen content of a ferrous carbidic product, using Auger Electron Spectroscopy (AES) technique, was 12.2 atomic % and the C/O ratio (atomic percent) was 2.4, an indication of the formation of a possible oxycarbide phase [15]. Due to the low titanium content of this alloy, the dark grain boundary precipitates were probably oxycarbide phases of chromium ($M_xO_yC_z$) where M can be Cr, Mo, and Ti combined to satisfy the stoichiometric relation.

Under the operating conditions of this stainless steel tube, brittle intergranular fracture can only occur in the presence of embrittling species like sulfur and phosphorus or a brittle phase at the grain boundaries. Carbides and oxides are known for their high fracture strength but low ductility. Probably a very important property is the strength of the bond between a precipitate and the alloy. Flaking and cracking of oxide scales due to thermal cycling is generally known, therefore cracking of this grain boundary precipitates in localized areas of high strain, due to the dent, should be expected. Crack propagation can occur by thermal cycling but probably more by vibratory stresses generated by the aircraft engine. In order to prevent this

---

[2]Hirth, J. P., private communication, Ohio State University, 1982.

form of failure, extra low carbon stainless steel should be used. This will prevent sensitization in the heat affected zone (HAZ) and possible extensive grain boundary precipitation.

High Strength Low Alloy (HSLA) Steel Equalizer Beam

An equalizer beam, part of a truck suspension, failed in service. This failure occurred at the rear axle locator, as can be seen in Fig. 1. The fractures were at the holes which accommodated the rear axle bushing restraining bolts. The effective cross-sectional area of the axle locator was, therefore, reduced at each of these holes. The inner surface of the broken piece was fairly worn (compare Fig. 3a and 3b), indicating it was supporting the axle load.

The fracture at the neck of the axle locator was mostly oxidized while the other fracture was bright and shiny, indicating a more recent fracture. The mostly oxidized fracture showed evidence of beach marks and many ratchet marks, characteristic of fatigue with multiple crack origins. The final fracture was not oxidized and was bright and shiny, therefore suggestive of a fatigue failure.

FIG. 3—(a) *Axle locator on equalizer beam after failure.* (b) *Surfaces of broken axle locator. Note wear on the locator surface caused by axle bearing.*

Fractographic examination revealed an oxide (iron) covered surface, as shown in Fig. 4a. Oxide removal was accomplished ultrasonically in an "Oakite 33" solution. The cleaned surface showed evidence of pitting and etching which revealed a pearlitic microstructure (see Fig. 4b). Figure 4c revealed evidence of fatigue striations which could not be attributed to microstructural etching because the striations are continuous across a grain

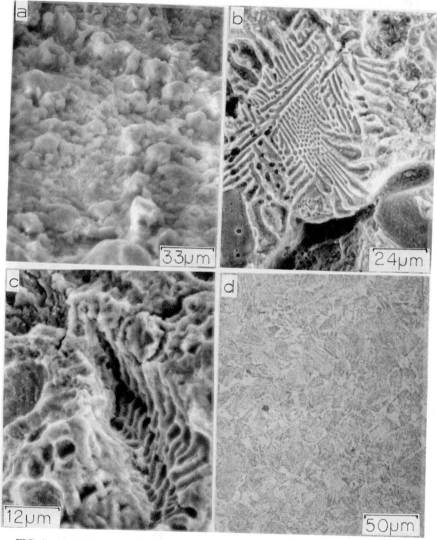

FIG. 4—(a) SEM micrograph of the oxidized fracture surface. (b) SEM micrograph of fracture surface after oxide removal. (c) Evidence of fatigue striations across a grain boundary on partially cleaned surface. (d) Microstructure of equalizer beam showing a pearlite and proeutectoid ferrite.

boundary. The microstructural information obtained from optical metallography, Fig. 4d, confirmed a pearlitic microstructure with proeutectoid ferrite at prior austenite grain boundaries. The striations observed across the grain boundary were, therefore, attributed to fatigue instead of to etching resulting from chemical attack.

Installation of the equalizer beam in the upside-down position exposed an area of smaller cross section to fluctuating stresses which were higher than the design stresses. This failure was, therefore, caused by improper installation of the equalizer beam.

HSLA Steel in Welded Roof Joists

The roof of a building collapsed after a heavy rain. Although the design stresses in the structure were compromised, it was established that the roof failure occurred well below the rated roof load. Examination of the wreckage revealed some fractured welds in the joints. These fractured joists were located around the area of highest loads.

The fracture surfaces were fairly rusted at the time of analysis, as indicated by the brownish appearance of one of the joists shown in Fig. 5a. Also shown in this photograph (arrow) is evidence of gross overlapping. Inspection of other welds indicates some undercutting and lack of penetration.

The extracted specimens were ultrasonically cleaned in a 50% solution of orthophosphoric acid for fractographic examination. Gas pores observed on the fracture surface are shown in Fig. 5b. Slag or oxide inclusions were also present in the fractured weld. An SEM micrograph of a spherical oxide inclusion is presented in Fig. 5c. This micrograph and its higher magnification (Fig. 5d) show two modes of failure, ductile and brittle, adjacent to one another. Microstructural information confirmed the presence of nonmetallic inclusions and gas pores (Fig. 6), as well as large ferrite grains and coarse pearlite at the fracture surface in the fusion zone (Fig. 6a). The observations just presented show that the welded joists which failed contained numerous weld defects.

Poor welding practice and quality control were responsible for this failure. The porosity observed indicates that wet electrodes, a source of hydrogen, were probably used in welding the joists. The dual mode of failure (Fig. 5) supports hydrogen embrittlement. Gross overlapping, undercutting, and lack of penetration should have been detected and reported, had it been inspected prior to or after painting.

Summary

Fracture surface oxides were removed successfully without any major damage to the underlying alloy using proprietary chemicals available com-

FIG. 5—(a) *Roof joist after failure showing oxidized fracture surface and a weld overlap* (arrow). (b) *SEM micrograph of fracture surface showing gas pores.* (c) *SEM micrograph of fracture showing a nonmetallic spherical inclusion. Note dual mode of failure.* (d) *Micrograph of fracture surface showing ductile and cleavage fracture.*

mercially and laboratory grade orthophosphoric acid. The oxide removal process was facilitated by ultrasonic effects.

Fractographic information was retrieved from the corroded fracture surfaces in the three case histories presented because the corrosion was not severe enough to destroy the fracture surfaces. Fractographic information retrieval was also facilitated by the oxide remover, which did not vigorously

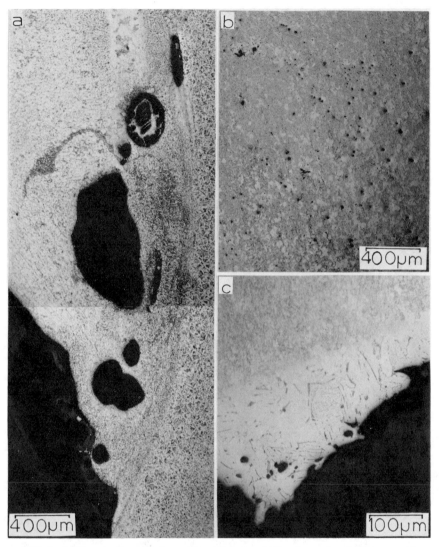

FIG. 6—(a) *Porosity, large ferrite grains, and coarse pearlite observed at the fracture surface* (*nital*). (b) *Weld containing gas pores and inclusions* (*nital*). (c) *Microporosity observed in the fusion zone* (*nital*).

attack the underlying alloy. In each of the three case histories, oxide removal revealed the following:

1. Intergranular fracture of a stabilized stainless steel aircraft manifold, adjacent to a weld, caused by poor material selection, improper welding, and heat-treating practice.

2. Beach marks and evidence of fatigue striations on a fractured HSLA steel equalizer beam caused by improper installation.

3. Porosity, slag inclusions, and evidence of hydrogen embrittlement in a welded HSLA steel roof joist caused by poor welding and quality control practices.

References

[1] Madeyski, A., *Pracktische Metallographie*, Vol. 17, 1980, pp. 598–607.

[2] Yuzawich, P. M. and Hughes, C. W., *Pracktische Metallaographie*, Vol. 15, 1978, pp. 184–195.

[3] Hughes, C. W. and Yuzawich, P. M., " Improved Technique for Removal of Oxide Scale from Fracture Surfaces of Ferrous Metals," presented at the Metals Congress, Philadelphia, PA, 1985.

[4] *SEM/TEM Fractographic Handbook*, Technical Report No. F33615-74-C-5004, prepared by McDonnell Douglass for the Air Force Materials Laboratory, G. F. Pittianto, V. Kerlins, A. Phillips, and M. A. Russo, Eds., prepared by MCIC, Dec. 1975, p. 6.

[5] *Metals Handbook*, Vol. 10, 8th ed., American Society for Metals, Metals Park, OH, 1975, pp. 10–29.

[6] Rapp, R. A., Colwell, J., and Ibidunni, A. O., "Corrosion Chemistry in Low-Oxygen Activity Atmospheres Characteristic of Gasified of Incompletely Combusted Coal," EPRI Report No. RP716-2, Electric Power Research Institute, Palo Alto, CA, 26 April 1979.

[7] Pettit, F. S., Goebel, J. A., and Goward, G. W., *Corrosion Science*, Vol. 9, 1963, p. 903.

[8] Page, R. A., Hack, J. E., and Brown, R. D., *Metallurgical Transactions A*, 1984, Vol. 15A, pp. 11–22.

[9] Fontana, M. G. and Greene, N. E., *Corrosion Engineering*, McGraw-Hill Book Co., NY, 1976, pp. 58–66.

[10] *Metals Handbook*, Vol. 3, 9th ed., American Society for Metals, Metals Park, OH, 1980, pp. 57–59.

[11] Hanneman, R. E. and Rao, P. in *Environment-Sensitive Fracture of Engineering Materials*, Z. A. Foroulis, Ed., The Metallurgical Society, Institute of Mining, Metallurgical and Petroleum Engineers, New York, 1979, pp. 153–177.

[12] Kubaschewski, O. and Alcock, C. B., *Metallurgical Thermochemistry*, 5th ed., Pergamon Press, Elmsford, NY, 1979.

[13] Seybolt, A. U., *Journal of the Electrochemical Society*, Vol. 107, No. 3, March 1960, pp. 147–156.

[14] *Inorganic Phases*, JCPDS, International Center for Diffraction Data, Swarthmore, PA, 1984, File No. 21-239.

[15] Ibidunni, A. O. and St. Pierre, G. R., "Carbon Deposition by the Decomposition of Carbon Monoxide on Reduced Iron Oxide," presented at the AIME Annual Conference, Los Angeles, CA, 1984, American Institute of Mining, Metallurgical and Petroleum Engineers, New York.

Metals II: Nonferrous Alloys

Srivathsan Venkataraman,[1] Theodore Nicholas,[2] and Noel E. Ashbaugh[3]

Micromechanisms of Major/Minor Cycle Fatigue Crack Growth in Inconel 718

REFERENCE: Venkataraman, S., Nicholas, T., and Ashbaugh, N. E., "**Micromechanisms of Major/Minor Cycle Fatigue Crack Growth in Inconel 718.**" *Fractography of Modern Engineering Materials: Composites and Metals, ASTM STP 948*, J. E. Masters and J. J. Au, Eds., American Society for Testing and Materials, Philadelphia, 1987, pp. 383–399.

ABSTRACT: Fatigue crack growth tests under major/minor cycle loading were conducted on Inconel 718 at 649°C using center-cracked tension specimens, $M(T)$. The loading waveform used consisted of a major cycle loading (1 Hz) and a hold at the major cycle maximum load, with a superimposed minor cycle loading (10 Hz) during the hold period. Crack growth data were obtained for a range of major and minor cycle stress intensities. Constant load amplitude tests were also run at various R-values to determine the effect of R on major/minor cycle fatigue. Detailed fractographic investigation was conducted on selected specimens using a scanning electron microscope to study the associated micromechanisms. Creep processes as indicated by an intergranular fracture mode were found to be dominant at lower minor cycle stress intensity ranges and higher values of R for the minor load. Transition to fatigue-dominant mechanisms as indicated by a transgranular fracture mode occurred at higher stress intensity ranges and lower R-values for the minor load. The minor cycle stress intensity range at which this transition occurred varied with the major cycle stress intensity range.

KEYWORDS: major/minor cycle, fatigue crack growth, nickel base superalloy, elevated temperature, fractography, micromechanism, load ratio (R), creep-fatigue interaction

Crack growth rate calculations for critical structural components in U.S. Air Force gas turbine engines under actual operating conditions are necessary for design and life management procedures. Programs such as Retirement-for-Cause [1] and the Air Force Engine Structural Integrity Program

[1]Resident National Research Council research associate, Materials Laboratory, Air Force Wright Aeronautical Laboratories, Wright-Patterson Air Force Base, Dayton, OH 45433.
[2]Senior scientist, Materials Laboratory, Air Force Wright Aeronautical Laboratories, Wright-Patterson Air Force Base, Dayton, OH 45433.
[3]Group leader, University of Dayton Research Institute, Dayton, OH 45429.

(ENSIP) require life prediction analysis based on crack growth rates calculated from initial flaw sizes using fracture mechanics concepts. Such calculations must be performed under typical engine-operating conditions. These operating conditions include variations in temperature, frequency, load ratio (R), and sustained load hold time. A typical loading profile experienced by an aircraft gas turbine engine consists basically of low-frequency loadings associated with thermal gradients and centrifugal forces and of a superimposed high-frequency loading associated with vibratory stresses.

In modern day, high-performance aircraft, the temperatures experienced by engine materials are ever increasing while stress amplitudes in the neighborhood of material yield strength are often experienced at various locations. As a result, low-frequency loading, also designated here as a major cycle loading, has a large enough stress amplitude to cause low-cycle fatigue failure. Additionally, this major cycle loading can be altered by superimposed high-frequency vibratory loading, designated here as minor cycle loading. It is important for life prediction analyses to be able to determine the effects of this high-frequency loading on the fatigue crack growth behavior of typical engine disk materials such as Inconel 718.

Fatigue crack growth under combined major/minor cycle loading has been investigated for several materials in the past. The loading consisted of a major load cycle with large amplitude and low frequency cycle and with a sustained load time at the maximum load. The minor load cycle, superimposed during the hold time, had low amplitude and high frequency. Powell et al. [2] have conducted a series of major/minor fatigue tests on Ti-6Al-4V to determine the threshold amplitude of the minor cycle stress intensity range at which the onset of the influence of the minor cycles occurred. Brown and Goodman [3] and Petrovich [4] have also studied the major/minor cycle fatigue crack growth in Inconel 718 sheet material over a wide range of major and minor cycle stress intensity ranges, hold times, and minor cycle frequencies. Both have proposed a linear summation model that predicts the effect of stress ratio, hold time, and minor cycle frequency on the crack growth rate.

Several investigators [5–7] have shown that the elevated temperature fatigue crack growth rate of nickel-base superalloys decreases with increasing frequency in the regime in which environmental and time-dependent deformation processes operate. Fractography in these studies revealed that this decrease in fatigue crack growth rate is accompanied by a change in fracture mode from intergranular to transgranular. In addition, Solomon and Coffin [8] and Weerasooriya [6] have shown that for test frequencies above a certain frequency the fatigue crack growth rate is independent of frequency.

The R-value has also been shown to influence the elevated temperature fatigue crack growth properties [9,10]. This is rationalized by two different mechanisms operable at the two extremes of R. At low R, the crack growth behavior is essentially cycle dependent, and mean stress has little influence on growth rate. At high R-values, however, mean stress becomes much more important as the mode of crack growth gradually changes toward purely time-dependent, sustained load crack growth. At high R-values, therefore, there are two contributions to the crack growth process, namely, the cyclic contribution due to stress amplitude and the time-dependent contribution due to mean load.

In major/minor cycle loading, there are potentially three contributions to the crack growth rate. First, low-frequency loading contributes a cyclic portion which may be anywhere from cycle-dependent at low-temperature to time-dependent at high temperatures and low frequencies. Second, the hold time at mean load of the minor cycles may contribute a time-dependent portion. Finally, the minor cycle has its contribution. The interaction of the major and minor load cycles on the fatigue crack growth behavior of Inconel 718 at elevated temperature has been studied in this investigation. This paper presents observations of the various crack growth micromechanisms operating under these conditions and their relationship to measured crack growth rates.

Experimental Procedure

A series of major/minor cycle fatigue crack growth tests were conducted on Inconel 718, a nickel-base superalloy. The composition and the heat treatment of this alloy is shown in Table 1. Center-cracked tension, $M(T)$, specimens of dimensions shown in Fig. 1 were machined out of the heat-treated sheet material of thickness 2.36 mm. The center hole and the starter notch were machined using an electrodischarge machining technique. Tests

TABLE 1—*Chemical composition and heat treatment for Inconel 718.*

CHEMICAL COMPOSITION			
Element	Weight %	Element	Weight %
Ni	50 to 55	Cr	17 to 21
Fe	Balance	Nb + Ta	4.75 to 5.5
Mo	2.8 to 3.3	Ti	0.65 to 1.15
Al	0.2 to 0.8	Co	1.0
C	0.08	Mn	0.35
Si	0.35	Cu	0.30
Ph	0.015	S	0.015
B	0.006		

HEAT TREATMENT
Solution at 968°C (1775°F) for 1 h
Air-cooled to 718°C (1325°F)
Held at 718°C (1325°F) for 8 h
Furnace-cooled at a rate of 38°C (100°F)/h to 621°C (1150°F).
Held at 621°C (1150°F) for 8 h
Air-cooled to room temperature

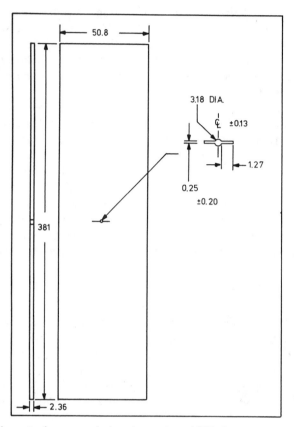

FIG. 1—*Schematic of center-cracked tension specimen*, M(T). *Dimensions are in millimeters.*

were carried out in an automated servohydraulic testing machine under computer control. The specimen was maintained at a constant temperature of 649°C in a resistance-heated oven equipped with quartz viewing ports through which optical crack length could be periodically monitored with the aid of a travelling microscope. Primary crack length data were obtained using an a-c electric potential technique [3]. Derived crack lengths were verified through optical surface crack length measurements.

Precracking was conducted following ASTM Method for Constant-Load-Amplitude Fatigue Crack Growth Rates Above 10^{-8} m/cycle (E 647-83). The loading waveform used for the major/minor cycle fatigue testing is shown in Fig. 2. It consisted of a 1-Hz, low-frequency major cycle loading with a 180-s hold at the maximum load. The R of the major cycle was maintained at 0.1. The 10-Hz minor cycle loading was then superimposed on the major cycle hold time load. Stress intensity was used as the controlling parameter throughout the tests. The stress intensity range for the major

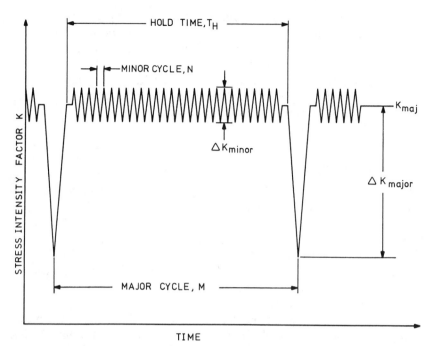

FIG. 2—*Schematic representation of the simple load spectrum used in major/minor cycle fatigue tests.*

cycle, ΔK major, was maintained constant during individual tests. This was done by decreasing the major cycle load as the crack length increased. Likewise, the minor cycle stress intensity range, ΔK minor, was initiated at a higher level and was decreased linearly with increasing crack length. The test parameters are shown in Table 2.

A series of constant load amplitude type tests was also carried out at the same 10 Hz frequency of the superimposed minor cycles. These tests used

TABLE 2—*Major/minor cycle fatigue test parameters.*

No.	ΔK Major, MPa$\sqrt{m}$	F Major, Hz	R Major	Hold Time, s	ΔK Minor, MPa$\sqrt{m}$	F Minor, Hz
1	30 to 90	1	0.1	0	0	0
2	40 to 110	1	0.1	600	0	0
3	15	1	0.1	180	10 to 30	10
4	20	1	0.1	180	5 to 20	10
5	25	1	0.1	180	6 to 30	10
6	30	1	0.1	180	0 to 25	10
7	35	1	0.1	180	0 to 25	10
8	40	1	0.1	180	0 to 17	10

NOTE: Maximum stress intensity for combined loading = $K_{\text{Major}}^{\text{Max}} + \frac{1}{2}\Delta K_{\text{Minor}}$.

center-cracked tension, $M(T)$, specimens at the test temperature of 649°C. Values for R for these tests were chosen between 0.1 and 0.9. These tests differed from the major/minor cycle tests in that there was no major load excursion and no variation in R during each test.

To understand the micromechanisms of major/minor cycle fatigue crack growth, detailed fractographic analyses were carried out on all fractured specimens using a scanning electron microscope.

Results and Discussion

The major/minor cycle fatigue crack growth rate data of Inconel 718 are shown in Fig. 3 in the form of da/dM versus ΔK minor relationships. The da/dM represents the crack growth rate per major cycle, and ΔK minor is defined in Fig. 2. The maximum stress intensity of the minor cycle is thus $K_{Major} + \frac{1}{2}\Delta K_{Minor}$. The minor cycle stress intensity range is used as the independent variable since the main interest is in determining the influence of the amplitude of the minor cycles on the fatigue crack propagation behavior of the major cycles. This can also be interpreted as an influence of R on the crack growth rate. Since the ΔK major level was maintained constant during each major/minor cycle fatigue experiment, the effect of minor cycle loading is more easily determined. As seen in Fig. 3, in the ΔK minor range above approximately 6 MPa$\sqrt{m}$ there is a continuous increase in the fatigue crack growth rate with increasing minor cycle stress intensity range. Since the mean stress intensity remains constant in each test, R decreases with increasing minor cycle amplitude. The minor cycle contribution to the fatigue crack growth process is predominant above approximately 6 MPa$\sqrt{m}$ for all values of ΔK major. Increase in the major cycle stress intensity range also resulted in increased crack growth rates. This is due to a combination of the cyclic contribution of the major cycle stress amplitude and the increase in the stress intensity at the hold time. The latter results in an increase in the sustained load crack growth rate contribution. Finally, a slight decrease in the crack growth rate was observed with an increasing minor cycle amplitude at the minor cycle stress intensity range between 2 and 6 MPa$\sqrt{m}$ for tests where the major cycle stress intensity range was above 35 MPa$\sqrt{m}$. A similar decrease in crack growth rate with minor cycle amplitude has also been reported by Petrovich [4] on the same material. An analogous phenomenon has also been observed by Venkiteswaran et al. [11], who reported a reduction in creep rate due to an applied fatigue cycle for precipitation hardened Inconel X-750. They attributed this reduction to the complex dislocation tangles and vacancy condensation along dislocation lines.

The data of Fig. 3 can be replotted as crack growth rate (da/dt) against

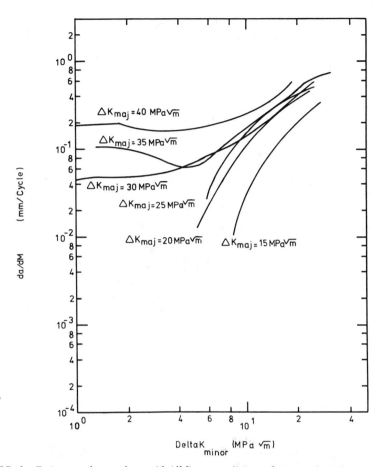

FIG. 3—*Fatigue crack growth rate* (da/dM) *versus minor cycles stress intensity range* (ΔK *minor*) *plots for Inconel 718 tested at 649°C and major cycle stress intensity range between 15 and 40 MPa*$\sqrt{m}$. R *minor decreases with increasing minor cycle stress intensity range.*

minor load R for constant values of ΔK major as shown in Fig. 4. Also shown in the figure are the points from sustained load crack growth tests, which are plotted as being equivalent to $R = 1$. Low ΔK minor levels corresponding to high R minor values appear to influence the crack growth behavior of the major cycle alone ($R = 1$). As shown in Fig. 4, as R decreases, that is, as the minor cycle amplitude is increased, a slight decrease in the crack growth rate results. The minimum growth rate occurs in the R minor range of 0.8 to 0.9 for ΔK major of 40 MPa$\sqrt{m}$. For the other two cases of ΔK major equal to 20 and 30 MPa$\sqrt{m}$, it is not as obvious where the minimum occurs because of insufficient data, but a minimum is

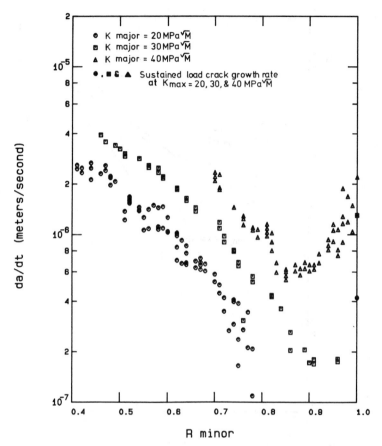

FIG. 4—*Fatigue crack growth rate* (da/dt) *versus minor load* R-*value* (R *minor*) *plots for Inconel 718 tested at 649°C.*

apparent by observing the values of *da/dt* for $R = 1$ (sustained load crack growth at the major cycle maximum stress intensity). An interaction between the fatigue and creep contributions to the crack growth process apparently takes place in this region to cause this behavior.

To establish a baseline for the major cycle behavior alone (without hold time), a specimen was tested under constant load range at a frequency of 1 Hz, $R = 0.1$, and ΔK major level from 30 to 110 MPa$\sqrt{m}$. Fractographic examination of this failed specimen tested with major cycles alone without hold times revealed two types of fracture morphology. First, in the low ΔK major region (approximately 40 MPa$\sqrt{m}$), an intergranular morphology was observed as shown in Fig. 5a. On the other hand, at higher ΔK major levels (approximately 90 MPa$\sqrt{m}$), a transgranular morphology was

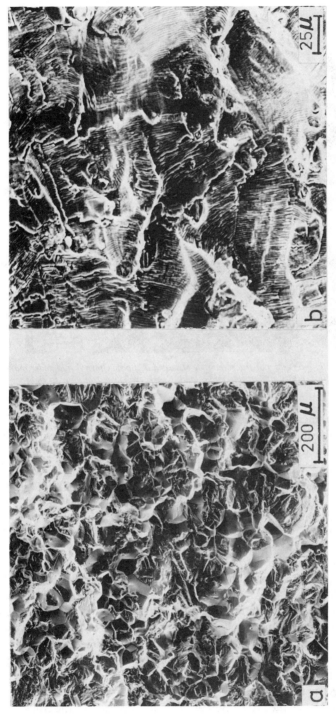

FIG. 5—Fracture morphologies observed in Inconel 718 specimen tested under major cycle loading alone. An intergranular fracture mode was observed at lower major cycle stress intensity ranges (a). On the other hand, a transgranular fracture mode was observed at higher major cycle stress intensity ranges (b). Fatigue striations were also found in the latter case. Crack growth is from left to right.

FIG. 6—*Totally intergranular fracture mode observed in the specimen tested under major cycle loading alone with a 600-s hold at maximum load. Crack growth is from left to right.*

observed as indicated in Fig. 5*b*. Fatigue striations were also visible in this region, and the spacing between the striations matched closely with the crack growth measured per major cycle.

Application of the hold time of 600 seconds at the maximum load of the major cycle resulted in an intergranular fracture morphology for the entire stress intensity range tested between 40 and 110 MPa$\sqrt{\text{m}}$ as indicated in Fig. 6. No transition to the transgranular fracture morphology at higher major cycle stress intensity range levels was observed with the application of the hold time. In such cases the crack growth is controlled by creep processes due to the major cycle hold load.

The change in fracture morphology can be seen by comparing results at two different values of ΔK minor. The major/minor cycle fatigue sample tested at the major cycle stress intensity range of 20 MPa$\sqrt{\text{m}}$ revealed an intergranular fracture morphology at lower ΔK minor levels (approximately 9 MPa$\sqrt{\text{m}}$) as shown in Fig. 7*a*. The corresponding R minor was 0.7. At higher ΔK minor levels (approximately 20 MPa$\sqrt{\text{m}}$), however, a mixed mode fracture morphology indicating a transition to the transgranular mode was observed as indicated in Fig. 7*b*. The R minor ratio for this ΔK minor level was 0.4. Hence the interaction of the minor cycles with the crack growth

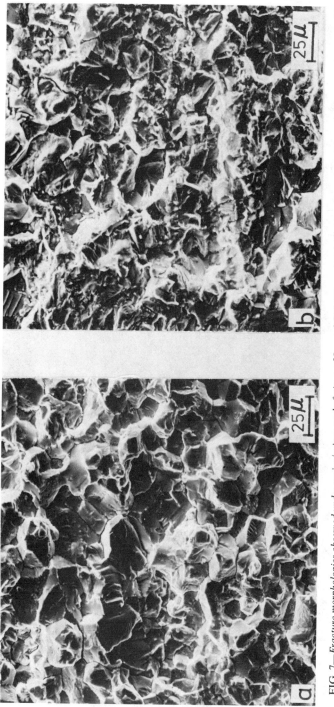

FIG. 7—Fracture morphologies observed under major cycle fatigue of Inconel 718. The major cycle stress intensity range for the test was 20 MPa√m. (a) Intergranular fracture morphology corresponding to a minor load R-value of 0.7. (b) Mixed mode (transgranular plus intergranular) fracture morphologies corresponding to a minor load R-value of 0.4. Crack growth is from left to right.

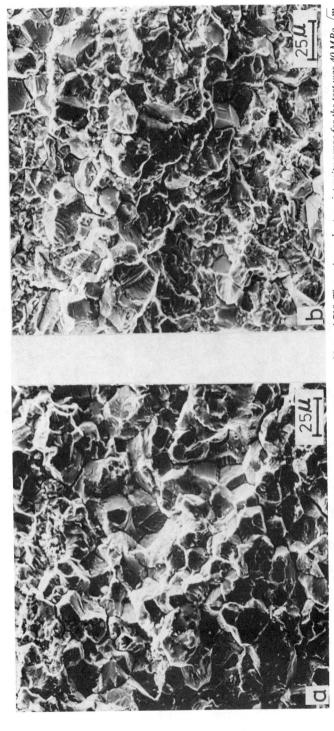

FIG. 8—*Fracture morphologies observed under major/minor cycle fatigue of Inconel 718. The major cycle stress intensity range for the test was 40 MPa√m. Fracture mode was intergranular for the entire test. The minor load R-values corresponding to the fractographs were (a) 0.8 and (b) 0.7. Crack growth is from left to right.*

process occurs at lower minor cycle R-values for these low values of ΔK major. On the other hand, for the test run with a major cycle stress intensity range of 40 MPa$\sqrt{\text{m}}$, the fracture morphology was totally intergranular for the entire minor cycle stress intensity range studied, as shown in Fig. 8a and 8b. The minor cycle R, however, was above 0.7 for the entire test. In this situation, the major cycle, which includes the 180-s hold time, appears to dominate the crack growth behavior as seen in Fig. 3. In fact, work by Weerasooriya and Nicholas [12] has shown that for hold times beyond approximately 10 s for this type of major cycle, the contribution of the sustained load portion of the cycle to crack growth exceeds that of the cyclic excursion for ΔK major = 40 MPa$\sqrt{\text{m}}$ and $R = 0.1$.

The fractographic analysis of major/minor cycle specimens indicates a strong influence of R on the crack growth mechanism. This can be illustrated by examining the fracture morphology at different locations in tests that were run under conditions of constant load amplitude, that is, constant values of minor cycle R. There are no major cycle excursions in these tests. For the constant R tests conducted, the fracture morphology was totally intergranular at R-values greater than 0.7 as illustrated in Fig. 9a, 9b, and 9c. This figure shows results for minor cycle $R = 0.8$. The experimental data for da/dN versus ΔK are shown in Fig. 9c along with the two points corresponding to the location of the fractographs shown. The mean loads corresponded to K-values ranging from 21 to 41 MPa$\sqrt{\text{m}}$. For tests conducted at R-values below 0.7, at low-stress intensity ranges, fracture morphology was again intergranular. At high-stress intensity ranges, however, a transgranular mode of failure was observed as shown in Fig. 10A and 10B. The stress intensity range at which this transition occurred decreased with decreasing R ratios due to an accompanying reduction in the mean stress effect.

Conclusion

1. Major/minor cycle fatigue crack growth data of Inconel 718 at 649°C indicated a strong minor cycle interaction at higher minor cycle stress intensities. A slight reduction in growth rate was also observed at lower minor cycle stress intensities and at R-values between 0.7 and 0.9.

2. The major/minor cycle stress intensity ranges as well as R minor had a strong influence in determining the operating micromechanism. At R-values over 0.7 the crack growth was mainly due to the creep (integranular fracture mode) contribution, while the fatigue (transgranular fracture mode) contribution was a major factor at R-values less than 0.7 and high minor cycle stress intensity ranges.

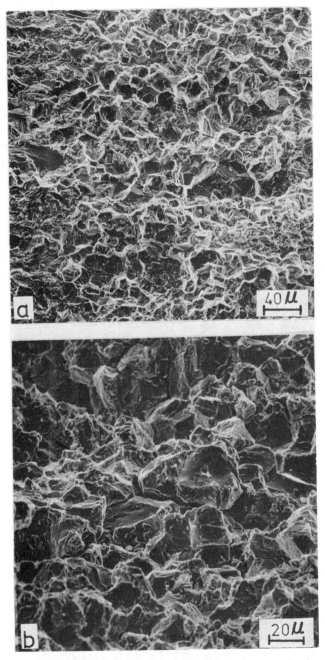

FIG. 9—*Fracture morphology observed in the Inconel 718 specimen tested under constant load amplitude with* R = 0.8 *and frequency* = 10 Hz. *Test parameters corresponding to the fractographs are indicated in the* da/dN *versus* ΔK *plot. Totally intergranular fracture mode was observed for the entire stress intensity range studied. Crack growth is from left to right.*

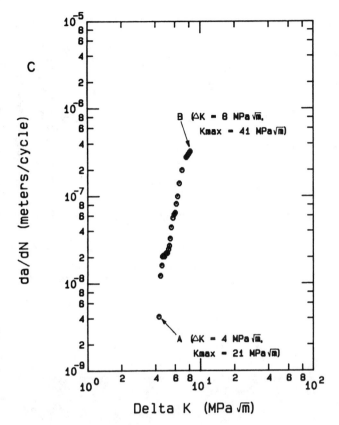

FIG. 9—*Continued.*

Acknowledgments

S. Venkataraman would like to thank the National Research Council for sponsoring his research work. The authors would like to express their appreciation to R. C. Goodman and A. M. Brown for their contribution to the experimental work covered in this paper. This work was supported by the U.S. Air Force under Project 2302P1.

References

[1] Larsen, J. M. and Nicholas, T., "Cumulative-Damage Modeling of Fatigue Crack Growth in Turbine Engine Materials," *Engineering Fracture Mechanics*, Vol. 22, No. 4, 1985, pp. 713–730.

[2] Powell, B. E., Duggan, T. V., and Ieal, R. H., "The Influence of Minor Cycles on Low Cycle Fatigue Crack Propagation," *International Journal of Fatigue*, Vol. 4, No. 1, 1982.

[3] Goodman, R. C. and Brown, A. M., "High Frequency Fatigue of Turbine Blade Materials," Report No. AFWAL-TR-82-4151, Materials Laboratory, Air Force Wright Aeronautical Laboratories, Dayton, OH, Oct. 1982.

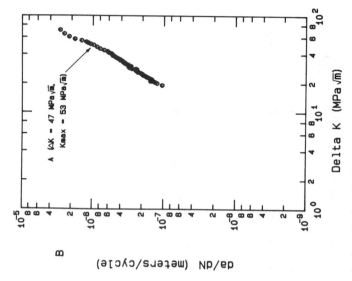

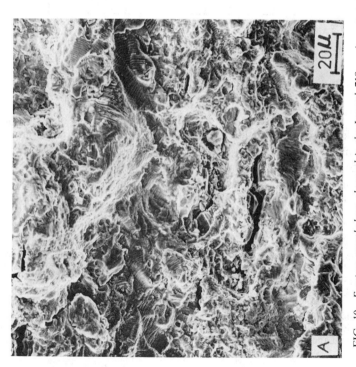

FIG. 10—*Fracture morphology observed in the Inconel 718 specimen tested under constant load amplitude with R = 0.1 and frequency = 10 Hz. Test parameters corresponding to the fractograph is indicated in the da/dN versus ΔK plot. Fracture mode was transgranular with vivid fatigue striations. Crack growth is from left to right.*

[4] Petrovich, A., "Interactive Effects of High- and Low-Frequency Loading of Fatigue," Report No. AFWAL-TR-85-4045, Materials Laboratory, Air Force Wright Aeronautical Laboratories, Dayton, OH, May 1985.

[5] Clavel, M. and Pineau, A., "Frequency and Wave Form Effects on the Fatigue Crack Growth Behavior of Alloy 718 at 298°K and 823°K," *Metallurgical Transactions*, Vol. 9A, 1978, p. 471.

[6] Weerasooriya, T., University of Dayton Research Institute, Dayton, OH, unpublished research. (Figure 14 of Ref 1.)

[7] Scarlin, R. B., "Effects of Loading Frequency and Environment on High Temperature Fatigue Crack Growth in Nickel-Base Alloys," proceedings of International Conference on Fracture, Waterloo, Canada, June 1977, *Fracture 1977*, Vol. 2, Pergamon Press, New York, p. 849.

[8] Solomon, H. D. and Coffin, L. F., Jr., "Effects of Frequency and Environment on Fatigue Crack Growth in A-286 at 100°F," in *Fatigue at Elevated Temperatures, ASTM STP 520*, 1973, p. 112.

[9] Nicholas, T., Laflen, J. H., and VanStone, R. H., "A Damage Tolerant Design Approach to Turbine Engine Life Prediction," AFWAL report, Air Force Wright Aeronautical Laboratories, Dayton, OH, 1985.

[10] Shahinian, P. and Sadananda, K., "Effects of Stress Ratio and Hold Time on Fatigue Crack Growth in Alloy 718," *Journal of Engineering Materials and Technology*, Vol. 101, July 1979, p. 224.

[11] Venkiteswaran, P. K., Ferguson, D. C., and Taplin, D. M. R., "Combined Creep-Fatigue Behavior of Inconel Alloy X-750," in *Fatigue at Elevated Temperatures, ASTM STP 520*, 1973, pp. 462–472.

[12] Weerasooriya, T. and Nicholas, T., "Hold-Time Effects in Elevated Temperature Fatigue Crack Propagation," Report No. AFWAL-TR-84-4184, Materials Laboratory, Air Force Wright Aeronautical Laboratories, Dayton, OH, March 1985.

Dale A. Meyn[1] and Robert A. Bayles[1]

Fractographic Analysis of Hydrogen-Assisted Cracking in Alpha-Beta Titanium Alloys

REFERENCE: Meyn, D. A. and Bayles, R. A., **"Fractographic Analysis of Hydrogen-Assisted Cracking in Alpha-Beta Titanium Alloys,"** *Fractography of Modern Engineering Materials: Composites and Metals, ASTM STP 948*, J. E. Masters and J. J. Au, Eds., American Society for Testing and Materials, Philadelphia, 1987, pp. 400–423.

ABSTRACT: The role of hydrogen in the subcritical crack growth of titanium alloys is a subject of intense interest to those who use these alloys for structural applications. The most widely accepted theories are based on the absorption of hydrogen (in the case of environmental hydrogen-assisted cracking), its diffusion to regions of high tensile stress, and the initiation of microcracks at hydrides precipitated in such regions. One of the difficulties with such a mechanism is the paucity of observations of titanium hydrides on fracture surfaces, such as those caused by stress corrosion cracking or inert-environment sustained load cracking caused by residual hydrogen impurity. Another difficulty has been the identity in microstructural crack paths between stress corrosion cracking in hydrogen-containing (saltwater, etc.) and hydrogen-free (carbon tetrachloride) environments.

Studies reported in this paper suggest that hydrogen is not unique in its ability to cause alpha cleavage, but may be a primary cause of alpha-beta interface cracking. Alpha cleavage also occurs in sustained load cracking, but is not, in the present studies, influenced by hydrogen content. Cracking in environmental hydrogen gas is primarily distinguished by alpha-beta interface cracking. The detailed fractographic character of such interface cracking appears to show that the microstructural locus of interface cracking is at the boundary between the alpha phase and the interface phase. The existence of this phase in bulk materials has recently been called into question, and it will be one purpose of this paper to discuss the role of hydrogen content on its formation and the effects of this phase on fracture mechanisms.

KEYWORDS: subcritical cracking, fractography, stress corrosion cracking, hydrogen embrittlement, titanium alloys, mechanical properties, fracture (materials), cracking mechanisms

[1]Code 6312, U.S. Naval Research Laboratory, Washington, DC, 20375-5000.

Hydrogen is considered a major factor in stress corrosion cracking (SCC), inert-environment sustained load cracking (SLC), and environmental hydrogen gas-induced cracking (HGC) of alpha-beta titanium alloys. For example, the exposure of such titanium alloys as Ti-6Al-4V or Ti-8Al-1Mo-1V under stress to hydrogen (H_2) gas causes rapid crack growth [1,2]. The alpha-beta alloys can also crack under sustained load in the absence of any active environment, and increasing alloy hydrogen contents usually causes increasing crack propagation rates [3–7]. Aqueous SCC in alpha-beta titanium alloys is distinguished by cleavage through the alpha phase. Intergranular cracking (IGC) or alpha-beta interface cracking (IFC) is seldom observed, even in microstructures having continuous or extensive beta in alpha grain boundaries. Because SLC with high internal hydrogen contents and HGC do cause IGC and IFC, doubt has existed about a hydrogen-induced cracking mechanism in the SCC case. Nelson [8] observed that as the hydrogen gas pressure was reduced to very low levels, the crack path shifted from IGC/IFC to transgranular in Ti-6Al-4V with continuous beta phase microstructure, thus simulating SCC. He proposed that SCC produces a very low effective hydrogen fugacity at the crack tip. Koch et al. [9] found electron diffraction evidence of titanium hydrides on aqueous SCC fracture surfaces of Ti-8Al-1Mo-1V, an indication that the cracking mechanism involved hydrogen generated by reactions between titanium and water.

On the other hand, it has always been difficult to explain the close similarity between aqueous SCC and SCC in organic solvents, notably carbon tetrachloride (CCl_4) [3,10–12], which has no source of hydrogen except for residual impurities such as water. Scully and Powell [28] hypothesized that residual water in the CCl_4 caused the generation of hydrogen, which then caused embrittlement. Vassel [12] has shown that residual tritiated water in CCl_4 apparently did not cause absorption of hydrogen, but that titanium dichloride ($TiCl_2$) was left on the fracture surface. This observation seems to point to an active species other than hydrogen for SCC in CCl_4.

The role of hydrogen in the SLC case is better understood. IGC/IFC increase with hydrogen content [7], and reflection high energy electron diffraction evidence for hydrides has been found on SLC fracture surfaces in one case [9], although fractographic observations of hydrides in the fracture surfaces or in the microstructure have not been made [7]. The existence of a monolithic, hydride-like interface phase (IFP) at the alpha-beta interface [13,14] in transmission electron microscope (TEM) specimens, however, leads to the hope that a direct microstructural link between hydrogen content and SLC behavior exists. This IFP has the same crystal structure (FCC) and lattice parameter as gamma titanium hydride, but is morphologically different, being a monolithic layer at the alpha-beta interface, rather than flakes or needles. Furthermore, there is evidence that IFP is a hydrogen-stabilized phase [15–17], although some investigators believe it is present only in thin foil TEM specimens, not in bulk [17].

It is the purpose of this paper to review the major fractographic aspects of the various kinds of static cracking modes attributed to hydrogen-assisted mechanisms, add some new results comparing SCC in aqueous solution with CCl_4 and some results on effects of hydrogen on the IFP, and then discuss the authors' viewpoint as to the role of hydrogen in the various forms of cracking.

Materials and Methods

The two titanium alloys used to provide examples of fractographic characteristics and the new results were Ti-8Al-1Mo-1V and Ti-6Al-4V. The Ti-8Al-1Mo-1V was received as plate, approximately 6.7 mm thick, in the mill-annealed condition, with microstructure as shown in Fig. 1A. The microstructure consists of elongated primary alpha with partly transformed beta on the alpha particle boundaries. Some specimens were annealed at 1339 K in vacuum and furnace-cooled to produce large beta grains and large Widmanstatten or plate-like alpha colony microstructure, Fig. 1B. Others were annealed at 1144 K in a vacuum and furnace-cooled to provide low-hydrogen-content material. The starting material had a hydrogen content of 41 ppm and 0.15% oxygen; vacuum annealing at 1339 K resulted in 21-ppm hydrogen and vacuum annealing at 1144 K produced 5-ppm hydrogen. The Ti-6Al-4V material, obtained from 26-mm alpha-beta processed plate, was used only for SLC experiments, and for the results to be described was annealed in vacuum at 1200 K for several hours, then furnace cooled, resulting in a hydrogen content of about 8 ppm. Specimens with higher hydrogen contents were produced by reannealing at 1200 K in H_2 gas for several hours, then furnace cooling. This information is summarized in Table 1. The resulting microstructures, shown in Fig. 2, were equiaxed or plate-like alpha particles with interparticle beta phase.

SLC specimens were compact tension type, slightly modified to an overall size of 50 by 50 mm, approximately, with sharp, shallow notches on the faces parallel to the fracture plane (side grooves) to increase the effective constraint in the plane of the crack and prevent excessive crack front curvature during propagation. The specimens were precracked by fatigue, then loaded to the desired load in a screw-thread loaded stress-rupture frame in air or in vacuum. As the crack grew, the load dropped because of the stiffness of the loading system, and this load drop was calibrated to provide a continuous indirect reading of crack depth. The stress intensity factor rose at a moderate rate during crack growth, enabling a crack growth rate versus stress intensity factor curve to be drawn from each specimen. More complete details are given elsewhere [7].

Crack growth in hydrogen gas was studied with small, single edge–notched specimens of Alloy A (see Table 1), using the same test configuration as for the SLC experiments. All tests were performed in a vacuum

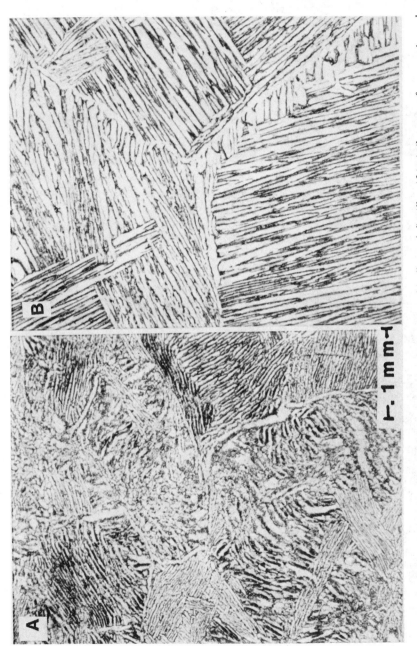

FIG. 1—Microstructure of Ti-8Al-1Mo-1V. Kroll's etch. (A) Alloy A: As received (mill annealed). Alloy A2 is similar except for coarsening and rounding of microstructural elements. (B) Alloy A1: Beta annealed.

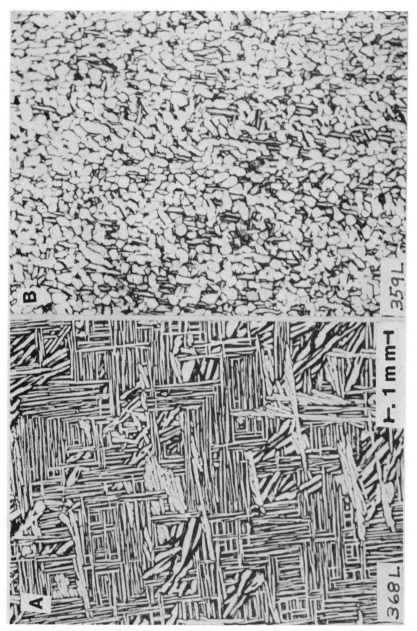

FIG. 2—Microstructure of Ti-6Al-4V, Alloy B, annealed at 1200 K. Krolls's etch. (A) Area of plate-like alpha. (B) Area of equiaxed alpha.

TABLE 1—*Properties of alloys studied in this work.*

Alloy		Heat Treatment	Yield Strength, MPa	Oxygen Content, %	Hydrogen Content, ppm
A	Ti-8Al-1Mo-1V	Mill annealed	924	0.15	41
A1	Ti-8Al-1Mo-1V	1339 K vacuum annealed	...	0.15	21
A2	Ti-8Al-1Mo-1V	1144 K vacuum annealed	...	0.15	5
B	Ti-6A1-4V	1200 K vacuum annealed	970	0.18	Varies

chamber, which was first evacuated to $< 100 \ \mu$Pa, then back-filled with ultra-pure hydrogen gas (10 ppm total impurity) at atmospheric pressure.

The stress corrosion cracking experiments comparing the saltwater and CCl_4 environments were performed using cantilever bend specimens. The specimens were first notched with an abrasive saw and then mounted vertically in the test machine. The load train is connected to an eccentric, and the specimen is fatigued. A computer system [18] monitors the load and stops the machine after the load has dropped 20%, producing a fatigue precrack of the desired length. A polytetrafluoroethylene (PTFE) (Teflon) cup is tightly fit to the specimen to hold the liquid environment around the crack. In the case of the CCl_4 experiments, a glove box surrounding the specimen is closed and purged with nitrogen gas for at least 12 h following the pre-cracking and throughout the subsequent slow strain rate experiment. The glove box is used to exclude atmospheric moisture from the test environment and also to protect workers from the CCl_4 vapors. The saltwater experiments are performed in lab air. In order to obtain the desired deflection rate, the load train was attached to a micrometer driven by a reduction gearbox. The deflection rate was chosen to produce fracture in a period of about 50 ks, which is about the time taken for the more usual tension slow strain rate tests. In future work the strain rate will be varied to find that which produces the highest sensitivity to the environments. This rate may not be the same for different environments [19]. The rate of increase of stress intensity (before crack growth begins) is about 1 MPa$\sqrt{m}$/ks. The data obtained from the experiment are in terms of load versus time. Calibrations of load as a function of deflection for various crack depths were made using sawed specimens, and the trend was found to be linear. Replotting this result showed that the crack depth was a linear function of the slope of the load versus deflection curve. After a slow strain rate experiment the initial (after precracking) and the final crack depths can be measured on the fracture surface, and the slope of the load versus deflection for the beginning and end of the experiment can be determined from the load versus time data and the constant deflection rate. This compliance calibration permitted the intermediate crack lengths to be calculated. Once crack lengths were determined, stress intensity factors

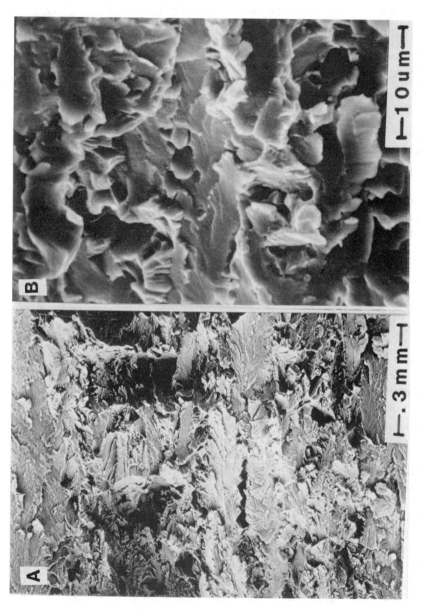

FIG. 3—*SCC of Alloy A in saltwater. (A) SEM fractograph, X100. (B) SEM fractograph. Note well-defined alpha cleavage, evidence of fluting. (C) Cross section showing transgranular secondary cracks. Kroll's etch.*

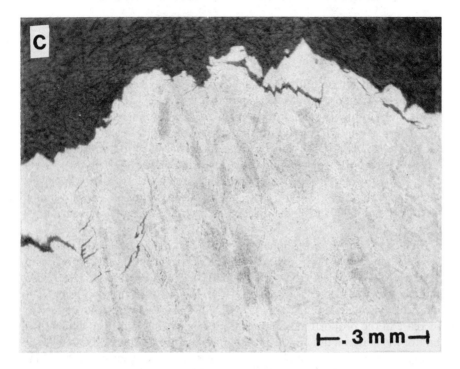

FIG. 3—*Continued.*

and crack growth rates were calculated. Due to the small specimen size relative to the size of the microstructural features and the crude calibration procedure, the stress intensity factor and crack growth rate results are considered to be rough values. Nevertheless, the crack growth calculations reflect expected trends and are useful to compare the crack growth kinetics for different environments.

Specimens were examined directly in a scanning electron microscope (SEM) or indirectly as two-stage (plastic primary, carbon-platinum secondary) replicas in a TEM. SEM specimens were cleaned by organic solvent soaking and were not coated. Some polished metallographic cross sections were looked at, unetched, in the SEM using primary back-scattered electron detection mode to visualize the beta phase through atomic number contrast. Others were examined by light microscopy after etching.

The interface phase examinations were undertaken on a 200-kV TEM, using foils thinned by jet polishing from both sides at 200 K in 5% sulfuric acid (H_2SO_4)-methanol electrolyte. All measurements and micrographs were made with the [0001] pole of the hexagonal close-packed alpha phase parallel to the incident electron beam.

Results

Review of Behavior and Fractography

SCC in aqueous solutions typically causes cleavage of the alpha phase in alpha-beta titanium alloys, such as pictured in Fig. 3. This cleavage in alloys moderately-to-highly susceptible to SCC looks very much like classic cleavage in brittle materials and is distinguished by typical cleavage steps, flat facets and flutes [20,21], and by the appearance of initiation at the facet edges. There is no sign of initiation occurring inside the cleavage facets. Crack growth rates in such susceptible alloys as Ti-8Al-1Mo-1V can be very high, over 100 μm/s [10,12]. SCC in organic solvents looks very similar [3,10–12], but in CCl$_4$, for example, crack rates are typically about ten times slower [12a].

SLC can occur in most alpha-beta alloys and usually causes alpha phase cleavage similar to SCC, except that the cleavage facets may show more evidence of microplasticity, such as small tearing elements (Fig. 4). Also, there is the appearance of initiation within the cleavage facets [4] similar to quasicleavage in tempered martensite [22]. Some investigators have noted an increase in the proportion of cleavage with increasing hydrogen content [6], but for the material used in the present investigation (Alloy B: Ti-6Al-4V) no change in cleavage morphology nor increase in prevalence could be found with increasing hydrogen content. The most marked change in SLC fracture mechanism with increasing hydrogen content is the appearance of IFC (Fig. 5). This feature has some characteristics of cleavage, but lacks true river-pattern cleavage steps. It is often marked by fatigue striation-like arrest marks and by a finely serrated or "shark's skin" texture. Another characteristic of IFC is that a part of the beta (or beta + IFP) layer may be seen on one matching fracture area, the remainder on the other (Fig. 6). This often creates the appearance of fracture along alternate boundaries of a glue layer. Figure 7 is a sketch of the geometrical relationships of the microstructural elements involved. A more complete discussion of IFC is given in Ref 7. The crack growth rates under SLC are far less than those of SCC, and no true Stage II, stress intensity, factor-independent behavior occurs. Growth rates increase very rapidly with hydrogen content, but the threshold stress intensity factor decreases only moderately [7].

HGC in the Ti-8Al-1Mo-1V alloy caused IGC and IFC (Fig. 8) [2]. The characteristics of the IFC are very similar to those under SLC conditions with relatively high hydrogen contents, but, in addition, titanium hydride flakes can be found on the fracture surfaces. In the Ti-8Al-1Mo-1V alloy (A1) the HGC rates were up to 5 μm/s or about twenty times lower than SCC rates in Ti-8Al-1Mo-1V [12a]. Nelson [23] has recorded rates in Ti-6Al-4V having acicular (continuous beta) microstructure of 100 μm/s in H$_2$ at 91 kPa ($\sim$ 1 atm), which compares favorably with SCC rates. At very low

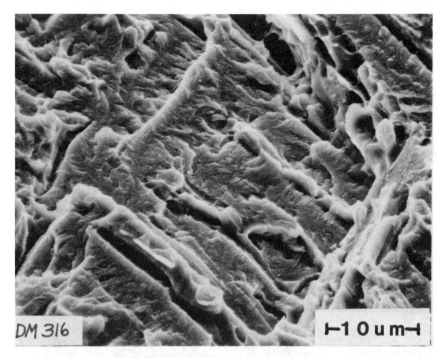

FIG. 4—*SEM fractograph of SLC cleavage in Alloy B containing 215-ppm hydrogen. Note lack of fluting versus Fig. 3.*

H_2 gas pressures or in alloys having the beta phase in isolated grains, only alpha cleavage occurred, similar to that caused by SCC or SLC [8]; crack rates were not recorded but probably were very low.

SCC in Saltwater and CCl_4 (New Results)

Stress corrosion cracking results in saltwater and in CCl_4 have been carefully compared to discover any differences. The experimental technique allowed the kinetics of crack reinitiation (from the fatigue precrack) and propagation to be examined qualitatively. Figure 9 shows the variation in load during slow straining of specimens in saltwater and CCl_4 as the load rises during constant rate deflection of the cantilever bend specimens. In saltwater some loss of linearity becomes evident as plastic strains at the crack tip become significant. More pronounced bending of the load versus time (deflection) line is observed, possibly due to initiation of some crack growth, then a sudden load drop caused by a spurt of crack growth (in some specimens more that one spurt is observed), and finally the rapid decay of load caused by fast, steady crack growth at rates of about 80 μm/s. Figure 9 shows similar stages of behavior for a specimen strained in CCl_4, except that

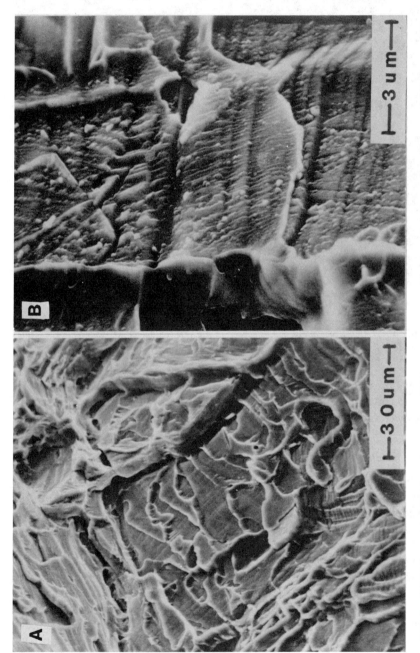

FIG. 5—SEM fractographs of SLC interface cracking. (A) Representative area. Note irregular arrest marks of "terraces." (B) Fine serrations or shark's skin texture on IFC facets.

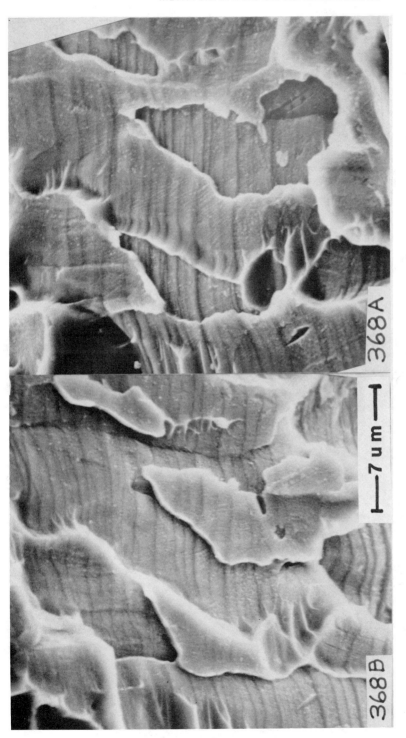

FIG. 6—SEM fractographs of precisely matched mating areas of IFC facets showing beta layer retained on one side, missing from the mating side.

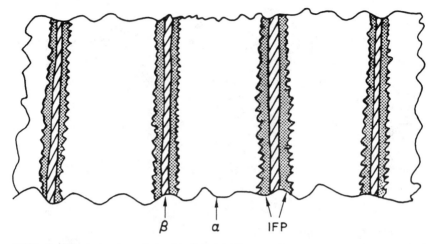

FIG. 7—*Sketch showing relationships of alpha, beta, and interface phases in a plate-like microstructure as in Fig. 1A and Fig. 2B.*

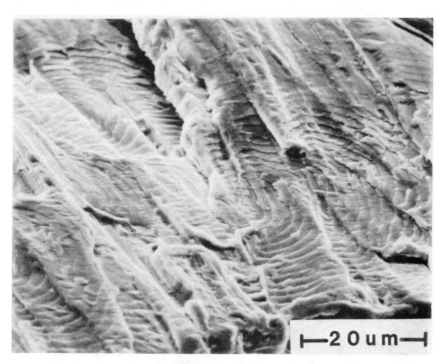

FIG. 8—*SEM fractograph of IFC caused by hydrogen gas in Alloy A1. Note regular, well-defined crack arrest marks or "terraces" in this large-scale microstructure. Serrations like those of Fig. 5B are visible at very high magnifications.*

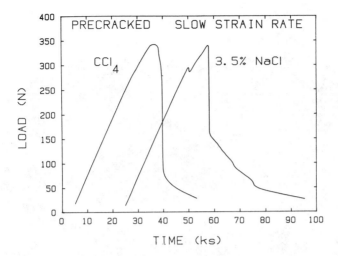

FIG. 9—*Load versus time records for SCC of precracked Alloy A specimens under slow displacement rates in labelled environments.*

the early spurt of crack growth is not observed and the onset and final decay of load due to steady crack growth is not as rapid, the steady state growth rate being about 30 μm/s. The total time of the test is about 50 ks. The fracture surfaces of several specimens fractured by this slow strain rate technique in saltwater and CCl_4 were examined by light optical and scanning electron microscopy. Cross sections of the fracture surfaces were also examined to see if qualitative differences existed. Figures 3 and 10 show typical examples of these examinations. No differences between the saltwater and CCl_4 fractures could be found, both had similar-appearing alpha cleavage, with similar proportions of microplastic tearing. Saltwater caused more secondary cracking, but little IGC or IFC was found in the SCC experiments in either environment.

Interface Phase and Hydrogen

The prevalence of IFC and its correlation with apparent hydrogen activity in the cracking process has been just pointed out without presenting detailed data about the exact microstructural locus of such cracking. In order to discuss this point and better appreciate how hydrogen might influence IFC behavior, it is useful to consider how hydrogen affects prevalence of the IFP in alpha-beta titanium alloys. Electropolished thin foils of Alloy B have been examined in a TEM. Two hydrogen contents were studied—8 ppm (weight) and 215 ppm (weight). Several areas had alpha grain orientations such that the $\langle 0001 \rangle$ pole of the alpha grains could be oriented parallel to the electron beam, ensuring that the alpha-beta interfaces, which are usually parallel to

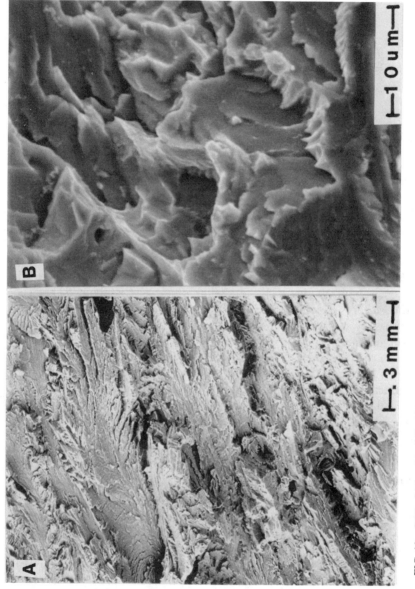

FIG. 10—*SCC of Alloy A in CCl₄.* (A) *SEM fractograph, low magnification.* (B) *SEM fractograph, high magnification.* (C) *Cross section showing transgranular secondary cracks. Kroll's etch.*

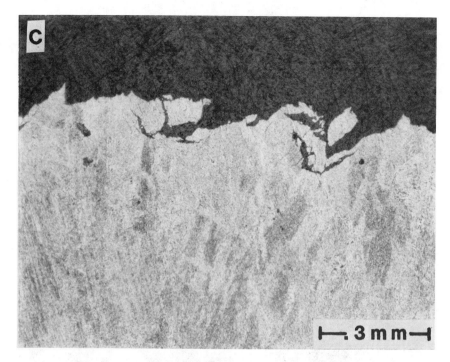

FIG. 10—Continued

{4150} planes [14], would be parallel to the beam. This made it possible to measure the true thickness of the phases at the interface. The results of such measurements are given in Table 2. There are two structures distinguishable from the alpha and beta phases, and both are called interface phase in Table 2. Figure 11 provides an example from a specimen of 215 ppm (weight) nominal hydrogen content. The phase immediately adjacent to the beta phase is of face centered cubic (FCC) crystal structure, and its lattice parameter is 0.414 nm in 8-ppm foils and 0.427 nm in 215-ppm foils. This structure and lattice parameter correspond to gamma titanium hydride [24]. The outer layer, between this FCC phase and the alpha phase, appears to be hexagonal close packed (HCP) structure identical to the alpha phase, but apparently faulted and of different contrast both in bright and dark field imaging from the alpha. Neither the beta phase nor the outer HCP IFP vary in lattice parameter or average thickness with bulk alloy hydrogen content, whereas the FCC IFP showed a 3% greater lattice parameter and a nearly three-fold greater average thickness in specimens of 215-ppm hydrogen content compared with 8 ppm.

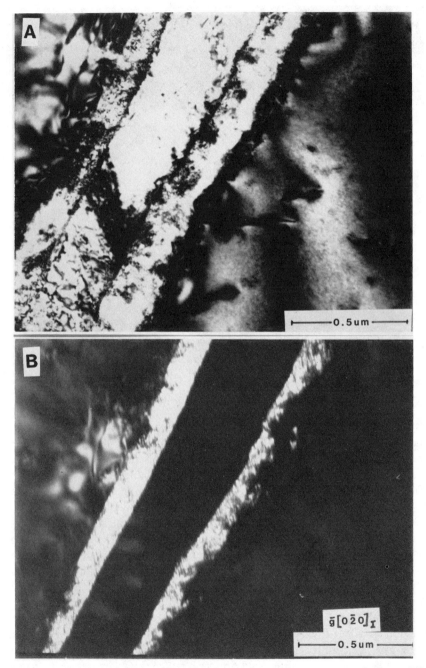

FIG. 11—*TEM micrographs of thin foils from Alloy B containing 215-ppm hydrogen.* (A) *Bright field image showing alpha (dark matrix), beta (wide white strip), and interface phase (strips flanking beta).* (B) *Dark field, highlighting FCC interface phase.* (C) *Diffraction pattern with origin of each spot indicated.*

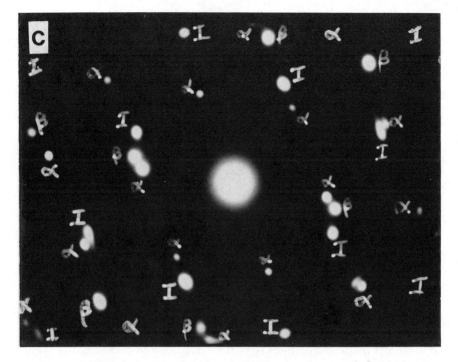

FIG. 11—*Continued.*

TABLE 2—*Thickness of interface phase versus hydrogen content in annealed Ti-6Al-4V.*

Hydrogen Content, ppm (wt)	Phase	Structure	Average Thickness, μm
8	BETA	BCC	7.8
8	IFP	FCC	2.6
8	IFP	HCP	8.8
215	BETA	BCC	11.6
215	IFP	FCC	7.2
215	IFP	HCP	7.1

Discussion

The overview of the three classes of static load cracking (SCC, SLC, and HGC) leaves the impression that hydrogen certainly causes IGC and IFC, but may not be the cause of alpha cleavage under all of the conditions considered. The observation that alpha cleavage under SCC conditions appears to start at the edge of the facet is an indication that the embrittling agent is not absorbed, but is a surface active agent as in the case of liquid metal cracking. Alpha cleavage occurring under SLC, on the other hand, initiates inside the border of the cleavage facet. This may signify the action of an absorbed embrittler, or it may be an intrinsic mode of creep-induced cracking in the alpha phase under static or slowly-applied stresses [25].

The tendency of large hydrogen fugacity to enhance intergranular or interface cracking is well documented, whereas there is less direct evidence that hydrogen causes alpha cleavage. Nelson [8], noting that cracking of Ti-6Al-4V changes from IGC and IFC to transgranular cracking at very low fugacities, suggests that SCC occurs primarily by alpha cleavage because the effective hydrogen fugacity during SCC is extremely low. This idea seems at variance with the high cleavage crack velocities caused by aqueous SCC. For example, in the present work, Stage II SCC rates for Alloy A (Ti-8Al-1Mo-1V) in salt water are about 80 μm/s. IGC or IFC growth rates for this alloy in H_2 gas at 100 kPa ($\sim$ 1 atm) were about 5 μm/s, or 10 times less. Nelson's data for acicular Ti-6Al-4V in H_2 gas at 91 kPa ($\sim$ 1 atm) showed IGC rates of 100 μm/s, about the same as for SCC of Alloy A and about 20 times greater than for H_2 gas cracking of Alloy A. Extrapolation of his data to 10 Pa H_2 gas pressure (at which transgranular cracking occurred) indicates crack rates about 10 nm/s, 10^5 times less than at atmospheric pressure, and 5000 times less than aqueous SCC rates for Alloy A. Based on these figures, it is evident that low hydrogen fugacity is not likely to cause cleavage crack growth rates of the order of those caused by aqueous SCC in acicular alpha-beta microstructures. In order to produce SCC-like crack velocities, fugacities sufficient to produce significant amounts of IFC and IGC seem necessary. At the very least, it can be said that if hydrogen is the embrittling species for aqueous transgranular SCC in acicular (more-or-less continuous beta phase) microstructures, then it is acting by a different mechanism from that in H_2 gas environments.

The SCC experiments in saltwater and CCl_4 were intended to show possible differences in crack growth behavior and microstructural crack growth path. Only a difference in behavior (for example, stop-start unsteady crack growth in the early stages of cracking in saltwater) was detected. No differences in the fracture surface morphology or the metallography of crack propagation could be detected. The observation by Hoffmann et al. [26], that transverse intergranular (IG) splitting could be seen in saltwater but not in CCl_4, was not reproduced for Ti-8Al-1Mo-1V. Both environments caused only transgranular primary and secondary cracking. It is possible that an alloy effect is occurring, since their work used alpha-beta annealed Ti-6Al-2Nb-1Ta-0.8Mo, which has significantly more beta phase and different phase chemistries. It was anticipated that if saltwater caused hydrogen-induced cracking while CCl_4 did not, then in the former case the fracture path should be at least partly along grain boundaries and alpha-beta interfaces, but this was not observed.

The start-stop early crack growth in saltwater may be caused by any of the following circumstances:

1. The embrittling agent or process is only intermittently present, or supplied at such a low rate as to require some time to build up to a sufficient level to embrittle. This would imply lower overall crack growth rates in saltwater than in CCl_4.

2. The agent is of a type requiring an incubation period because of the mechanism for embrittlement, such as absorption by a diffusion process. This implies that the agents for saltwater and CCl_4 are different and, if absorption is involved, that rates in saltwater should be lower.

3. The agent or process is microstructurally selective, requiring a particular local microstructure or grain orientation to act. Again, it must be different from that for CCl_4.

4. The agent is supplied so copiously or is so active in the saltwater case as to cause excessive microbranching, reducing the effective sharpness of the crack and hindering crack growth at low stress intensity levels.

5. The saltwater environment promotes intermittent shielding of the crack tip region from the embrittlement process, as by alternating formation and disruption of a passive film.

The hypotheses 1, 2, and 3 are unable to explain the results. The observations that no difference in crack growth morphology could be found between saltwater and CCl_4 (the cleavage facets show no signs of internal initiation and no IGC or IFC are observed) and that crack rates are very high make a hydrogen absorption-based embrittlement mechanism unattractive. Hypothesis 4 is attractive since secondary cracking is observed on the saltwater fracture surfaces. Case 5 is also attractive since evidence abounds that a common element in SCC processes is the repetitive, alternating formation and disruption of a passive layer at the tip of the crack [27,28]. The specific mechanisms whereby cracks are thus propagated are not clear, but it is reasonably well established that embrittlement reactions are facilitated by localized disruption of a protective film at the crack tip by a combination of accumulated strain and chemical activity. The film is reformed partially or wholly until again disrupted. If the environment is such as to promote rapid reformation of relatively strong or rupture-resistant films, then greater strains are required to disrupt them, and the consequent increment of crack growth may be greater. If the environment is such that film repair is slow or film strength is low, then very small accumulated strain may be sufficient to disrupt the thinner or less resistant film, leading to apparently continuous crack growth. In particular, it may be that CCl_4 produces weak, thin films because of low oxidizing potential and because it somehow solubilizes the passive film, although there are no reports that it can do so in the same way as methanol [12a], for example. Saltwater, on the other hand, undoubtedly has greater oxidizing activity despite the presence of chloride ion, and probably promotes formation of thicker, stronger films. Hypotheses 4 and 5 need not be exclusive; these processes may work together.

The present experimental results and the review of previous information do not strongly support a hypothesis that aqueous SCC and CCl₄ SCC are caused by different mechanisms, or that the mechanism of aqueous SCC involves hydrogen, despite the already quoted studies of Nelson [8] and of Koch et al. [9] and others [29,30] not discussed here. One serious reason for invoking an embrittling species for SCC is that it is difficult, perhaps impossible, to conceive of an anodic dissolution mechanism which can produce such high crack growth rates as have been observed, occurring by brittle cleavage cracking with no evidence of corrosion on the cleavage facets. A recent article by Jones [27] offers a third possibility—that anodic corrosion occurring in material exposed by disruption of the passive film at the crack tip may trigger a cleavage increment through effects on the dislocation reactions thus catalyzed. In this case, the chemical identity of the environment, except so far as it facilitates the passive film formation-disruption-repair cycle, is unimportant, and no specific embrittling species needs to be hypothesized. This would be especially helpful for the CCl₄ environment, where attempts to identify such species have been unfruitful.

The mechanisms for alpha-beta IFC and IGC for the cases of HGC and SLC are easier to hypothesize, because in both cases the growth rates and proportion of alpha-beta IFC and IGC increase with hydrogen fugacity. The very high growth rates achieved raise questions about the viability of a mechanism involving absorption of hydrogen and precipitation of an embrittling hydride, but diffusion along an interface "pipeline" is likely to be rapid, compared with that within alpha grains, for example. The observed increase in thickness of the FCC alpha-beta IFP with hydrogen content of SLC specimens seen in thin foil TEM specimens may provide a microstructural link with the effect of hydrogen on SLC kinetics and crack path morphology. Considerable controversy exists as to the origin of this IFP, especially as to whether it exists in bulk materials or is only present in very thin foils used for TEM examination [17]. It is probable, however, that it is a hydrogen stabilized phase, if not a hydride [15–17]. The techniques used for producing the thin foils in the present investigation have been shown to inject hydrogen into the foils [17]. Nonetheless, the three-fold greater thickness of the FCC IFP in foils from specimens of 215-ppm hydrogen content versus 8-ppm hydrogen content is an indication that part of the FCC IFP measured was present in the bulk, at least in the 215-ppm specimens. An additional indicator is the topography of the IFP facets (Figs. 5,6). The finely serrated or shark's skin appearance corresponds approximately to the serrated interface between the alpha phase and the interface phase [7]. This appearance has also been attributed to fine zig-zag slip-band cracking within the FCC IFP [16]. In either case, there is indirect evidence that the IFP exists in the bulk, and that it is involved in the mechanism whereby increasing hydrogen contents cause alpha-beta IFC and IGC.

The mechanism for alpha cleavage under SLC conditions is less clear. It has been reported in one investigation that the proportion of alpha cleavage increases with hydrogen content for one alloy and microstructural condition [6]. However, in the present case no consistent effect of hydrogen content on prevalence of alpha cleavage could be demonstrated for observations at the same level of stress intensity factor. It was found that the initial region of crack growth ahead of the fatigue precrack had more and better defined cleavage at high hydrogen contents than at low hydrogen contents, but this was judged to arise from the lower stress intensity factor thresholds for the higher hydrogen contents [7]. Nelson's observations [8] tend to support the hypothesis that hydrogen can cause transgranular cracking, but the crack path morphology (cleavage, for example) was not described for the acicular (continuous beta) microstructure. For the equiaxed (continuous alpha) microstructure, cracking appeared to be at least partly alpha cleavage [1], but there is a possibility that very similar cracking could be obtained in air or vacuum, judging from the lack of effect of hydrogen gas pressure on crack propagation in that microstructure [8]. The alpha cleavage plane is normally on, or within twelve angular degrees of, the (0001) basal plane of the HCP structure [11], and it has been shown that titanium hydrides can form on these planes [31], so, conceptually at least, hydrogen should be capable of causing or triggering alpha cleavage. More specific and conclusive studies on the effects of hydrogen fugacity on crack propagation mechanisms would be helpful in resolving the questions implicit in this discussion.

Concluding Remarks

This examination of the role of hydrogen in static cracking of alpha-beta titanium alloys has not been comprehensive, but was intended to show what information can be gained from fractographic and metallographic correlations with cracking behavior, and to demonstrate that there are problems with the full acceptance of hydrogen as a universal embrittling agent. It is clear that hydrogen is a major enabler of IGC and IFC, although it is not perfectly certain that simple hydrogen absorption, hydride growth, hydride cracking models apply to all cases. Further study of the FCC interface phase is needed to ascertain whether it is a hydride, exists in bulk material, and is involved in IGC and IFC mechanisms. The role of hydrogen in alpha cleavage, especially for aqueous SCC, is not securely established, despite evidence of hydride precipitation on the (0001) or near-(0001) cleavage planes [31]. Questions about the high crack growth rates, the initiation site of such cleavage, and the similarity to SCC in a hydrogen-free environment (CCl_4) need to be answered satisfactorily. The finding of hydrides on transgranular

SLC and aqueous SCC fracture surfaces [9] is an indication that titanium-hydrogen reactions take place during or after crack growth, but more specific knowledge of when the hydrides form and where in the microstructure is needed. The mechanism whereby CCl_4 causes SCC cleavage is virtually completely mysterious. The close similarity, indeed the essential identity, of crack path morphology between SCC caused by CCl_4 and saltwater observed in the present study points strongly to the possibility that a singular embrittling agent (such as hydrogen or even chloride ion) for SCC need not be hypothesized, but that some chemical-species-independent process such as outlined by Jones [27] is the cause of alpha cleavage under SCC conditions. Such a process can also explain the minor differences in kinetic behavior between CCl_4 and saltwater as a difference in passivation characteristics in the two media.

Acknowledgments

H. C. Wade assisted with portions of the experimental work. This work was partially supported by the Naval Air Systems Command, Washington, DC.

References

[1] Nelson, H. G., William, D. P., and Stein, J. E., *Metallurgical Transactions*, Vol. 3, Feb. 1972, pp. 469–475.
[2] Meyn, D. A., *Metallurgical Transactions*, Vol. 3, Aug. 1972, pp. 2302–2305.
[3] Sandoz, G. in *Fundamental Aspects of Stress Corrosion Cracking*, R. W. Staehle, A. J. Forty, and D. Van Rooyen, Eds., National Association of Corrosion Engineers, Houston, TX, 1969, pp. 684–690.
[4] Meyn, D. A., *Metallurgical Transactions*, Vol. 5, Nov. 1974, pp. 2405–2414.
[5] Boyer, R. R. and Spurr, W. F., *Metallurgical Transactions*, Vol. 9A, Jan. 1978, pp. 23–29.
[6] Sastry, S. M. L., Lederich, R. J., and Rath, B. B., *Metallurgical Transactions*, Vol. 12A, Jan. 1981, pp. 83–94.
[7] Meyn, D. A. in *Environmental Degradation of Engineering Materials in Hydrogen*, M. R. Louthan, Jr., R. P. McNitt, and R. D. Sisson, Jr., Eds., Virginia Polytechnic Institute, Blacksburg, VA, 1981, pp. 383–392.
[8] Nelson, H. G. in *Hydrogen in Metals*, I. M. Bernstein and A. W. Thompson, Eds., American Society for Metals, Metals Park, OH, 1974, pp. 445–464.
[9] Koch, G. H., Bursle, A. J., Liu, R., and Pugh, E. N., *Metallurgical Transactions*, Vol. 12A, Oct. 1981, pp. 1833–1843.
[10] Beck, T. R. and Blackburn, M. J., *AIAA Journal*, Vol. 6, Feb. 1968, pp. 326–332.
[11] Meyn, D. A. and Sandoz, G., *Transaction of The Metallurgical Society of AIME*, Vol. 245, June 1969, pp. 1253–1258.
[12] Blackburn, M. J., Smyrl, W. H., and Beck, T. R. in *Stress Corrosion Cracking in High Strength Steels and in Titanium and Aluminum Alloys*, B. F. Brown, Ed., Naval Research Laboratory, Washington, DC, 1972, Chapter 5, pp. 245–363.
[12a] Vassel, A. in *Environment-Sensitive Fracture of Engineering Materials*, Z. A. Foroulis, Ed., TMS-AIME, 1979, pp. 277–283.
[13] Rhodes, C. G. and Williams, J. C., *Metallurgical Transactions*, Vol. 6A, Aug. 1975, pp. 1670–1671.

[14] Rhodes, C. G. and Paton, N. E., *Metallurgical Transactions*, Vol. 10A, Feb. 1979, pp. 209-216.
[15] Banerjee, D. and Arunchalam, V. S., *Acta Metallurgica*, Vol. 29, 1981, pp. 1685-1694.
[16] Moody, N. R., Greulich, F. A., and Robinson, S. L., *Metallurgical Transactions*, Vol. 15A, Oct. 1984, pp. 1955-1958.
[17] Banerjee, D., Rhodes, C. G., and Williams, J. C. in *Titanium Science and Technology*, G. Lütjering, U. Zwicker, and W. Bunk, Eds., Deutsche Gesellschaft für Metallkunde e. V., Oberusel, West Germany, 1985, pp. 1597-1604.
[18] Meyn, D. A., Moore, P. G., Bayles, R. A., and Denney, P. E. in *Automated Test Methods for Fracture and Fatigue Crack Growth, ASTM STP 877*, W. H. Cullen, R. W. Landgraf, L. Kaisand, and J. H. Underwood, Eds., American Society for Testing and Materials, Philadelphia, PA, 1985, pp. 27-43.
[19] Scully, J. C. in *Stress Corrosion Cracking—The Slow Strain-Rate Technique, ASTM STP 665*, G. M. Ugianski and J. H. Payer, Eds., American Society for Testing and Materials, Philadelphia, 1979, pp. 237-253.
[20] Aitchison, I. and Cox. B., *Corrosion*, Vol. 28, March 1972, pp. 83-87.
[21] Meyn, D. A. and Brooks, E. J. in *Fractography and Materials Science, ASTM STP 733*, L. N. Gilbertson and R. D. Zipp, Eds., American Society for Testing and Materials, Philadelphia, PA, 1981, pp. 5-31.
[22] Beachem, C. D., *Transactions of ASME: Journal of Basic Engineering*, Series D, Vol. 87, 1965, pp. 299-306.
[23] Nelson, H. G., *Metallurgical Transactions*, Vol. 7A, May 1976, pp. 621-627.
[24] Jaffe, L. D., *Journal of Metals*, Vol. 206, 1956, pp. 861-865.
[25] Williams, D. N., *Metallurgical Transactions*, Vol. 5, Nov. 1974, pp. 2351-2358.
[26] Hoffmann, C. L., Judy, R. W., Jr., and Rath, B. B. in *Titanium Science and Technology*, G. Lütjering, U. Zwicker, and W. Bunk, Eds., Deutsche Gesellschaft für Metallkunde e. V., Oberusel, West Germany, 1985, pp. 2495-2502.
[27] Jones, D. A., *Metallurgical Transactions*, Vol. 16A, June 1985, pp. 1133-1141.
[28] Scully, J. C. and Powell, D. T., *Corrosion Science*, Vol. 10, 1970, pp. 719-733.
[29] Thompson, A. W., *Materials Science and Engineering*, Vol. 43, 1980, pp. 41-46.
[30] Bomberger, H. B., Jr., Meyn, D. A., and Fraker, A. C. in *Titanium Science and Technology*, G. Lütjering, U. Zwicker, and W. Bunk, Eds., Deutsche Gesellschaft für Metallkunde e. V., Oberusel, West Germany, 1985, pp. 2435-2454.
[31] Paton, N. E. and Spurling, R. A., *Metallurgical Transactions*, Vol. 7A, Nov. 1976, pp. 1769-1774.

Narayanaswami Ranganathan,[1] *Bernard Bouchet,*[2] *and Jean Petit*[3]

Fractographic Aspects of the Effect of Environment on the Fatigue Crack Propagation Mechanism in a High-Strength Aluminum Alloy

REFERENCE: Ranganathan, N., Bouchet, B., and Petit, J., **"Fractographic Aspects of the Effect of Environment on the Fatigue Crack Propagation Mechanism in a High-Strength Aluminum Alloy,"** *Fractography of Modern Engineering Materials: Composites and Metals, ASTM STP 948*, J. E. Masters and J. J. Au, Eds., American Society for Testing and Materials, Philadelphia, 1987, pp. 424–446.

ABSTRACT: Fatigue crack propagation tests were conducted in three different environments, air, vacuum, and high purity nitrogen (N_2) [3 ppm water vapor (H_2O) + 1 ppm oxygen (O_2)], on a high-strength aluminum alloy, 2024-T351. From tests conducted at a load ratio of 0.5, a typical behavior in the N_2 environment is brought into light. At moderate ΔK-values, the crack growth behavior in N_2 is similar to that observed in vacuum. As ΔK decreases, the crack growth curve shifts towards the one obtained in air showing a progressive environmental effect occurring in the presence of traces of water vapor.

The fractographic observations for these tests show that the environmental effect observed in the crack growth curves is accompanied by changes in the fracture surface morphology.

The observed results are discussed considering an energy-based model and by taking into account a discontinuous crack growth mechanism at low growth rates.

KEYWORDS: fatigue crack propagation, specific energy, fractography, striations, coarse slip steps, discontinuous crack growth, crack growth model

The fatigue crack growth rate is strongly influenced by an active environment [1]. In ambient air, the crack growth rate, da/dN, can be as much as ten times higher than that obtained in vacuum at comparable values of the amplitude of the stress intensity factor, ΔK. This difference is accentuated

[1]Maître de conferences, E.N.S.M.A., Poitiers, France 86034.
[2]Ingénieur de recherche, E.N.S.M.A., Poitiers, France 86034.
[3]Directeur de recherche, E.N.S.M.A., Poitiers, France 86034.

near threshold conditions [1,2]. The active species, with respect to aluminum alloys, is considered to be essentially water vapor molecules [3–6].

Recently, with the development of the crack closure concepts [7], attempts have been made to rationalize the crack growth behavior in different environments with respect to the effective stress intensity factor range, ΔK_{eff}. It has been shown that the crack growth rates in aluminum alloys of the 2XXX and 7XXX series can be described by power law relationships in terms of ΔK_{eff}, depending upon the environment [2,5]. This approach adds a valuable contribution to the understanding of the crack growth mechanisms involved, but at the present state of knowledge no unified model exists in the literature which can take into account different crack closure effects such as plasticity-induced closure, roughness-induced closure, closure due to a Mode II component, etc., and hence it is difficult to correlate unambiguously the ΔK_{eff} value to the external loading condition.

In the past few years the energy-based approach developed by Weertman [8] has attracted considerable interest. In this model, the crack growth rate is related to ΔK and to the energy to create a unit surface U. Attempts to study the validity of this model have met with some success, and techniques have been developed to measure the quantity U [9–11]. These studies are limited to high ΔK-values, and very little data exist concerning the measurement of the specific energy at low growth rates approaching the threshold.

Recently a technique has been presented to measure the hysteretic energy during a loading cycle, and the evolution of the specific energy has been quantified in three environments—air, vacuum, and high purity nitrogen (N_2)—on high-strength aluminum Alloy 2024-T351 over a large ΔK range [12].

In this study it was shown that the da/dN-ΔK and the U-ΔK relationships are strongly influenced by the environment.

The aim of this paper is to understand the crack growth mechanisms involved in order to explain the observed changes due to the environment in the evolutions of the crack growth rate and the specific energy with respect to ΔK on the basis of microfractographic observations of the fracture surfaces and micro-optical observations of the crack profiles.

The observed results are compared to available data in the literature, and broad lines of a crack growth model are developed.

Experimental Details

The composition and mechanical properties of the studied alloy are given in Tables 1 and 2.

The grain structure shows the presence of unrecrystallized grains of dimensions 15 by 15 by 12 μm and recrystallized grains of dimensions 190 by 70 by 60 μm.

The crack growth behavior was studied near threshold and at high growth

TABLE 1—*Nominal composition, alloy 2024-T351.*

Element	Si	Fe	Cu	Mn	Mg	Cr	Zn	Ti	Al
% weight	0.1	0.22	4.46	0.66	1.5	0.01	0.04	0.02	Remaining

TABLE 2—*Average mechanical properties.*

Yield Strength	Tensile Strength	% Elongation (gage length, 30 mm)	Cyclic Yield Strength
300 MPa	502 MPa	11	500 MPa

rates (by means of constant amplitude tests) at two load ratios R of 0.5 and 0.1 in vacuum, air, and high purity N_2 [3 ppm water (H_2O) + 1 ppm oxide (O_2)].

In this paper only the crack growth behavior at an R value of 0.5 is discussed in detail.

An environmental chamber mounted on a servohydraulic machine was used in this study in which a vacuum of the order of 10^{-3} Pa could be obtained [13]. The relative humidity was typically 50% during tests in air. The tests in N_2 were conducted at a pressure of about 12.4×10^4 Pa. The loading frequency was typically 35 Hz.

Modified compact tension specimens shown in Fig. 1 were used in this study in which the crack opening displacement δ is measured under the loading axis [14]. The specimens were 75 mm wide and 10 mm thick and were so machined to have crack growth parallel to the rolling direction (TL orientation).

At points of interest during the tests, the load P versus differential opening displacement δ' diagrams were traced using an X-Y plotter at a frequency of 0.2 Hz; δ' is given by

$$\delta' = G(\delta - \alpha P) \tag{1}$$

where

α = the specimen compliance with the crack fully open, and
G = the electronic amplification factor (G = 20 to 50).

From these diagrams the hysteretic energy dissipated per cycle Q was measured.

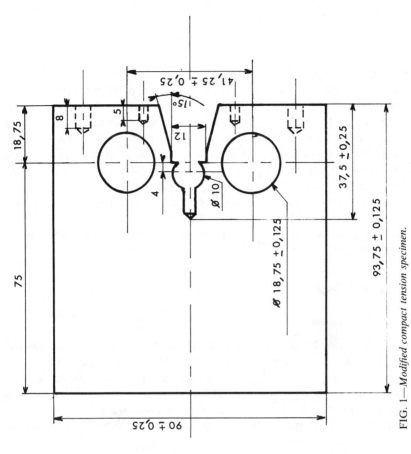

FIG. 1—*Modified compact tension specimen.*

The specific energy, U is defined as

$$U = \frac{Q}{2b(da/dN)} \tag{2}$$

b is equal to the specimen thickness.

The evolution of energy parameters at an R value of 0.5 for threshold tests in the three environments are presented here with complementary results obtained at an R value of 0.1. Complete results are given in Refs 12 and 15.

After the tests the specimen surfaces were examined to characterize the evolution of the crack path and the slip bands accompanying the crack, using an optical microscope at magnifications varying from X200 to X1000.

The specimens were then broken open and the fracture surfaces were examined in scanning electron microscope JSM-50A at magnifications varying from X300 to X25 000.

Two-stage tungsten-carbon replicas were also made to illustrate specific fractographic details.

Experimental Results

da/dN-ΔK *Relationship*

Figure 2 shows the *da/dN*-ΔK relationships for the tests conducted in the three different environments for threshold tests at an R value of 0.5.

From this figure it can be seen that at moderate ΔK values ($\Delta K >$ 9 MPa$\sqrt{m}$) the crack growth rates are the highest in air, while in vacuum and in the N_2 environment the crack growth rates are similar. For lower ΔK-values the crack growth curve in N_2 begins to shift towards the one in air.

This transition occurs at a *da/dN* value of approximately 5×10^{-6} mm/cycle, and in the range $3.5 < \Delta K < 5.5$ MPa$\sqrt{m}$ the crack growth rate is almost independent of ΔK.

For lower ΔK values, the crack growth behavior in air and N_2 are similar, leading to much lower values of the stress intensity factor amplitude at threshold, ΔK_{th}, in these two environments with respect to that in vacuum.

This behavior confirms the occurrence near threshold of a marked environmental effect in the nitrogen environment containing only traces of H_2O in the studied alloy, which is comparable to previously reported results in the aluminum alloys of the 2XXX and 7XXX series [3,5,16].

Energy Analysis

Figure 3 shows the relationship between U and ΔK in the range $2.5 < \Delta K < 30$ MPa$\sqrt{m}$.

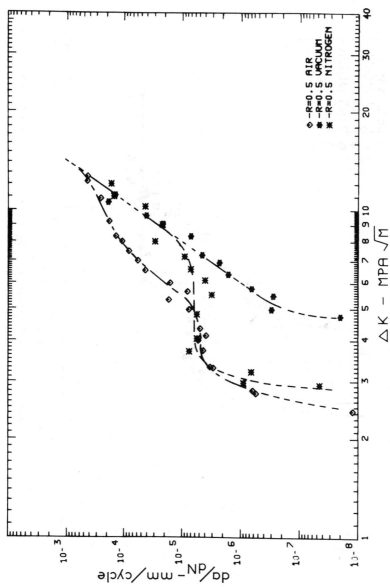

FIG. 2—da/dN-ΔK relationships in the three environments at R = 0.5.

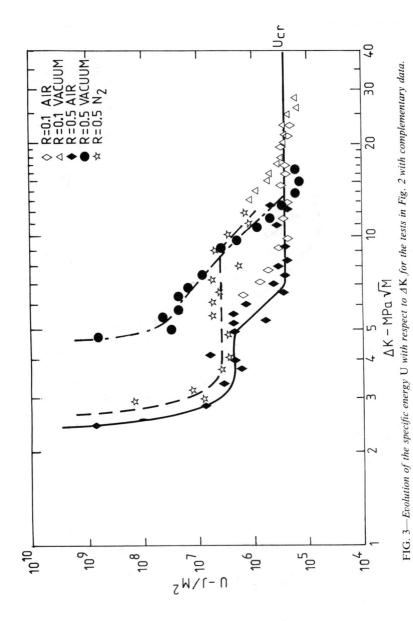

FIG. 3—*Evolution of the specific energy U with respect to ΔK for the tests in Fig. 2 with complementary data.*

In this figure energy measurements include the tests in Fig. 2 as well as constant ΔP tests in air at $R = 0.1$ and in vacuum at $R = 0.1$ and 0.5 as reported in Refs 12 and 15.

These results can be summarized as follows:

1. For ΔK values higher than a critical value called ΔK_{cr}, the specific energy reaches a constant level called U_{cr}. Considering the experimental scatter ($\simeq 15\%$), no significant influence of R ratio and environment is noted on the value of U_{cr}, which is estimated at $2.6 \pm 0.6 \times 10^5$ J/m^2. This value is comparable to results obtained by Liaw et al. [17] using the strain gage technique for the 2024-T4 aluminum alloy.

2. The value of ΔK_{cr}, on the other hand, appears to be dependent on the R ratio and the environment. At $R = 0.1$, ΔK_{cr} in air is on the order of 10 MPa$\sqrt{m}$, while in vacuum the value of this parameter is about 18 MPa$\sqrt{m}$. Also, at $R = 0.5$, ΔK_{cr} is about 6.5 MPa$\sqrt{m}$ in air and approximately equal to 14 MPa$\sqrt{m}$ in vacuum, that is, at much lower levels than the corresponding values at $R = 0.1$. In the nitrogen environment at $R = 0.5$, the plateau corresponding to U_{cr} cannot be observed based on available data (Fig. 3). However, on the basis of the crack growth behavior and on the basis of limited data of U measurements in the range $9 < \Delta K < 12$ MPa$\sqrt{m}$, it can be expected that the behavior in this environment would be similar to that observed in vacuum for ΔK-values higher than 9 MPa$\sqrt{m}$.

3. For a particular environment and load ratio, the specific energy increases in the range $\Delta K < \Delta K_{cr}$.

4. In air at an R ratio of 0.5 (Fig. 3), the U-ΔK relationship indicates the existence of a second plateau in a range of ΔK between 3.5 and 5.5 MPa$\sqrt{m}$, with a specific energy level of 2×10^6 J/m^2. This plateau level is about 10 times as high as the U_{cr}-value.

In N$_2$ at $R = 0.5$ (Fig. 3), the experimental data suggest that the U-ΔK relationship is comparable to the one observed in vacuum for $\Delta K > 9$ MPa$\sqrt{m}$ and a similar second plateau is observed in the range of ΔK between 3.5 and 9 MPa$\sqrt{m}$. This second plateau is not observed in vacuum.

The ΔK range for this plateau in N$_2$ corresponds to the one observed in Fig. 2 where the transitional behavior exists in the ΔK-da/dN relationship. Thus the plateau behavior at a U level of 2×10^6 J/m^2 is associated to a characteristic environmental effect. The specific energy in air and N$_2$ is much lower than that in vacuum for comparable ΔK-values as can be seen in Fig. 3, in the range 5.5 to 9 MPa$\sqrt{m}$. The significance of this behavior is discussed in the latter part of this paper.

5. Near threshold, the specific energy U increases asymptotically for all the experimental conditions studied.

Fractographic and Micro-Optical Observations

Crack Growth in Vacuum—In the range $6.5 < \Delta K < 12$ MPa$\sqrt{\text{m}}$, the fracture surfaces are characterized by very ductile transgranular features with dimples formed by decohesion at the second phase particle interface with the matrix (Fig. 4a) and by flatter regions characterizing a slip mechanism where coarse slip steps can be observed at high magnifications (Fig. 4b). As ΔK decreases, this second kind of feature becomes predominant. The corresponding crack path is slightly wavy, and evidence of slip bands can be observed on either side of the crack (Fig. 4c). The density and length of slip bands appear to decrease as ΔK decreases.

Near threshold very large crystallographic facets can be observed (Fig. 5a) which are typical of the Stage I crack growth where only one slip system is considered to be active [5,18]. Similar results have been obtained in underaged and peak-aged 7075 aluminum alloys [5,18] and is associated with low crack growth rates and high threshold levels. Coarse slip steps are observed in adjacent areas (Fig. 5b). The crack profile (Fig. 5c) shows the occurrence of very large crack deflections and branching consistent with very rough fracture surfaces. The presence of debris due to the rubbing of the crack faces can also be noted in Fig. 5c.

The general evolution of the fractographic features is similar to the results obtained by Kirby and Beevers [19] in a 2L93 aluminum alloy but for the presence of crystallographic facets near threshold.

Crack Growth in Air—In the ΔK range $7.5\ \Delta K < 12$ MPa$\sqrt{\text{m}}$, the fracture surfaces are characterized by transgranular features with dimples similar to the ones in vacuum and flatter areas (Fig. 6a).

In the flatter areas, two kinds of striations can be observed side by side. One of a ductile aspect (Fig. 6b) and another of a more brittle nature with sharper edges (Fig. 6c). This result is similar to that reported by Nix and Flower on a 7010 aluminum alloy [20].

In this ΔK range, the crack profile is characterized by the presence of well-formed slip bands which look like microcracks on either side of the crack (Fig. 6d).

In the range $3.5 < \Delta K < 7.5$ MPa$\sqrt{\text{m}}$, the main fractographic feature is the presence of transgranular pseudocleavage facets (Fig. 7a). As ΔK decreases, these facets have a more brittle appearance. At high magnification, coarse slip steps can be observed (Fig. 7b). There is also evidence of the crack front bypassing second phase particles, and locally the slip step spacing is altered (Fig. 7c), a result previously reported by Feeney et al. on the same alloy [21].

In this range the crack profile shows that crack branching is predominant, and evidence of secondary cracking can also be found (Fig. 7d).

Near threshold, the fracture surfaces are characterized by similar pseudocleavage features as just noted (Fig. 8a) and the existence of striation-like

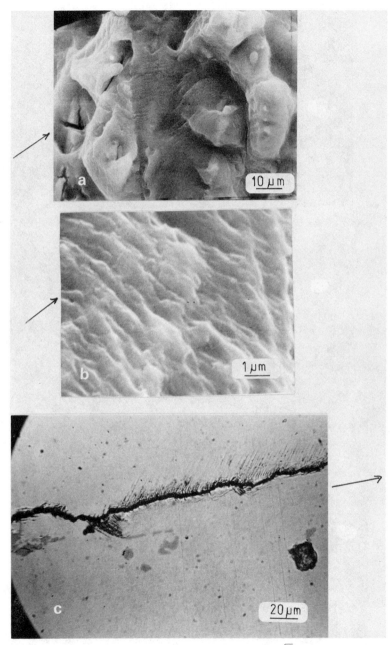

FIG. 4—*Typical fractographs and crack profile, $\Delta K > 6.5\,MPa\sqrt{m}$ vacuum*: (a) *Dimples and flat areas* $\Delta K \simeq 11.11\,MPa\sqrt{m}$; (b) *Coarse slip steps* $\Delta K = 9\,MPa\sqrt{m}$; (c) *Slip bands* $\Delta K \simeq 7\,MPa\sqrt{m}$.

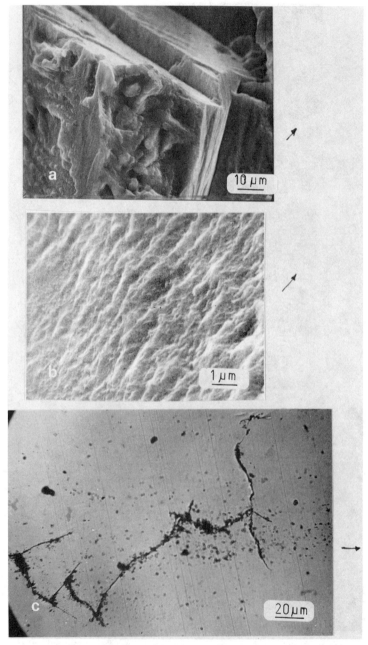

FIG. 5—*Fractography and crack profile near threshold in vacuum:* (a) *Crystallographic facets near threshold,* $\Delta K \simeq 4.70\ MPa\sqrt{m}$; (b) *Coarse slip steps* $\Delta K \simeq 4.80\ MPa\sqrt{m}$; (c) *Large crack deflections and debris on the surface* $\Delta K \simeq 4.80\ MPa\sqrt{m}$.

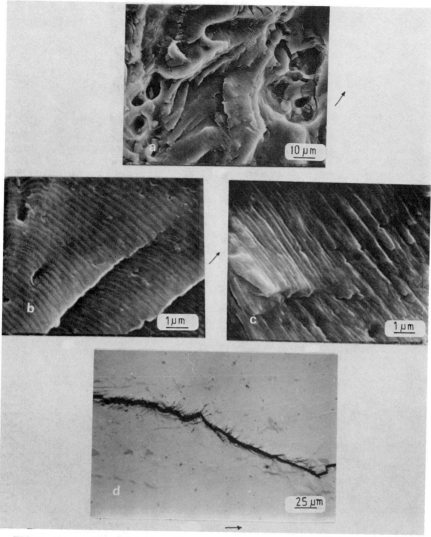

FIG. 6—*Fractography and crack profiles in the range* $\Delta K > 7.5\,MPa\sqrt{m}$, *in air*: (a) *Dimples and striations,* $\Delta K > 12\,MPa\sqrt{m}$; (b) *Ductile striations,* $\Delta K \simeq 12\,MPa\sqrt{m}$; (c) *"Brittle" striations* $\Delta K \simeq 12\,MPa\sqrt{m}$; (d) *Slip band structure* $\Delta K \simeq 12\,MPa\sqrt{m}$.

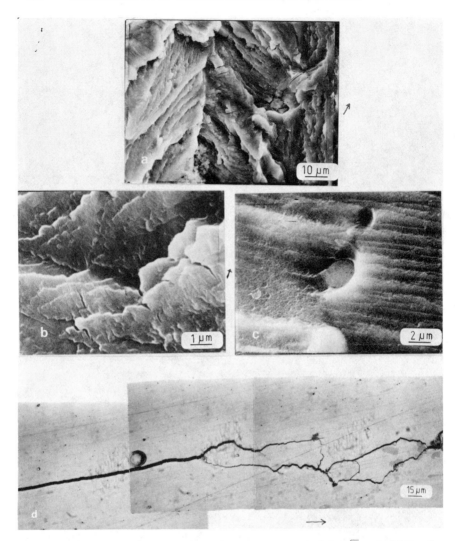

FIG. 7—*Typical fractographs and crack profiles, $3.5 < \Delta K < 7.5\, MPa\sqrt{m}$ in air*: (a) *Pseudo-cleavage facets $\Delta K \simeq 6.5\, MPa\sqrt{m}$*; (b) *Coarse slip steps $\Delta K \simeq 5.3\, MPa\sqrt{m}$*; (c) *Effect of a second phase particle: note modification of slip step spacing $\Delta K \simeq 3.6\, MPa\sqrt{m}$*; (d) *Secondary cracks observed on the surface $4.4 < \Delta K < 5.4\, MPa\sqrt{m}$.*

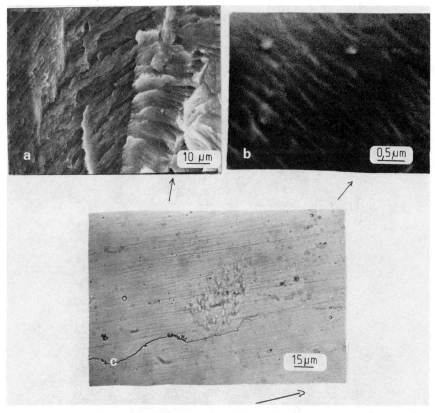

FIG. 8—*Observations near threshold in air*: (a) *Pseudocleavage facets*, $\Delta K \simeq 2.44\ MPa\sqrt{m}$; (b) *Striation-like steps*, $\Delta K \simeq 2.44\ MPa\sqrt{m}$; (c) *Crack profile*, $\Delta K \simeq 2.44\ MPa\sqrt{m}$.

steps at high magnification (Fig. 8*b*). The crack path is slightly wavy with very little crack branching (Fig. 8*c*).

Crack Growth in Nitrogen—In the ΔK range higher than 9 MPa$\sqrt{m}$, the fracture surface appearance (Fig. 9*a*) is generally similar to that obtained in vacuum (Fig. 4*a*) with the presence of dimples and coarse slip steps hardly visible (Fig. 9*b*). The corresponding crack path is wavy with slip bands accompanying the crack (Fig. 9*c*).

In the transition range, pseudocleavage facets become predominant (Fig. 10*a*) and the fracture surface is similar to the ones in air (Fig. 7*a*). The coarse slip steps become very distinct (Fig. 10*b*), and, as in air, the crack profile shows some branching (Fig. 10*c*).

These fractographic features illustrate the occurrence of a marked environmental effect in the transition range as previously observed on several other aluminum alloys [*3,16*].

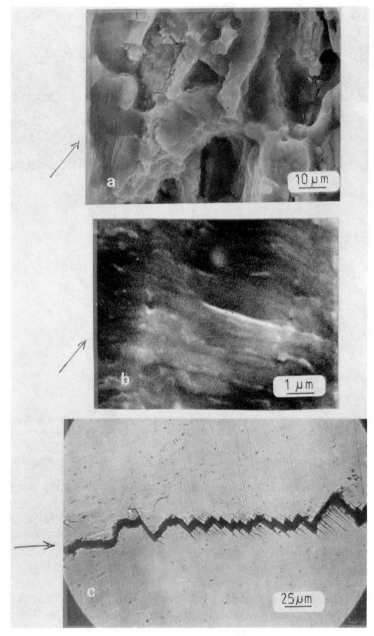

FIG. 9—*Fractographs and crack profile, $\Delta K \simeq 9\,MPa\sqrt{m}$ in N_2: (a) Ductile features, dimples and flatter areas, $\Delta K \simeq 12\,MPa\sqrt{m}$. (b) Ill-formed coarse slip steps, $\Delta K \simeq 10.25\,MPa\sqrt{m}$. (c) Slip bands: $\Delta K \simeq 11\,MPa\sqrt{m}$.*

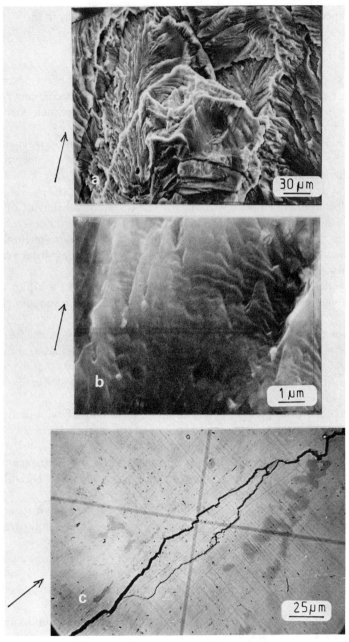

FIG. 10—*Fractographs and crack profile in the transition range in* N_2: (a) *Pseudocleavage facets,* $\Delta K \simeq 6 \, MPa\sqrt{m}$; (b) *Well-formed slip steps,* $\Delta K \simeq 6 \, MPa\sqrt{m}$; (c) *Secondary cracks,* $\Delta K \simeq 3.6 \, MPa\sqrt{m}$.

In the range $\Delta K < 5.5$ MPa$\sqrt{m}$ and near threshold, the fracture surface appearance and crack profiles in N_2 is very similar to that in air.

Discussion

It has been shown in this study that the environmental effect observed in the da/dn-ΔK relationship is accompanied by changes in specific energy levels and by changes in the fracture surface morphology and the main features of the crack profiles as observed on the surface.

Comparing ambient air and vacuum (considered here as an inert environment), the environmental effect is attributed to water vapor adsorption and subsequent hydrogen gas (H_2) embrittlement [5,22].

The N_2 atmosphere used in this study (containing traces of H_2O) is considered to be an intermediate environment. The transitional behavior observed in this atmosphere for a crack growth rate of 5×10^{-6} mm/cycle can be attributed to a gradual appearance of this particular environmental effect which depends on diverse parameters such as water vapor partial pressure, frequency, temperature, etc. [4,5,23].

For a threshold test in N_2 for ΔK values greater than 9 MPa$\sqrt{m}$, the specific energy and da/dN values are comparable to that in vacuum (Figs. 2 and 3).

For lower ΔK values, the just-mentioned transitional behavior leads to a decrease in specific energy in N_2 with respect to vacuum (Fig. 3). For a ΔK of 6 MPa$\sqrt{m}$, the U value in N_2 is about 20 times lower than that in vacuum.

According to Tetelman [24], the surface energy of H_2 assisted crack growth in iron (Fe) can be 20 times lower than the true surface energy.

It has been shown previously that similar transitional behavior in the crack growth curves in N_2 environment is obtained for comparable crack growth rates in certain steels and aluminum alloys [5].

Based on these observations, the behavior in the N_2 atmosphere (with traces of H_2O) can be associated to a gradual appearance of a H_2 embrittlement effect.

As regards fractographic aspects, it has been shown here that in general the fracture surface features in N_2 and in vacuum are similar for ΔK value > 9 MPa$\sqrt{m}$.

In N_2, for lower ΔK values, the development of pseudocleavage facets and the presence of secondary cracks in the transition range show that the fracture morphology in this environment becomes gradually similar to the one in air.

These results substantiate previously observed results on aluminum alloys of the 2XXX and 7XXX series [3,5,16].

Another important result obtained in this study is the presence of regular steps on the fracture surface right up to the threshold in all the environments studied.

Figures 11*a* and 11*b* show the differences in the appearance of classical striations observed in air and the coarse slip steps described in works of Wanhill [25], Ranganathan [26], and Davidson and Lankford [27].

The basic difference between the striations and coarse slip steps is that the striations are very regular with well-marked fronts, while the coarse slip steps are more chaotic in nature and their fronts are ill-defined as can be seen by comparing Figs. 11*a* and 11*b*. Also, the striation spacing can be correlated with the macroscopic growth rates for moderate da/dN values ($da/dN > 10^{-4}$ mm/cycle in aluminum alloys) [15,25,28], while the distance between the coarse slip steps shows very little change with respect to the macroscopic growth rate [15,25].

To illustrate this point, the distance between striations/coarse slip step spacing is compared to da/dN values in Fig. 12. The results obtained in Ref 28 on the same alloy are shown here for comparison.

It can be seen here that only for $da/dN > 10^{-4}$ mm/cycle in air is there a correspondence between the striation spacing and the macroscopic growth rate. The value of ΔK for which this correspondence is first observed is about 7.5 MPa$\sqrt{\text{m}}$ (at $R = 0.5$), which is comparable to the ΔK_{cr} value defined from the U-ΔK relationship for these experimental conditions [12,15].

In vacuum for tests conducted at $R = 0.1$, for da/dN values of about 10^{-3} mm/cycle, the coarse slip step spacing coincides with the macroscopic growth rate, and the corresponding value of ΔK is also comparable to ΔK_{cr} for these experimental conditions [12,15].

In general for $\Delta K < \Delta K_{cr}$, the slip step spacing is greater than the microscopic da/dN values, suggesting that the crack growth in this range is a result of a cumulative damage built up over a large number of cycles, resulting in a step-wise growth mechanism [15,19,27,29].

The average number of cycles, N_F, for the crack to advance a distance equal to a coarse slip step spacing, p, defined as [15,27,29]

$$N_F = p/(da/dN) \tag{3}$$

The increase of U for $\Delta K < \Delta K_{cr}$ can now be expressed in terms of a power law relationship in terms of N_F [12,15]

$$U = U_o(N_F)^\alpha \tag{4}$$

where U_o is equal to U_{cr} in air and in vacuum for the $da/dN > 10^{-5}$ mm/cycle. Correspondingly, the estimated value of α is equal to 0.79 in air and for low N_F values in vacuum, and it is equal to 0.44 for high N_F values near threshold in vacuum [15]. In N_2 the relationship between U and N_F confirms the transitional behavior described in this study, associated to the progressive appearance of the effect of water vapor just described and which could be taken into account by a change in u_o due to hydrogen embrittlement as well as in air for ΔK lower than 5.5 MPa$\sqrt{\text{m}}$.

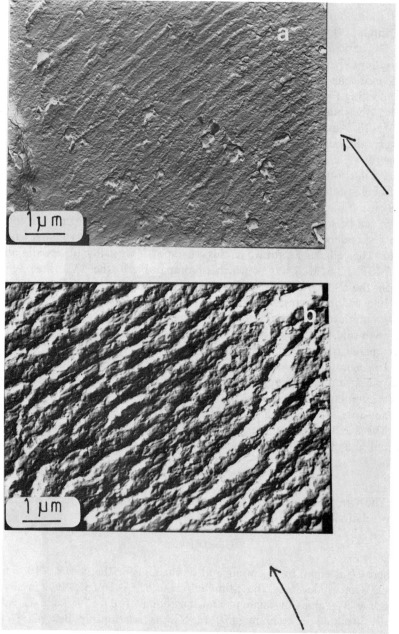

FIG. 11—*Examples of striations and coarse slip steps*: (a) *Striations in air,* $\Delta K = 14\,MPa\sqrt{m}$; (b) *Coarse slip steps in vacuum* $\Delta K = 15.7\,MPa\sqrt{m}$.

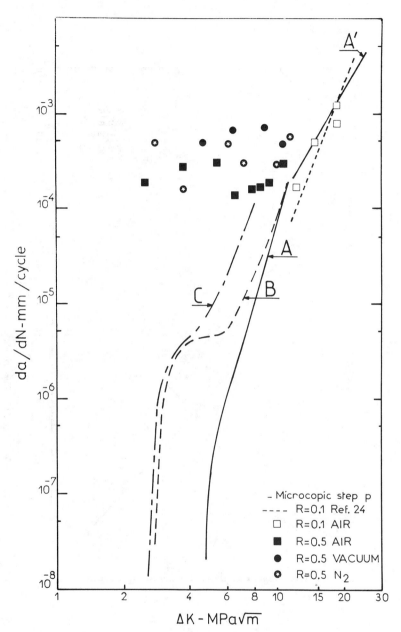

FIG. 12—*Comparison between striation/coarse slip step spacing with macroscopic growth rates*: A *Mean* da/dN-ΔK *curve* R = 0.5 *vacuum*; B *Mean* da/dN-ΔK *curve* R = 0.5 N_2; C *Mean* da/dN-ΔK *curve* R = 0.5 *air*; A′ da/dN-ΔK *curve at high* ΔK *values in air and vacuum at* R = 0.1.

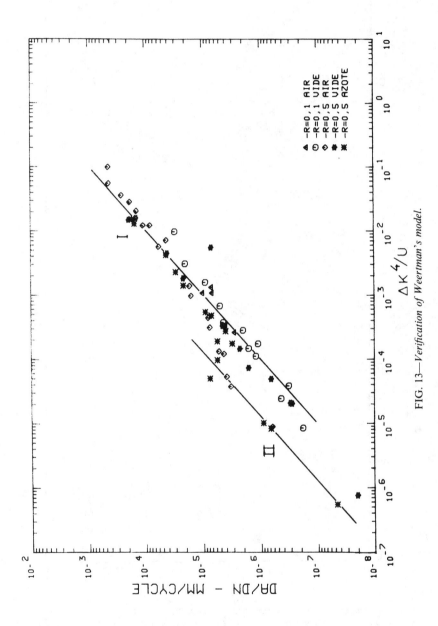

FIG. 13—*Verification of Weertman's model.*

Equation 4 is similar to the one proposed by Halford [30] between accumulated energy and the number of cycles to failure in low cycles fatigue tests.

Based on this analysis, the relationship between da/dN, ΔK, and U proposed by Weertman [8] can be generalized. In its orignal form it was proposed that

$$\frac{da}{dN} = \frac{A \, \Delta K^4}{\mu \sigma_o^2 U}$$

(5)

where

A = a constant = 0.259,
μ = the shear modulus = 27600 MPa, and
σ_o = the flow strength of the material.

Weertman proposed that the value of U should be a constant, while in the present study it has been shown that such a situation exists only for $\Delta K > \Delta K_{cr}$.

Considering that variations on U represent changes in intrinsic material behavior, the relationship between da/dN and $\Delta K^4/U$ is given in Fig. 13. It can be seen in this figure that for $da/dN > 5 \times 10^{-6}$ mm/cycle there is a unique relationship with a slope of 1 independent of the environment (Line I).

For lower growth rates for tests in air and N_2, a second straight line (Line II) could be obtained, but this relationship is approximate considering the limited number of experimental points.

Taking the flow stress to be equal to the cyclic yield strength ($\sigma_o \simeq 500$ MPa), the value of the Constant A for Line I is equal to 6.9×10^{-2}, which is lower than the theoretical estimate of Weertman [8,15].

Conclusions

1. The environmental effect observed in the evolutions of the crack growth rate and the specific energy with respect ΔK are accompanied by changes in the fracture surface morphology.

2. Evidence of step-wise crack growth at low growth rates is confirmed and is consistent with a discontinuous crack advance over a microscopic step after every N_F cycle defined in this study.

3. A relationship between the specific energy U and N_F is proposed which is similar to that obtained in low-cycle fatigue, which suggests the occurrence of an accumulated damage process during N_F cycles.

4. Weertman's model can describe the crack propagation in the studied rate range with a unique relationship independent of the environment for da/dN values $> 5 \times 10^{-6}$ mm/cycle.

References

[1] Hudson, C. M. and Seward, S. K., *Engineering Fracture Mechanics*, Vol. 8, 1976, pp. 315–332.

[2] Petit, J. et al., in *Proceedings*, 3rd European Congress on Fracture, *Fracture and Fatigue*, J. C. Radon, Ed., Pergamon Press, Oxford, U.K., 1980, pp. 320–337.

[3] Petit, J. and Zeghloul, A., in *Proceedings*, ISFT conference, *Fatigue Thresholds*, J. Backlund et al., Eds., EMAS Publications, Warley, United Kingdom, 1981, pp. 563–579.

[4] Achter, M. R., *Scripta Metallurgica*, Vol. 2, 1968, pp. 525–528.

[5] Petit, J., *Fatigue Crack Growth Threshold Concepts*, D. Davidson and S. Suresh, Eds., TMS, AIME symposium, Philadelphia, American Institute of Mining, Metallurgical and Petroleum Engineers, New York, 1984, pp. 3–24.

[6] Bouchet, B., thesis, Dr es sciences, Poitiers, France, 1976.

[7] Elber, W., *Damage Tolerance in Aircraft Structure*, ASTM STP 486, ASTM, Philadelphia, 1971, pp. 230–242.

[8] Weertman, J., *International Journal of Fracture Mechanics*, Vol. 10, 1974, pp. 125–130.

[9] Ikeda, S. et al., *Engineering Fracture Mechanics*, Vol. 9, 1977, pp. 123–136.

[10] Gross, T. D. and Weertman, J., *Metallurgical Transactions*, Vol. 13A, 1982, pp. 2165–2172.

[11] Liaw, P. K. et al., *Fatigue of Engineering Materials and Structures*, Vol. 3, 1980, pp. 59–74.

[12] Ranganathan, N. et al., *Strength of Metals and Alloys*, Vol. 2, H. J. McQueen et al., Eds., proceedings of Conference ICSMA7, Montreal, Canada, 1985, Pergamon Press, Oxford, U.K., 1985, pp. 1267–1272.

[13] Petit, J. et al., *Review de Physique Appliquée*, Vol. 15, 1980, pp. 919–923.

[14] Landes, J. D. and Begley, J. A., *Fracture Analysis*, ASTM STP 560, American Society for Testing and Materials, Philadelphia, 1974, pp. 170–186.

[15] Ranganathan, N., thesis, Docteur ès Sciences, ENSMA, Poitiers, France, 1985.

[16] Zeghloul, A., thesis, Docteur Ingénieur, ENSMA, Poitiers, France, 1980.

[17] Liaw, P. K. et al., *Metallurgical Transactions*, Vol. 12A, 1981, pp. 49–55.

[18] Renaud, P., thesis, Docteur Ingénieur, ENSMA, Poitiers, France, 1982.

[19] Kirby, B. R. and Beevers, C. J., *Fatigue of Engineering Materials and Structures*, Vol. 1, 1979, pp. 289–303.

[20] Nix, K. J. and Flower, H. M., *Materials, Experimentation and Design in Fatigue*, F. Sherraft and J. B. Sturgeon, Eds., Westbury House, Surrey, England, 1981, pp. 117–121.

[21] Feeney, J. A. et al., *Metallurgical Transactions*, Vol. 1A, 1970, pp. 1741–1757.

[22] Westwood, A. R. C. and Ahearn, J. S., *Physical Chemistry of the Solid State: Application to Metals and Their Compounds*, proceedings of the 37th International Meeting of the Society de Chimie Physique, Paris, Elsevier Science Publishers, Netherlands, 1984, pp. 65–88.

[23] Wei, R. P. et al., *Metallurgical Transactions*, Vol. 11A, 1980, pp. 151–158.

[24] Tetelman, A. S., *Hydrogen in Metals*, J. M. Bernstein and A. W. Thompson, Eds., American Society of Metals, Metals Park, OH, 1974, pp. 17–49.

[25] Wanhill, R. J. H., *Metallurgical Transactions*, Vol. 6A, 1975, pp. 1587–1596.

[26] Ranganathan, N., thesis, Docteur Ingénieur, ENSMA, Poitiers, France, 1979.

[27] Davidson, D. L. and Lankford, J., *Fatigue of Engineering Materials and Structures*, Vol. 7, No. 1, pp. 29–39.

[28] Bignonnet, thesis, MSC, University de Montreal, Canada, 1979.

[29] Davidson, D. L. and Lankford, J., *Fatigue of Engineering Materials and Structures*, Vol. 6, pp. 241–256.

[30] Halford, G. R., *Journal of Materials*, Vol. 1, 1966, pp. 3–13.

Subject Index

Author Index